AF352943

Evaluation and Prevention of Mycotoxin Contamination and Toxicological Effects

Evaluation and Prevention of Mycotoxin Contamination and Toxicological Effects

Editors

Houda Berrada
María José Ruiz Leal

Basel • Beijing • Wuhan • Barcelona • Belgrade • Novi Sad • Cluj • Manchester

Editors
Houda Berrada
University of Valencia
Burjassot, Valencia
Spain

María José Ruiz Leal
University of Valencia
Burjassot, Valencia
Spain

Editorial Office
MDPI
St. Alban-Anlage 66
4052 Basel, Switzerland

This is a reprint of articles from the Special Issue published online in the open access journal *Toxins* (ISSN 2072-6651) (available at: https://www.mdpi.com/journal/toxins/special_issues/evaluation_mycotoxin).

For citation purposes, cite each article independently as indicated on the article page online and as indicated below:

Lastname, A.A.; Lastname, B.B. Article Title. *Journal Name* **Year**, *Volume Number*, Page Range.

ISBN 978-3-7258-0885-4 (Hbk)
ISBN 978-3-7258-0886-1 (PDF)
doi.org/10.3390/books978-3-7258-0886-1

Contents

Article

Genotoxicity of 12 Mycotoxins by the SOS/umu Test: Comparison of Liver and Kidney S9 Fraction

Maria Alonso-Jauregui [1], Elena González-Peñas [2], Adela López de Cerain [1] and Ariane Vettorazzi [1,*]

[1] MITOX Research Group, Department of Pharmacology and Toxicology, School of Pharmacy and Nutrition, Universidad de Navarra, 31008 Pamplona, Spain; malonso.17@alumni.unav.es (M.A.-J.); acerain@unav.es (A.L.d.C.)

[2] MITOX Research Group, Department of Pharmaceutical Technology and Chemistry, School of Pharmacy and Nutrition, Universidad de Navarra, 31008 Pamplona, Spain; mgpenas@unav.es

* Correspondence: avettora@unav.es

Abstract: Liver S9 fraction is usually employed in mutagenicity/genotoxicity in vitro assays, but some genotoxic compounds may need another type of bioactivation. In the present work, an alternative S9 fraction from the kidneys was used for the genotoxicity assessment of 12 mycotoxins with the SOS/umu test. The results were compared with liver S9 fraction, and 2–4 independent experiments were performed with each mycotoxin. The expected results were obtained with positive controls (4-nitroquinoline-N-oxide and 2-aminoanthracene) without metabolic activation or with liver S9, but a potent dose-dependent effect with 4-nitroquinoline-N-oxide and no activity of 2-aminoanthracene with kidney S9 were noticed. Aflatoxin B1 was genotoxic with metabolic activation, the effect being greater with liver S9. Sterigmatocystin was clearly genotoxic with liver S9 but equivocal with kidney S9. Ochratoxin A, zearalenone and fumonisin B1 were negative in all conditions. Trichothecenes were negative, except for nivalenol, 3-acetyldeoxynivalenol, 15-acetyldeoxynivalenol, T-2 and HT-2 toxins, which showed equivocal results with kidney S9 because a clear dose-response effect was not observed. Most of the mycotoxins have been assessed with kidney S9 and the SOS/umu test for the first time here. The results with the positive controls and the mycotoxins confirm that the organ used for the S9 fraction preparation has an influence on the genotoxic activity of some compounds.

Keywords: genotoxicity; liver S9; kidney S9; in vitro; bioactivation

Key Contribution: 4-nitroquinoline-N-oxide showed a high dose-related genotoxic activity with kidney S9. Trichothecenes were assessed for the first time by the SOS/umu test with kidney S9. The genotoxic activity of the positive controls and seven mycotoxins was different depending on the origin of the S9 fraction.

Citation: Alonso-Jauregui, M.; González-Peñas, E.; López de Cerain, A.; Vettorazzi, A. Genotoxicity of 12 Mycotoxins by the SOS/umu Test: Comparison of Liver and Kidney S9 Fraction. *Toxins* **2022**, *14*, 400. https://doi.org/10.3390/toxins14060400

Received: 18 May 2022
Accepted: 7 June 2022
Published: 10 June 2022

Publisher's Note: MDPI stays neutral with regard to jurisdictional claims in published maps and institutional affiliations.

1. Introduction

Mycotoxins are secondary metabolites produced by fungi with several toxicological effects to humans and animals [1]. Genotoxicity and carcinogenicity are important toxicological endpoints due to the possible chronic exposure to mycotoxins [2]. A prioritization strategy based on the genotoxic potential of 12 mycotoxins has been performed using in silico approaches and the SOS/umu test [3]. The mycotoxins selected were aflatoxin B1 (AFB1), sterigmatocystin (STER), ochratoxin A (OTA)—all of which are produced by *Aspergillus* and *Penicillium* species—and 9 mycotoxins produced by *Fusarium* species: zearalenone (ZEA), fumonisin B1 (FB1), nivalenol (NIV), deoxynivalenol (DON), 3-acetyldeoxynivalenol (3ADON), 15-acetyldeoxynivalenol (15ADON), T-2 (T-2), HT-2 (HT-2) and fusarenon-X (F-X). In the SOS/umu test performed with and without metabolic activation, AFB1 and STER were classified as genotoxic dependent on metabolic activation, whereas OTA, ZEA and FB1 were nongenotoxic [3]. The S9 mixture for the SOS/umu test was prepared with a commercial post-mitochondrial supernatant 9000 g fraction (S9) from rodent livers at 8%.

The study design of the in vitro genotoxicity assays must consider the use of an exogenous metabolic activation fraction for chemical testing [4,5]. Rat liver S9 fraction is the standardly used in the genotoxicity assays; it can be obtained from different rat strains and after different treatments with enzyme inducers, but some inconsistencies have been found [6]. Moreover, there is no a clear statement on the rat strain, enzyme concentration (mg protein/mL) and the most convenient treatment for the induction. This lack of homogeneity can generate variable genotoxic responses during in vitro assessments [6]. To overcome this problem, some alternatives have been proposed, such as the biotechnological animal free "ewoS9R" [7]. On the other hand, the International Council for Harmonisation of Technical Requirements for Pharmaceuticals for Human Use (ICH) guideline for the genotoxicity assessment of pharmaceuticals (ICH S2R1) recommends the use of alternative methods/systems for metabolic activation when negative results are obtained in vitro and further testing is considered [5]. One of the reasons for explaining the failure of these strategies for detecting genotoxic carcinogens is concerning metabolism and the different biotransformation pathways of chemicals in cells of different tissues. Therefore, the International Workshop on Genotoxicity Test (IWGT) Strategy Expert Group also recommended the use of alternative systems/tissues/species for genotoxicity testing in vitro [8]. The Organization for Economy and Co-operation and Development (OECD) bacterial reverse mutation assay guideline also considers different patterns for the metabolic activation of chemicals as a cause for missing some mutagens with this assay [4]. Different organs than liver have been considered for the S9 preparation (e.g., kidney and spleen) [9]. Human kidney microsomes exerted a similar response to a liver S9 fraction in the biotransformation of peptides [10]. The kidney has a prominent role in the toxicity of numerous drugs, environmental pollutants and natural substances [11]. In addition, several organic substances are nephrotoxic just after being transported into tubular cells. For instance, citrinin induces proximal tubular necrosis [12]. OTA has been shown to be negative in mutagenicity assays [13]. However, it was mutagenic when using a mice kidney microsomal fraction [14]. Therefore, the use of a kidney S9 fraction in genetic toxicology testing may be relevant for some compounds.

In this work, the in vitro genotoxic effect of 12 mycotoxins and the influence of the tissue origin of the metabolic activation fraction has been studied using two S9 fractions. On the one hand, is the standard liver S9 fraction from induced rats and on the other, is a kidney S9 fraction from uninduced rats. For that purpose, the medium-throughput test SOS/umu, previously used to prioritize mycotoxins based on their genotoxicity [3], has been selected. It has been demonstrated to have a high concordance with the regulatory accepted Ames test (TG OECD 471 [4]) [15,16]. The SOS/umu was selected because it has the following advantages with respect to the Ames test: (i) the results are obtained the same day of the experiment (1 day); (ii) the amount of test compound is lower; and (iii) it allows for the assessment of a maximum of six different compounds in one experiment. In summary, there is a significant reduction in material expenses and workload [17]. In the present study, the SOS/umu test has been performed with each of the mycotoxins and with both S9 fractions in 2–4 independent assays. Negative and positive controls were included in all the assays.

2. Results

The results of the negative control (C−) from eight independent experiments are presented in Figure 1. Seven replicates were generated in each experiment. The average and standard deviation of each of the replicates are shown in the figure. The absence of toxicity and genotoxicity is confirmed in each condition.

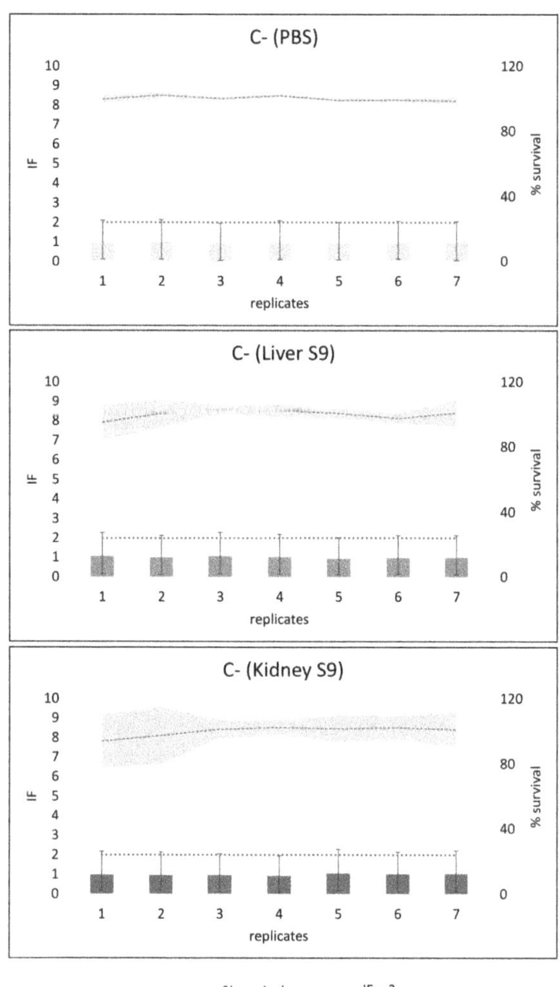

Figure 1. Negative control results in the SOS/umu test without metabolic activation (PBS) or with metabolic activation from liver S9 or kidney S9. Average and standard deviation of each of 7 replicates per experiment from a total of 8 experiments are presented. The bars represent the induction factor (IF) and the grey line the bacterial survival as percentage. The standard deviation (SD) between experiments is presented with the SD bars or with the soft grey area for the survival. Results were considered non-toxic if survival was >80%. Values were considered genotoxic if inductor factor (IF) value was ≥2 at non-toxic concentrations. The red line has been depicted to indicate IF = 2.

The results of the positive controls 4-nitroquinoline-N-oxide (4NQO) and 2-aminoanthracene (2AA) without metabolic activation (PBS) and with metabolic activation from liver (liver S9) or from kidney (kidney S9) are presented in Figure 2. 4NQO without metabolic activation has IF values greater than 2 at high concentrations (0.63–2.5 µg/mL) with a concentration–response tendency. In the presence of liver S9, a positive result was only observed at the highest concentration (2.5 µg/mL), but with kidney S9, a strong genotoxic effect with a concentration–response trend was registered. Moreover, the highest concentration had IF values around 6 (Figure 2). IF variability was greater with kidney S9 (max % CV: 62.9) than with liver S9 (max % CV: 50.6). Considering 2AA, it was negative without metabolic activation and with kidney S9. However, it was positive with liver S9, showing a concentration–response tendency and reaching IF values around 4 (Figure 2). IF variability was greater with kidney S9 (max % CV: 66.7) than with liver S9 (max % CV: 45.8).

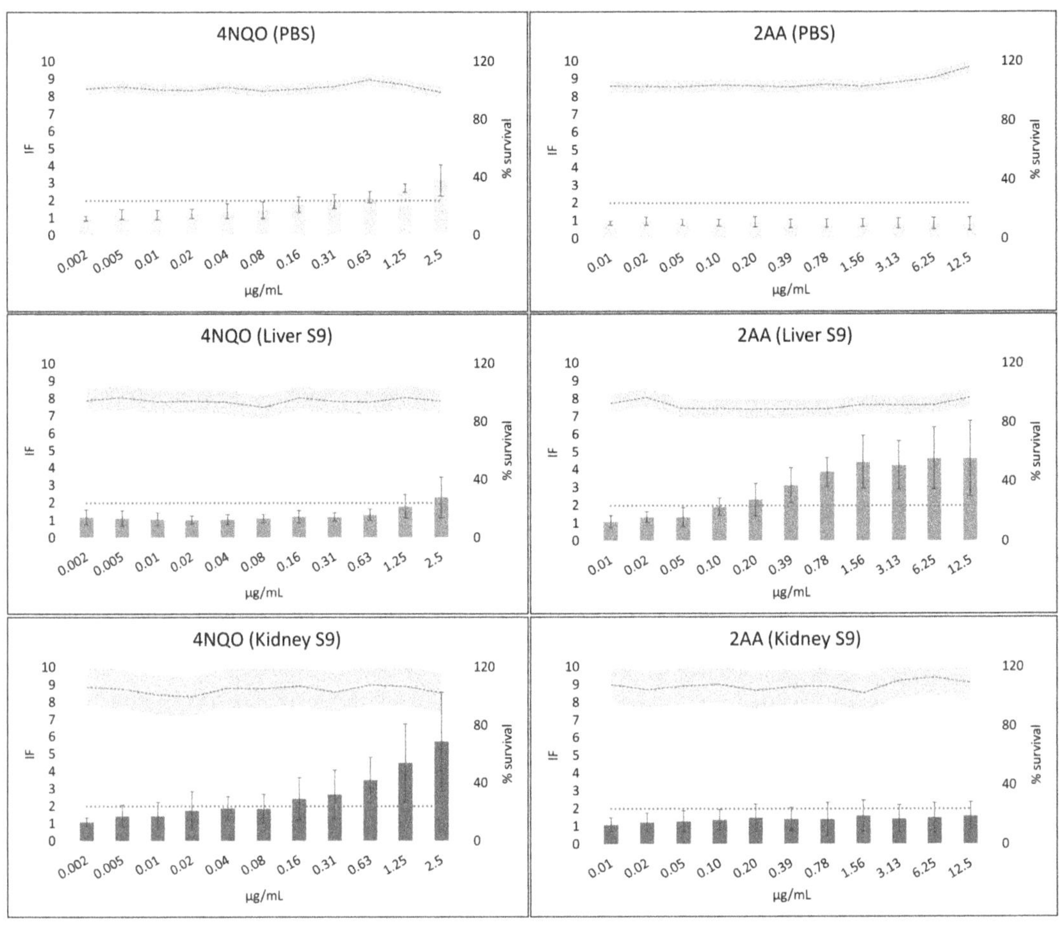

Figure 2. Positive controls results of the SOS/umu test without metabolic activation (PBS) or with metabolic activation from liver S9 or kidney S9. Average and standard deviation from 8 independent experiments are presented. The bars represent the inductor factor (IF) and the grey line the bacterial survival as percentage. The standard deviation (SD) is presented with the SD bars or with the soft grey area for the survival. Concentrations were considered non-toxic if survival was >80%. A compound was considered genotoxic if inductor factor (IF) value was ≥2 at non-toxic concentrations. The red line has been depicted to indicate IF = 2.

The results of AFB1 and STER are shown in Figure 3 and Figure S1 of Supplementary Materials. There was a weak genotoxic tendency without metabolic activation for both mycotoxins (see Figure S1). AFB1 showed a clear concentration–response tendency with both metabolic fractions, although the effect was more pronounced with liver S9. The bacterial survival decreases at the highest concentrations. However, AFB1 is still genotoxic and non-toxic at the lowest concentrations (see Figure 3). IF variability was greater with liver S9 (max % CV: 65.6) than with kidney S9 (max % CV: 39.9). STER gave IF values greater than two, and a concentration–response was evident with liver S9. However, equivocal results were observed with kidney S9: no genotoxicity in three out of four experiments. IF values of day 1 were above two with a weak concentration–response (see Figure 3). IF variability was greater with kidney S9 (max % CV: 87.2) than with liver S9 (max % CV: 47.4).

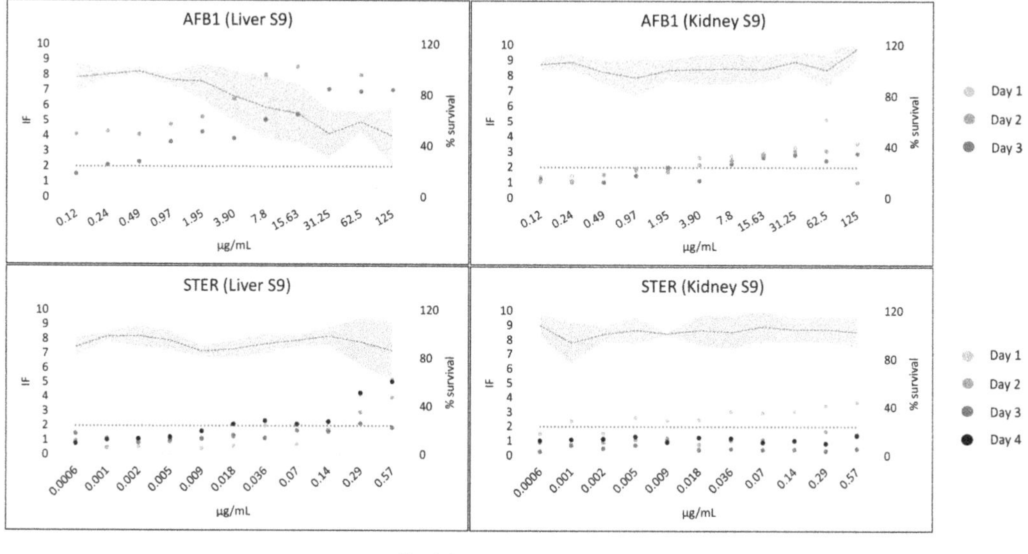

Figure 3. AFB1 and STER results in the SOS/umu test with metabolic activation from liver S9 or kidney S9. The number of independent experiments of AFB1 are 2 for the liver S9 or 3 for the kidney S9. The number of independent experiments of STER are 4 in both conditions. The dots represent the inductor factor (IF) of each individual experiment and the grey line the mean bacterial survival as percentage. The standard deviation of the survival is presented with the soft grey area. Concentrations were considered non-toxic if survival was >80%. A compound was considered genotoxic if inductor factor (IF) value was ≥2 at non-toxic concentrations. The red line has been depicted to indicate IF = 2. Data from two assays with liver S9 (days 3 (AFB1) and 4 (STER)) were published in [3]. Percentage survival values above 120 were corrected to 120%.

The results of DON, F-X, NIV, 3ADON, 15ADON, T-2 and HT-2 without metabolic activation were all negative (see Figures S2 and S4 of Supplementary Materials). The results with metabolic activation are shown in Figure S3 (DON, F-X) and 4 (NIV, 3ADON, 15ADON, T-2, HT-2). DON and F-X were negative with metabolic activation. The highest IF values obtained were 1.94 for DON (kidney S9, 431 µg/mL) and 1.88 for F-X (kidney S9, 1 µg/mL). There was no concentration–response tendency for any mycotoxin.

NIV, 3ADON, 15ADON, T-2 and HT-2 were all negative with liver S9 (Figure 4). However, NIV, 3ADON and T-2 gave IF values above two in three out of four experiments, and 15 ADON and HT-2 gave IF values above two in two out of four experiments with kidney S9. Nonetheless, there was not a clear concentration–response in any of the experiments (see Figure 4). IF variability was greater with kidney S9 (max % CV of 3ADON: 60.1; 15ADON: 51.8; T-2: 47.3; HT-2: 54.9) than with liver S9 (max % CV of 3ADON: 49.3; 15ADON: 42.8; T-2: 40.7; HT-2: 36). For NIV, liver S9 variability was greater (max % CV 54.2) than with kidney S9 (max % CV 51.3).

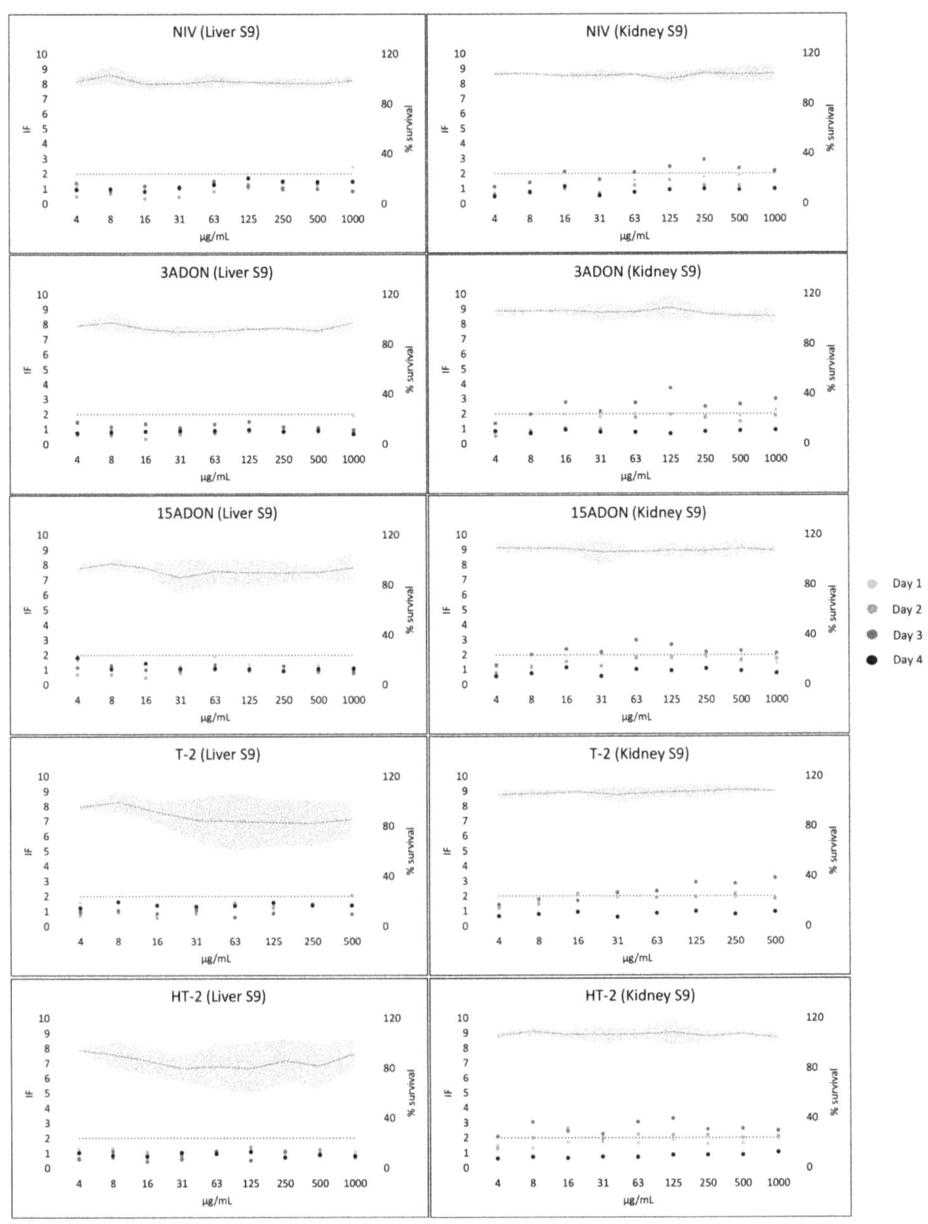

Figure 4. NIV, 3ADON, 15ADON, T-2 and HT-2 results in the SOS/umu test with metabolic activation from liver S9 or kidney S9. The number of independent experiments was 4 for all conditions. The dots represent the inductor factor (IF) of each individual experiment and the grey line the mean bacterial survival as percentage. The standard deviation of the survival is presented with the soft grey area. Concentrations were considered non-toxic if survival was >80%. A compound was considered genotoxic if inductor factor (IF) value was ≥ 2 at non-toxic concentrations. The red line has been depicted to indicate IF = 2. Data from two assays with liver S9 (days 2 and 3) were published in [3].

The results of OTA, ZEA and FB1 are shown in Figures S5 and S6 of Supplementary Materials. All the mycotoxins were negative both with liver or kidney S9 and without metabolic activation.

3. Discussion

Metabolic activation is a pivotal aspect in genetic toxicology testing, due to metabolic deficiencies of in vitro experimental systems. The enzyme preparations from mammalian cells are added to in vitro systems for simulating in vivo xenobiotic bioactivation [8]. Subcellular preparations are obtained by the centrifugation of tissue homogenates. Any tissue could be applied [18]. However, liver is the most widely used because the enzymes from the cytochrome P450 system are located mainly in the endoplasmic reticulum of hepatic cells. There are two types of subcellular preparations. On the one hand, S9 fraction is obtained with one centrifugation step at 9000 g; it contains the microsomal and the cytosolic phases. The microsomal fraction requires an ultracentrifugation at 100,000 g to separate the cytosolic soluble phase from the microsomes [18]. The most widely used S9 fraction is obtained from rodent livers treated with inducing agents [8]. Liver homogenates were recommended as the most convenient tissue in the Ames test [19]. When validating the SOS assays as screening tests, liver S9 was also the most widely used [15,20–25]. However, some procarcinogens could need another type of metabolic preparation to show their biological effects [4,5]. For example, an Ames protocol modification was the inclusion of a hamster S9 fraction for the assessment of azo compounds [26–28]. In addition, the use of a different organ other than liver for preparing the S9 fraction may be recommended in some circumstances. For instance, OTA was assessed with a mice kidney microsomal fraction because the kidney is the target organ of OTA genotoxicity in rodents [14]. Moreover, the kidney is one of the tissues responsible of the metabolic conversion of fusarenon-X into nivalenol [29] and T-2 toxin transformation into HT-2 [30,31]. The present work analyzes the mutagenic potential of 12 mycotoxins with a standard liver S9 fraction and a kidney S9 fraction with the SOS/umu test. Most of the mycotoxins are assessed with kidney S9 and the SOS/umu test for the first time here. This assay is usually performed once in the context of drug screening, but in the present study, 2–4 independent experiments have been carried out for confirming the results and analyzing the variability of the test.

4NQO is standardly used as positive control without metabolic activation [32]. According to our results, it had a weak genotoxic response in this condition. However, it was genotoxic with both metabolic fractions, which was not expected. Moreover, the genotoxicity with kidney S9 was greater than without metabolic activation. 4NQO is quickly reduced into 4-hydroxyaminoquinolone 1-oxide (4HAQO), a carcinogenic metabolite [33]. The suspected enzyme responsible of the reduction was located in the supernatant fraction of rat liver and kidney, but with a higher specific activity in liver [34]. However, we registered greater genotoxicity with kidney S9. The carcinogenic activity of 4HAQO was higher than 4NQO in mice. Consequently, it was suggested that 4NQO acts through its reduced form, 4HAQO [34]. Therefore, in our case, 4NQO's metabolic conversion with both S9 fractions and the possible presence of 4HAQO could explain the genotoxic response obtained with metabolic activation.

2AA is probably the most widely used mutagen [35] and seems to be activated by hepatic preparations from most animal species (rat, mice, hamster, pig and human) [36], which agrees with the results obtained with liver S9. However, it was negative with kidney S9. Just one author assessed the mutagenicity of 2AA with the Ames test and kidney S9 from oyster toadfish. The addition of the kidney S9 increased the mutagenicity of 2AA [37]. The different species and S9 concentration in the metabolic activation mixture could be the cause of the discrepancy.

AFB1 is a genotoxic and carcinogenic compound. It forms DNA adducts and induces gene mutations and chromosomal aberrations. The mutagenicity in *Salmonella* tester strains TA 98 and TA 100 was 1000 times higher with metabolic activation than without bioactivation [38]. AFB1 has been used as positive control for liver S9 in SOS test [32,39,40].

The results with metabolic activation in our study are also greater than without, with a more pronounced response with liver S9 than with kidney S9. Just one author explored the mutagenicity of AFB1 with the Ames test and kidney S9 from oyster toadfish. The addition of the kidney S9 increased the mutagenicity of AFB1 [37].

STER is an intermediate of the aflatoxins biosynthesis pathway with a comparable genotoxic potential than AFB1 [41]. Moreover, STER produced a greatest response than AFB1 in SOS/umu test with liver microsomes from human and rat [42]. In most of the SOS tests it is also genotoxic [40,43]. In our study STER was positive with liver S9 but equivocal with kidney S9. In all cases, it was less genotoxic than AFB1. Nonetheless, the assessment with kidney S9 has been conducted for the first time.

Regarding trichothecenes, NIV, F-X, DON, 3ADON and T-2 were assessed with the Ames test without and with liver S9. In all cases, mutagenic activity was not found [44–49]. DON was also negative with SOS chromotest [45]. In the present study, all were negative without and with liver S9. In some cases, the genotoxicity assessment with SOS/umu test was carried out for the first time (15ADON and HT-2) [50,51]. Moreover, the genotoxicity assessment with S9 fraction of kidney origin was carried out for the first time for all trichothecenes. DON and F-X were also negative with kidney S9; however, NIV, 3ADON, 15ADON, T-2 and HT-2 produced equivocal results.

Ochratoxin A is suspected of being one of the etiological agents of Balkan Endemic Nephropathy (BEN) and was associated with urinary track tumors in humans [52]. Since its last revision in 2020, the uncertainty of the mode of action for kidney carcinogenicity has increased [13]. However, it is known that oxidative stress is involved [53]. OTA mutagenicity assessed with *Salmonella* tester strains was consistently negative [13], which agrees with our results obtained in all conditions. OTA induced a weak response in SOS chromotest, but without a concentration-response trend, which was considered a negative outcome [40]. OTA induced SOS repair activity with SOS chromotest without metabolic activation. However, the concentration tested was cytotoxic [54]. A mixture of OTA and OTB (85% and 15%) induced a weak mutagenic response with a liver S9 fraction in the Ames test. However, in the same experiment, OTA alone was negative [48]. Therefore, in most cases, OTA was found as negative with and without metabolic activation in the SOS test [43,55], which agrees with our results.

However, OTA was mutagenic when using a mice renal microsomal fraction in 1535, 1538 and 98 strains with the Ames test. NADP and arachidonic acid were used as cofactors, and in both cases, OTA was positive in the three strains [14]. The type of preparation (microsomal fraction) and the species (mice) are different from our study, as is the use of arachidonic acid, which could explain the different results obtained with the kidney fraction in the present work. The mutagenicity of 1535 was higher with this cofactor, which would support the hypothesis according to which prostaglandin synthase would be the key element for oxidative metabolism of genotoxic compounds in mouse kidney [14].

Regarding zearalenone, negative results were observed in the Ames test without and with (liver S9) metabolic activation [44,48,56]. In addition, negative outcomes were obtained with SOS tests [39,40,43,55,57], which agrees with our results. In just one case, ZEA was able to induce SOS repair in *Escherichia coli*, but the genotoxic concentration (1.5 mM) was also toxic (IC50 1.45 mM) [58]. The assessment with kidney S9 has been performed for the first time.

FB1 was related with renal carcinogenicity in rats, probably not mediated by a genotoxic mode of action [59]. The mutagenic potential studied with the Ames test was negative without and with liver S9 [45,60,61]. Knasmüller et al. 1997 [45] assessed FB1 genotoxicity with *Escherichia coli* PQ37 in the SOS chromotest and was found to be negative, which is in agreement with our results [45]. In the present study, FB1 reached IF values of 1.99 with kidney S9 (31 and 63 µg/mL in day 1) but without a dose–response tendency (see Figure S6). The genotoxicity assessment with kidney S9 has been performed for the first time.

In all compounds tested, the IF variability was high (CV > 30%). In addition, it was higher with kidney S9 than with liver S9, except with NIV and AFB1, which showed higher variability with liver S9 than with kidney S9. In spite of this, the SOS/umu test can be considered a reproducible assay. This is very evident with the positive controls results. In seven out of eight experiments, it could be concluded that 4NQO was genotoxic with kidney S9, and in eight out of eight experiments, it was genotoxic with PBS. For 2AA, in eight out of eight experiments, it was genotoxic with liver S9. AFB1, was also genotoxic in all experiments with liver S9 and kidney S9. The common aspect observed in all of them was the concentration–response tendency. For NIV, 3ADON and T-2, IF values above two were registered in three out of four experiments with kidney S9; 15ADON and HT-2 in two out of four experiments; and STER in one out of four experiments. However, a concentration–response tendency was not observed in any of those cases. It seems that the high variability with kidney S9 would affect the reproducibility of these compounds under this condition. Therefore, these mycotoxins were classified as equivocal. In summary, the recommendation could be, depending on the compound, that the assay could be performed once, e.g., if there is a clear positive (with a concentration–response tendency) or negative response (IF values < 1.5). However, when analyzing equivocal compounds (IF values around 2), it could be advisable to perform at least two independent experiments. In addition, the Ames test is the gold standard assay for mutagenicity screening [62]. The SOS/umu test has a high concordance with this assay [15,16]. Therefore, it would be advisable to obtain confirmatory results with the miniatured version of the Ames test using a kidney S9 of the mycotoxins with equivocal results obtained in this work with kidney S9 (sterigmatocystin and trichothecenes type A and B). The miniaturized version of the Ames test has a high concordance with the standard version [63]. Only in the case of still obtaining equivocal results could the standard version be performed.

4. Conclusions

The SOS/umu test can be considered a reproducible test. Consistent results have been obtained with negative (PBS) and positive (4NQO and 2AA) controls. Thus, this assay could be used not only for screening but also for genotoxic characterization as a first step. The organ used for the S9 fraction preparation has an influence on the genotoxic activity of some compounds, including 4NQO and 2AA and some mycotoxins. Thus, the use of S9 fraction from kidney tissue may be advisable in some cases.

5. Materials and Methods

5.1. Chemicals and Reagents

DMSO and Na_2CO_3 were purchased from PanReac AppliChem (Barcelona, Spain). Bactotryptone for TGA medium was obtained from Bectone Dickinson (Madrid, Spain) and dextrose and NaCl were from PanReac AppliChem (Barcelona, Spain). Ampicillin, ONPG (2-nitrophenyl- β-D-galactopyranoside), the B-buffer ingredients (Na_2HPO_4 $2H_2O$, NaH_2PO_4 H_2O, $MgSO_4$ $7H_2O$, sodium dodecyl sulfate, β-mercaptoethanol) in which the ONPG was dissolved, PBS ingredients (Na_2HPO_4 $2H_2O$, NaH_2PO_4 H_2O) and the positive controls 2-aminoanthracene (2AA) and 4-nitroquinoline-N-oxide (4NQO) were purchased from Sigma-Aldrich (Darmstadt, Germany). The KCl for B-buffer was from PanReac AppliChem (Barcelona, Spain). Ingredients for the S9 mix preparation were obtained from Sigma-Aldrich (Darmstadt, Germany)—phosphate buffer (NaH_2PO_4 H_2O, Na_2HPO_4 $2H_2O$), glucose-6-phosphate and NADP solutions—or from PanReac AppliChem (Barcelona, Spain)—saline solution ($MgCl_2$ $6H_2O$, KCl).

5.2. Experimental System

The genetically modified *Salmonella typhimurium* 1535/pSK1002 used in the SOS/umu test was purchased from the German Collection for microorganisms and Cell cultures (DSMZ 9274) (Berlin, Germany). The bacterium contains a plasmid in which two genes are fused, one involved in DNA repair and the other accountable for β-galactosidase activity. Therefore, SOS repair activity was monitored through β-galactosidase induction activity determined with spectrophotometric measures.

5.3. Mycotoxins

All mycotoxins were purchased in powder form from Sigma Aldrich (Darmstadt, Germany), dissolved in DMSO or water and maintained at $-20\,^{\circ}$C until use. The reference commercial number and CAS number is indicated in each case: AFB1 (A6636; CAS: 1162-65-8), STER (S3255; CAS:10048-13-2), DON (D0156; CAS: 51481-10-8), F-X (33438; CAS:23255-69-8), NIV (32929; CAS: 23282-20-4), 3ADON (A6166; CAS:50722-38-8), 15ADON (32928; CAS: 88337-96-6), T-2 toxin (33947; CAS: 21259-20-1), HT-2 toxin (T4138; CAS: 26934-87-2), OTA (O1877; CAS: 303-47-9), ZEA (Z2125; CAS: 17924-92-4) and FB1 (F1147; CAS: 116355-83-0).

5.4. Rat Tissue Fractions and S9 Mix Preparation

Two rat S9 fractions were used, one obtained from liver and the other one from kidney. The kidney S9 was purchased from Tebu-bio (Barcelona, Spain), and the liver S9 was obtained from Trinova Biochem (Giessen, Germany). Both were extracted from Sprague Dawley male rats of 5 to 8 weeks of age. Liver S9 was obtained from rats treated with Aroclor 1254 as enzyme inductor. The kidney S9 was extracted from rats not treated with any inducing agent. The sterility of kidney S9 fraction was verified in our laboratory.

The S9 mixtures at 8% with each one of the fractions were prepared just before the assay by adding the S9 fraction to the following solution: phosphate buffer (0.2 M pH 7,4 NaH_2PO_4 H_2O, Na_2HPO_4 H_2O), glucose-6-phosphate (1 M), NAPD (0.1 M) and saline solution (1.6 M $MgCl_2$ $6H_2O$, 0.4 M KCl). The S9 mixture was filtered through a 0.45 µ filter.

5.5. Replicates and Number of Independent Experiments

The number of independent experiments for the positive and negative controls was 8 in all conditions. The number of replicates for the negative control in each experiment was 7. In the case of positive controls and mycotoxins, technical replicates were not included in each experiment. The number of independent experiments for the mycotoxins are between 2 and 4. If the conclusion among the experiments was the same (no genotoxic/genotoxic), there were two independent experiments. If there was no concordance in the conclusion of the experiment, it was repeated up to 4 times. Some experiments of the PBS and liver S9 fraction conditions were already published [3] (day 1 of FB1; day 3 of AFB1, OTA, ZEA, DON and F-X; day 4 of STER; days 2 and 3 of NIV, 3ADON, 15ADON, T-2 and HT-2).

5.6. SOS/umu Assay

The SOS/umu test was performed according to the method proposed by [20,24], with some modifications. The procedure of the assay and the stock solution preparation of the mycotoxins and positive controls are described in [3].

The bacteria were incubated overnight at 37 $^{\circ}$C in 100 mL TGA medium supplemented with ampicillin (50 µg/mL), with slight orbital shaking (155 rpm) from 15 to 17 h until an optimal orbital density (OD_{600} from 0.5 to 1.5) was reached. Then, the overnight culture was diluted with fresh TGA medium (not supplemented with ampicillin) and incubated for 2 h at 37 $^{\circ}$C with orbital shaking (155 rpm) in order to obtain a log-phase bacterial growth culture (OD_{600} from 0.15 to 0.4).

The test was performed in the absence (PBS) and presence of an external metabolic activation system (8% of rat S9 mix from liver or kidney S9 fraction). In each test, negative and positive controls were included.

The test procedure was as follows: first, mycotoxins were dissolved at their respective maximum concentrations. Then, 11 serial half dilutions in DMSO (or water for FB1) of mycotoxins and positive controls were prepared in a 96-well plate (plate A). The final volume in each well was 10 µL. The wells on the last row of the plate contained the negative control (DMSO or water). Afterwards, 70 µL water was added to each well. At this point, each well was checked in order to detect any precipitation of the mycotoxins. Then, the S9 mixes were prepared.

Thereafter, in another 96-well plate (plate B), 10 μL S9 mix or 10 μL PBS, were added, followed by 25 μL of the concentrations of the different mycotoxins previously prepared (plate A). Finally, 90 μL/well of exponentially growing bacteria suspension was added to each well. Then, the plates were incubated for 4 h at 37 °C with orbital shaking (500 rpm). After the incubation period, A_{600} was measured to evaluate toxicity as follows:

$$\% \text{ Survival} = \frac{A_{600} \text{ for each concentration tested}}{\text{Average } A_{600} \text{ for negative control}} \times 100 \tag{1}$$

Afterwards, for the determination of β-galactosidase activity, 30 μL/well of treatment plates (plates B) were transferred to new wells (plates C) containing 150 μL/well of ONPG solution for the enzymatic reaction. A total of 0.9 mg/mL in B-buffer was prepared according to [24] for an enzymatic reaction. Plates C were incubated for 30 min at 28 °C with orbital shaking (500 rpm) in the dark. After the incubation period, the reaction was stopped by adding 120 μL/well of Na_2CO_3 (1M). Finally, A_{420} was measured and β galactosidase activity was determined as follows:

β-galactosidase activity relative units (RU):

$$RU = \frac{A_{420} \text{ for each concentration tested}}{A_{600} \text{ for each concentration tested}} \tag{2}$$

Additionally, induction factor (IF)

$$IF = \frac{RU \text{ for each concentration tested}}{\text{Average RU for negative control}} \tag{3}$$

where average β-galactosidase RU for negative control:

$$RU = \frac{\text{Average} A_{420} \text{ for negative control}}{\text{Average } A_{600} \text{ for negative control}} \tag{4}$$

IF values above or equal to two were considered as positive when the bacterial survival is above 80%, following the criteria of [20]. IF values below 1.5 were considered negative. Finally, IF among 2 and 1.5 were considered equivocal.

Supplementary Materials: The following are available online at https://www.mdpi.com/article/10.3390/toxins14060400/s1, Figure S1. AFB1 and STER results in the SOS/umu test without metabolic activation (PBS). Figure S2. DON and F-X results in the SOS/umu test without metabolic activation (PBS) (*n* = 3 experiments). Figure S3. DON and F-X results in the SOS/umu test with metabolic activation from liver S9 or kidney S9 (*n* = 2 experiments). Figure S4. NIV, 3ADON, 15ADON, T-2 and HT-2 results in the SOS/umu test without metabolic activation (PBS) (*n* = 4 experiments). Figure S5. OTA, ZEA and FB1 results in the SOS/umu test without metabolic activation (PBS). Figure S6. OTA, ZEA and FB1 results in the SOS/umu test with metabolic activation from liver S9 and kidney S9 (*n* = 2 experiments).

Author Contributions: Conceptualization, A.L.d.C. and A.V.; methodology, M.A.-J. and A.V.; writing—original draft preparation, M.A.-J., A.V. and A.L.d.C.; writing—review and editing, M.A.-J., A.V., A.L.d.C. and E.G.-P., funding acquisition, A.V., A.L.d.C. and E.G.-P. All authors have read and agreed to the published version of the manuscript.

Funding: This work was funded by the Spanish "Ministerio de Economía, Industria y Competitividad, Agencia Estatal de Investigación" (AGL2017-85732-R) (MINECO/AEI/FEDER, UE). M.A.-J. thanks the "Asociación de Amigos" (University of Navarra) and the Spanish Government for the predoctoral grants (PRE2018-083527)".

Institutional Review Board Statement: Not applicable.

Informed Consent Statement: Not applicable.

Data Availability Statement: Data are contained within the article.

Conflicts of Interest: The authors declare no conflict of interests in the results.

References

1. Bennett, J.W.; Klich, M. Mycotoxins. *Clin. Microbiol. Rev.* **2003**, *16*, 497–516. [CrossRef] [PubMed]
2. Vettorazzi, A.; López de Cerain, A. Mycotoxins as Food Carcinogens. In *Environmental Mycology in Public Health*; Elsevier: Amsterdam, The Netherlands, 2016; pp. 261–298, ISBN 9780124114715.
3. Alonso-Jauregui, M.; Font, M.; González-Peñas, E.; López de Cerain, A.; Vettorazzi, A. Prioritization of Mycotoxins Based on Their Genotoxic Potential with an In Silico-In Vitro Strategy. *Toxins* **2021**, *13*, 734. [CrossRef] [PubMed]
4. OECD. Test. In *OECD Guidelines for the Testing of Chemicals, Section 4*; OECD: Paris, France, 2020; ISBN 9789264071247.
5. ICH. *ICH Guideline S2 (R1) on Genotoxicity Testing and Data Interpretation for Pharmaceuticals Intended for Human Use*; ICH: Geneva, Switzerland, 2008; Volume 2, pp. 1–28.
6. Yoshikawa, K.; Nohmi, T.; Miyata, R.; Ishidate, M.; Ozawa, N.; Isobe, M.; Watabe, T.; Kada, T.; Kawachi, T. Differences in Liver Homogenates from Donryu, Fischer, Sprague-Dawley and Wistar Strains of Rat in the Drug-Metabolizing Enzyme Assay and the *Salmonella*/Hepatic S9 Activation Test. *Mutat. Res. Mol. Mech. Mutagen.* **1982**, *96*, 167–186. [CrossRef]
7. Brendt, J.; Crawford, S.E.; Velki, M.; Xiao, H.; Thalmann, B.; Hollert, H.; Schiwy, A. Is a Liver Comparable to a Liver? A Comparison of Different Rat-Derived S9-Fractions with a Biotechnological Animal-Free Alternative in the Ames Fluctuation Assay. *Sci. Total Environ.* **2021**, *759*, 143522. [CrossRef] [PubMed]
8. Ku, W.W.; Bigger, A.; Brambilla, G.; Glatt, H.; Gocke, E.; Guzzie, P.J.; Hakura, A.; Honma, M.; Martus, H.-J.; Obach, R.S.; et al. Strategy for Genotoxicity Testing—Metabolic Considerations. *Mutat. Res. Toxicol. Environ. Mutagen.* **2007**, *627*, 59–77. [CrossRef] [PubMed]
9. Shahin, M.M.; Chopy, C.; Mayet, M.J.; Lequesne, N. Mutagenicity of Structurally Related Aromatic Amines in the *Salmonella*/Mammalian Microsome Test with Various S-9 Fractions. *Food Chem. Toxicol.* **1983**, *21*, 615–619. [CrossRef]
10. Zvereva, I.; Semenistaya, E.; Krotov, G.; Rodchenkov, G. Comparison of Various In Vitro Model Systems of the Metabolism of Synthetic Doping Peptides: Proteolytic Enzymes, Human Blood Serum, Liver and Kidney Microsomes and Liver S9 Fraction. *J. Proteom.* **2016**, *149*, 85–97. [CrossRef]
11. Barnett, L.M.A.; Cummings, B.S. Nephrotoxicity and Renal Pathophysiology: A Contemporary Perspective. *Toxicol. Sci.* **2018**, *164*, 379–390. [CrossRef]
12. Berndt, W.O. The Role of Transport in Chemical Nephrotoxicity. *Toxicol. Pathol.* **1998**, *26*, 52–57. [CrossRef]
13. EFSA (European Food Safety Authority); Schrenk, D.; Bodin, L.; Chipman, J.K.; del Mazo, J.; Grasl-Kraupp, B.; Hogstrand, C.; Hoogenboom, L.R.; Leblanc, J.; Nebbia, C.S.; et al. Risk Assessment of Ochratoxin A in Food. *EFSA J.* **2020**, *18*, e06113. [CrossRef]
14. Obrecht-Pflumio, S.; Chassat, T.; Dirheimer, G.; Marzin, D. Genotoxicity of Ochratoxin A by *Salmonella* Mutagenicity Test after Bioactivation by Mouse Kidney Microsomes. *Mutat. Res. Toxicol. Environ. Mutagen.* **1999**, *446*, 95–102. [CrossRef]
15. Yasunaga, K.; Kiyonari, A.; Oikawa, T.; Abe, N.; Yoshikawa, K. Evaluation of the *Salmonella* umu Test with 83 NTP Chemicals. *Environ. Mol. Mutagen.* **2004**, *44*, 329–345. [CrossRef] [PubMed]
16. Reifferscheid, G.; Heil, J. Validation of the SOS/Umu Test Using Test Results of 486 Chemicals and Comparison with the Ames Test and Carcinogenicity Data. *Mutat. Res. Toxicol.* **1996**, *369*, 129–145. [CrossRef]
17. McDaniels, A.E.; Reyes, A.L.; Wymer, L.J.; Rankin, C.C.; Stelma, G.N. Comparison of the *Salmonella* (Ames) Test, Umu Tests, and the Sos Chromotests for Detecting Genotoxins. *Environ. Mol. Mutagen.* **1990**, *16*, 204–215. [CrossRef]
18. Parmentier, Y.; Bossant, M.-J.; Bertrand, M.; Walther, B. In Vitro Studies of Drug Metabolism. In *Comprehensive Medicinal Chemistry II*; Elsevier: Amsterdam, The Netherlands, 2007; pp. 231–257.
19. Maron, D.M.; Ames, B.N. Revised Methods for the *Salmonella* Mutagenicity Test. *Mutat. Res. Mutagen. Relat. Subj.* **1983**, *113*, 173–215. [CrossRef]
20. Oda, Y.; Nakamura, S.; Oki, I.; Kato, T.; Shinagawa, H. Evaluation of the New System (Umu-Test) for the Detection of Environmental Mutagens and Carcinogens. *Mutat. Res. Mutagen. Relat. Subj.* **1985**, *147*, 219–229. [CrossRef]
21. Oda, Y.; Yamazaki, H.; Watanabe, M.; Nohmi, T.; Shimada, T. Development of High Sensitive Umu Test System: Rapid Detection of Genotoxicity of Promutagenic Aromatic Amines by *Salmonella typhimurium* Strain NM2009 Possessing High O-Acetyltransferase Activity. *Mutat. Res. Mutagen. Relat. Subj.* **1995**, *334*, 145–156. [CrossRef]
22. Oda, Y.; Funasaka, K.; Kitano, M.; Nakama, A.; Yoshikura, T. Use of a High-Throughputumu-Microplate Test System for Rapid Detection of Genotoxicity Produced by Mutagenic Carcinogens and Airborne Particulate Matter. *Environ. Mol. Mutagen.* **2004**, *43*, 10–19. [CrossRef]
23. Hamer, B.; Bihari, N.; Reifferscheid, G.; Zahn, R.K.; Müller, W.E.; Batel, R. Evaluation of the SOS/Umu-Test Post-Treatment Assay for the Detection of Genotoxic Activities of Pure Compounds and Complex Environmental Mixtures. *Mutat. Res. Toxicol. Environ. Mutagen.* **2000**, *466*, 161–171. [CrossRef]
24. Reifferscheid, G.; Heil, J.; Oda, Y.; Zahn, R.K. A Microplate Version of the SOS/Umu-Test for Rapid Detection of Genotoxins and Genotoxic Potentials of Environmental Samples. *Mutat. Res. Mutagen. Relat. Subj.* **1991**, *253*, 215–222. [CrossRef]
25. Whong, W.-Z.; Wen, Y.-F.; Stewart, J.; Ong, T. Validation of the SOS/Umu Test with Mutagenic Complex Mixtures. *Mutat. Res. Lett.* **1986**, *175*, 139–144. [CrossRef]
26. Prival, M.J.; Mitchell, V.D. Analysis of a Method for Testing Azo Dyes for Mutagenic Activity in *Salmonella typhimurium* in the Presence of Flavin Mononucleotide and Hamster Liver S9. *Mutat. Res. Mutagen. Relat. Subj.* **1982**, *97*, 103–116. [CrossRef]

27. Prival, M.J.; Bell, S.J.; Mitchell, V.D.; Peiperl, M.D.; Vaughan, V.L. Mutagenicity of Benzidine and Benzidine-Congener Dyes and Selected Monoazo Dyes in a Modified *Salmonella* Assay. *Mutat. Res. Toxicol.* **1984**, *136*, 33–47. [CrossRef]
28. Prival, M.J.; Davis, V.M.; Peiperl, M.D.; Bell, S.J. Evaluation of Azo Food Dyes for Mutagenicity and Inhibition of Mutagenicity by Methods Using *Salmonella typhimurium*. *Mutat. Res. Toxicol.* **1988**, *206*, 247–259. [CrossRef]
29. Poapolathep, A.; Sugita-Konishi, Y.; Doi, K.; Kumagai, S. The Fates of Trichothecene Mycotoxins, Nivalenol and Fusarenon-X, in Mice. *Toxicon* **2003**, *41*, 1047–1054. [CrossRef]
30. Schuhmacher-Wolz, U.; Heine, K.; Schneider, K. Report on Toxicity Data on Trichothecene Mycotoxins HT-2 and T-2 Toxins. *EFSA Support. Publ.* **2010**, *7*, 65E. [CrossRef]
31. SCF. *Opinion of the Scientific Panel on Food on Fusarium Toxins. Part 5: T-2 Toxin and HT-2 Toxin*; SCF: Brussels, Belgium, 2001; Volume 5.
32. Nakamura, S.; Oda, Y.; Shimada, T.; Oki, I.; Sugimoto, K. SOS-Inducing Activity of Chemical Carcinogens and Mutagens in *Salmonella typhimurium* TA1535/PSK1002: Examination with 151 Chemicals. *Mutat. Res. Lett.* **1987**, *192*, 239–246. [CrossRef]
33. Matsushima, T.; Kobuna, I.; Fukuoka, F.; Sugimura, T. Metabolism of 4-Nitroquinoline 1-Oxide. IV. Quantitative Determination of In Vivo Conversion of 4-Nitroquinoline 1-Oxide in Subcutaneous Tissue of Rat. *Gan* **1968**, *59*, 247–250.
34. Shirasu, Y. Comparative Carcinogenicity of 4-Nitroquinoline 1-Oxide and 4-Hydroxyaminoquinoline 1-Oxide in 3 Strains of Mice. *Exp. Biol. Med.* **1965**, *118*, 812–814. [CrossRef]
35. Ayrton, A.D.; Neville, S.; Ioannides, C. Cytosolic Activation of 2-Aminoanthracene: Implications in Its Use as Diagnostic Mutagen in the Ames Test. *Mutat. Res. Mol. Mech. Mutagen.* **1992**, *265*, 1–8. [CrossRef]
36. Phillipson, C.E.; Ioannides, C. Activation of Aromatic Amines to Mutagens by Various Animal Species Including Man. *Mutat. Res. Toxicol.* **1983**, *124*, 325–336. [CrossRef]
37. Guobaitis, R.J.; Ellingham, T.J.; Maddock, M.B. The Effects of Pretreatment with Cytochrome P-450 Inducers and Preincubation with a Cytochrome P-450 Effector on the Mutagenicity of Genotoxic Carcinogens Mediated by Hepatic and Renal S9 from Two Species of Marine Fish. *Mutat. Res. Mutagen. Relat. Subj.* **1986**, *164*, 59–70. [CrossRef]
38. IARC. Some Traditional Herbal Medicines, Some Mycotoxins, Naphtalene and Styrene. *Monogr. Eval. Carcinog. Risks Hum.* **2002**, *82*, 169–522.
39. Auffray, Y.; Boutibonnes, P. Induction of SOS Function InEscherichia Coli by Some Mycotoxins. *Toxic. Assess.* **1988**, *3*, 371–378. [CrossRef]
40. Krivobok, S.; Olivier, P.; Marzin, D.R.; Seigle-Murandi, F.; Steiman, R. Study of the Genotoxic Potential of 17 Mycotoxins with the SOS Chromotest. *Mutagenesis* **1987**, *2*, 433–439. [CrossRef]
41. Theumer, M.G.; Henneb, Y.; Khoury, L.; Snini, S.P.; Tadrist, S.; Canlet, C.; Puel, O.; Oswald, I.P.; Audebert, M. Genotoxicity of Aflatoxins and Their Precursors in Human Cells. *Toxicol. Lett.* **2018**, *287*, 100–107. [CrossRef]
42. Baertschi, S.W.; Raney, K.D.; Shimada, T.; Harris, T.M.; Guengerich, F.P. Comparison of Rates of Enzymic Oxidation of Aflatoxin B1, Aflatoxin G1, and Sterigmatocystin and Activities of the Epoxides in Forming Guanyl-N7 Adducts and Inducing Different Genetic Responses. *Chem. Res. Toxicol.* **1989**, *2*, 114–122. [CrossRef]
43. Sakai, M.; Abe, K.-I.; Okumura, H.; Kawamura, O.; Sugiura, Y.; Horie, Y.; Ueno, Y. Genotoxicity of Fungi Evaluated by SOS Microplate Assay. *Nat. Toxins* **1992**, *1*, 27–34. [CrossRef]
44. Wehner, F.C.; Marasas, W.F.O.; Thiel, P.G. Lack of Mutagenicity to *Salmonella* Typhimurium of Some Fusarium Mycotoxins. *Appl. Environ. Microbiol.* **1978**, *35*, 659–662. [CrossRef]
45. Knasmüller, S.; Bresgen, N.; Kassie, F.; Mersch-Sundermann, V.; Gelderblom, W.; Zöhrer, E.; Eckl, P.M. Genotoxic Effects of Three Fusarium Mycotoxins, Fumonisin B1, Moniliformin and Vomitoxin in Bacteria and in Primary Cultures of Rat Hepatocytes. *Mutat. Res. Toxicol. Environ. Mutagen.* **1997**, *391*, 39–48. [CrossRef]
46. Takakura, N.; Nesslany, F.; Fessard, V.; Le Hegarat, L. Absence of In Vitro Genotoxicity Potential of the Mycotoxin Deoxynivalenol in Bacteria and in Human TK6 and HepaRG Cell Lines. *Food Chem. Toxicol.* **2014**, *66*, 113–121. [CrossRef] [PubMed]
47. Ueno, Y. Mode of Action of Trichothecenes. *Pure Appl. Chem.* **1977**, *49*, 1737–1745. [CrossRef] [PubMed]
48. Kuczuk, M.H.; Benson, P.M.; Heath, H.; Wallace Hayes, A. Evaluation of the Mutagenic Potential of Mycotoxins Using *Salmonella* Typhimurium and Saccharomyces Cerevisiae. *Mutat. Res. Mutagen. Relat. Subj.* **1978**, *53*, 11–20. [CrossRef]
49. EFSA (European Food Safety Authority). Scientific Opinion on Risks for Animal and Public Health Related to the Presence of Nivalenol in Food and Feed. *EFSA J.* **2013**, *11*, 3262. [CrossRef]
50. EFSA (European Food Safety Authority). Scientific Opinion on the Risks for Animal and Public Health Related to the Presence of T-2 and HT-2 Toxin in Food and Feed. *EFSA J.* **2011**, *9*, 2481. [CrossRef]
51. Knutsen, H.K.; Alexander, J.; Barregård, L.; Bignami, M.; Brüschweiler, B.; Ceccatelli, S.; Cottrill, B.; Dinovi, M.; Grasl-Kraupp, B.; Hogstrand, C.; et al. Risks to Human and Animal Health Related to the Presence of Deoxynivalenol and Its Acetylated and Modified Forms in Food and Feed. *EFSA J.* **2017**, *15*, 04718. [CrossRef]
52. Bach, P.H.; Gregg, N.J.; Delacruz, L. Relevance of a Rat Model of Papillary Necrosis and Upper Urothelial Carcinoma in Understanding the Role of Ochratoxin A in Balkan Endemic Nephropathy and Its Associated Carcinoma. *Food Chem. Toxicol.* **1992**, *30*, 205–211. [CrossRef]
53. Rašić, D.; Mladinić, M.; Želježić, D.; Pizent, A.; Stefanović, S.; Milićević, D.; Konjevoda, P.; Peraica, M. Effects of Combined Treatment with Ochratoxin A and Citrinin on Oxidative Damage in Kidneys and Liver of Rats. *Toxicon* **2018**, *146*, 99–105. [CrossRef]

54. Malaveille, C.; Brun, G.; Bartsch, H. Structure-Activity Studies in E. Coli Strains on Ochratoxin A (OTA) and Its Analogues Implicate a Genotoxic Free Radical and a Cytotoxic Thiol Derivative as Reactive Metabolites. *Mutat. Res. Mol. Mech. Mutagen.* **1994**, *307*, 141–147. [CrossRef]
55. Reiss, J. Detection of Genotoxic Properties of Mycotoxins with the SOS Chromotest. *Naturwissenschaften* **1986**, *73*, 677–678. [CrossRef]
56. Bartholomew, R.M.; Ryan, D.S. Lack of Mutagenicity of Some Phytoestrogens in the *Salmonella*/Mammalian Microsome Assay. *Mutat. Res. Toxicol.* **1980**, *78*, 317–321. [CrossRef]
57. Auffray, Y.; Boutibonnes, P. Evaluation of the Genotoxic Activity of Some Mycotoxins Using Escherichia Coli in the SOS Spot Test. *Mutat. Res. Toxicol.* **1986**, *171*, 79–82. [CrossRef]
58. Ghédira-Chékir, L.; Maaroufi, K.; Zakhama, A.; Ellouz, F.; Dhouib, S.; Creppy, E.; Bacha, H. Induction of a SOS Repair System in Lysogenic Bacteria by Zearalenone and Its Prevention by Vitamin E. *Chem. Biol. Interact.* **1998**, *113*, 15–25. [CrossRef]
59. Müller, S.; Dekant, W.; Mally, A. Fumonisin B1 and the Kidney: Modes of Action for Renal Tumor Formation by Fumonisin B1 in Rodents. *Food Chem. Toxicol.* **2012**, *50*, 3833–3846. [CrossRef] [PubMed]
60. Park, D.L.; Rua, S.M.; Mirocha, C.J.; Abd-Alla, E.-S.A.M.; Weng, C.Y. Mutagenic Potentials of Fumonisin Contaminated Corn Following Ammonia Decontamination Procedure. *Mycopathologia* **1992**, *117*, 105–108. [CrossRef] [PubMed]
61. Aranda, M. Assessment of In Vitro Mutagenicity in *Salmonella* and In Vivo Genotoxicity in Mice of the Mycotoxin Fumonisin B1. *Mutagenesis* **2000**, *15*, 469–471. [CrossRef]
62. Escobar, P.A.; Kemper, R.A.; Tarca, J.; Nicolette, J.; Kenyon, M.; Glowienke, S.; Sawant, S.G.; Christensen, J.; Johnson, T.E.; McKnight, C.; et al. Bacterial Mutagenicity Screening in the Pharmaceutical Industry. *Mutat. Res. Mutat. Res.* **2013**, *752*, 99–118. [CrossRef]
63. Diehl, M.S.; Willaby, S.L.; Snyder, R.D. Comparison of the Results of a Modified Miniscreen and the Standard Bacterial Reverse Mutation Assays. *Environ. Mol. Mutagen.* **2000**, *36*, 72–77. [CrossRef]

Article

Occurrence of Mycotoxins and Toxigenic Fungi in Cereals and Application of Yeast Volatiles for Their Biological Control

Asma Alkuwari, Zahoor Ul Hassan, Randa Zeidan, Roda Al-Thani and Samir Jaoua *

Environmental Science Program, Department of Biological and Environmental Sciences, College of Arts and Sciences, Qatar University, Doha 2713, Qatar; asma.alkuwarii@gmail.com (A.A.); zahoor@qu.edu.qa (Z.U.H.); rzedan@qu.edu.qa (R.Z.); rathani@qu.edu.qa (R.A.-T.)
* Correspondence: samirjaoua@qu.edu.qa; Tel.: +974-440-34536

Abstract: Fungal infections in cereals lead to huge economic losses in the food and agriculture industries. This study was designed to investigate the occurrence of toxigenic fungi and their mycotoxins in marketed cereals and explore the effect of the antagonistic yeast *Cyberlindnera jadinii* volatiles against key toxigenic fungal strains. *Aspergillus* spp. were the most frequent contaminating fungi in the cereals, with an isolation frequency (Fr) of 100% in maize, followed by wheat (88.23%), rice (78.57%) and oats (14.28%). Morphological and molecular identification confirmed the presence of key toxigenic fungal strains in cereal samples, including *A. carbonarius*, *A. flavus*, *A. niger*, *A. ochraceus* and *A. parasiticus*. Aflatoxins (AFs) were detected in all types of tested cereal samples, with a significantly higher level in maize compared to wheat, rice, oats and breakfast cereals. Ochratoxin A (OTA) was only detected in wheat, rice and maize samples. Levels of mycotoxins in cereals were within EU permissible limits. The volatiles of *Cyberlindnera jadinii* significantly inhibited the growth of *A. parasiticus*, *A. niger* and *P. verrucosum*. The findings of this study confirm the presence of toxigenic fungi and mycotoxins in cereals within the EU permissible limits and the significant biocontrol ability of *Cyberlindnera jadinii* against these toxigenic fungi.

Keywords: toxigenic fungi; ochratoxin A; aflatoxins; wheat; rice; maize; oats; breakfast cereals; biocontrol; food safety

Key Contribution: Mycotoxins were detected within the permissible limits in cereals marketed in Qatar. *Cyberlindnera jadinii*, an antagonistic yeast, significantly inhibits key toxigenic fungi in cereals.

Citation: Alkuwari, A.; Hassan, Z.U.; Zeidan, R.; Al-Thani, R.; Jaoua, S. Occurrence of Mycotoxins and Toxigenic Fungi in Cereals and Application of Yeast Volatiles for Their Biological Control. *Toxins* **2022**, *14*, 404. https://doi.org/10.3390/toxins14060404

Received: 19 May 2022
Accepted: 7 June 2022
Published: 13 June 2022

Publisher's Note: MDPI stays neutral with regard to jurisdictional claims in published maps and institutional affiliations.

1. Introduction

Throughout the world, dietary starch and proteins are mainly obtained from cereals [1]. Other nutritional components of cereals include fiber, non-starch carbohydrates, lipids, minerals and vitamins [2]. Because of their high nutritive values, good health effects and their availability, cereals have been an essential source of human food for millions of years [3]. In the year 2022, estimated cereal production is 2799 million tons, with a high proportion of coarse grains, wheat, maize and rice [4]. However, cereal crops are prone to several biotic and abiotic stressors, among which fungal infections are considered a major biotic factor, rendering deceased and/or low quality cereals [5]. Mold contamination of the cereals occurs at different stages of their growth, processing and preservation [6]. These infections can be categorized as pathogenic, for infections leading to plant diseases and low productivity, and toxigenic, for infections resulting in the accumulation of toxic metabolites affecting productivity. In both cases, the quality and quantity of the cereals is compromised, leading to huge annual economic losses in the agricultural sector worldwide [7].

Fungi, particularly those from the genera *Aspergillus*, *Penicillium* and *Fusarium*, might accumulate, during target infection, secondary metabolites known as mycotoxins. Among the 300 known mycotoxins, aflatoxins (AFs), synthesized by *A. flavus* and *A. parasiticus*; ochratoxin A (OTA), synthesized by *A. carbonarius*, *A. ochraceus*, *A. westerdijkiae*, and some strains of *A. niger*; and Zearalenone (ZEN), synthesized by *F. graminearum* and *F. culmorum*, are widely studied due to their significant health effects [8]. In 2004, Abdulkadar and others [9] detected AFs in 3 out of 5 basmati rice samples in the range of 0.14–0.24 µg/kg, while the other varieties of rice were free from AF contamination. The presence of OTA, DON and ZEN were also confirmed in the rice samples in their study. Similarly, wheat and corn samples were found contaminated with *Fusarium* mycotoxins (such as DON and ZEN) only. More recently, *Aspergillus*, *Penicillium* and *Fusarium* fungi and their toxins were detected in cereals intended for animal feed in the state of Qatar [10,11].

A huge range of toxic effects of mycotoxins are reported in people exposed to cereals contaminated with mycotoxins. Depending on the nature of the mycotoxin and their exposure dose, the effects may be mild gastrointestinal, growth retardation, immunosuppression, teratogenicity, mutagenicity, etc. [12]. In keeping toxicity data, world food regulatory bodies have set regulatory limits for some mycotoxins. However, in recent years, there are increasing reports of the (simultaneous) occurrence of multiple mycotoxins in the food chain [13,14]. The presence of several mycotoxins in food commodities is due to either (a) the contamination of a commodity with different fungal species, (b) the production of more than one mycotoxin by a single fungus, or (c) the mixing of cereals contaminated with different mycotoxins. Although the individual mycotoxin levels are within the permissible limits, the synergistic or additive toxic effects may pose huge risks to the exposed communities [15–17]. This developing dilemma of the co-occurrence of fungal toxins in food and the setting of regulatory limits by the food and health regulatory authorities can be understood better through adequate mycotoxin interaction data.

Conventionally toxigenic fungi are identified on the basis of their colony morphology, such as their size, shape, color, sporulation, etc. Such a method is not only time consuming but also requires expertise in the field of fungal identification [18]. Recently, species-specific PCR primers targeting ITS regions have been designed for the accurate and early identification of fungal species [19,20]. This molecular identification technique allows foe the differentiation of closely related fungal species, which is relatively difficult to achieve through morphological identification. Moreover, the amplification of cluster genes involved in mycotoxin synthesis pathways can be a helpful tool in differentiating potentially toxigenic fungal strains from non-toxigenic ones [10,21].

Although it is impossible to prevent fungal infection and mycotoxin accumulation in cereals, it is essential to minimize the levels of toxins in food. There are several chemical and physical approaches to inhibit the growth and spread of fungi in cereal crops and their products [22]. Fungicides are effective in preventing fungal growth, but the accumulation of synthetic chemicals in the food chain renders these cereals unacceptable for the consumer. Currently, consumers prefer green-labeled, minimally proceeded and high-quality products, driving the food industry towards bio-preservation, which, in fact, safeguards nutritive quality and organoleptic characteristics. Several in vitro studies reported successful application of friendly yeast and bacterial strains against toxigenic fungi [23–25]. The volatile organic compounds (VOCs) of *Lachancea thermotolerans*, identified mainly as 2-phenylethanol, significantly inhibited *A. parasiticus*, *P. verrucosum* and *F. graminearum* in artificial media and suppressed their potential to synthesize mycotoxins. Additionally, the VOCs of *Lachancea thermotolerans* inhibited the germination and spread of *F. oxysporum* spores inoculated on tomato leaf surfaces [23]. In another study [25], a yeast, *Kluyveromyces marxianus* QKM-4, isolated from a local dairy product (laban), inhibited several fungal species from the genera *Aspergillus*, *Penicillium* and *Fusarium*. Moreover, this strain was able to inhibit OTA synthesis by *A. carbonarius* and *P. verrucosum* by 98.7% and 99.6%, respectively. Microbial volatiles and diffusible organic compounds are being tested for their antifungal activities to replace the application of chemical pesticides on food crops.

This study was designed to investigate the presence of toxigenic fungi and mycotoxins in cereals marketed in Qatar. A PCR-based approach using specific primers was applied due to its suitability for identifying toxigenic fungi from cereal foods. Finally, an antagonistic yeast strain was applied for the biocontrol of toxigenic fungi. The novelty of this study consists in the validation of molecular techniques for the early and reliable identification of toxigenic mycobiota in cereals and in the potential application of *Cyberlindnera jadinii* in the protection of cereals, particularly during their storage, from the toxigenic fungal growth and synthesis of mycotoxins.

2. Results and Discussion

2.1. Fungal Contamination of Cereal Samples

Fungal contamination of cereals showed the highest contamination in maize, followed by wheat, rice and oats. All the samples of breakfast cereals were free from fungal contamination, which can be attributed to the impact of heat treatment during processing, killing the fungal spores. Another reason could also be the quality of the cereal, which may be augmented by the fact that there was no OTA detected in the breakfast cereal samples (Section 2.1). The presence of AFs in all cereal samples and the absence of fungal communities together suggest the killing of fungal spores during processing. The average contamination of the grain samples with at least one fungal colony was 7.5% (oats), 52.7% (maize), 16.1% (rice) and 23.2% (wheat). The contamination of cereal grains with mycotoxigenic fungi has been reported by several researchers [10,21,26]. In a similar, study conducted on animal feed cereals marketed in Qatar [18], there were higher percentages of grains contaminated with mycotoxigenic fungi in the animal feed samples. Relatively lower contamination was noticed in the present study, which could be associated with the good quality of the cereals for human consumption. Damaged grains, either by insects or due to other physical factors, can also lead to fungal infections and mycotoxin contamination. In the present study, all the cereals were of good quality and apparently free from any damage.

There are several reports on the prevalence of mycotoxigenic fungi in food cereals. *Aspergilli* is the most important food spoilage fungi. In Iran [27], *A. flavus* was the major producer of AFs in marketed rice and other cereals. In the Romanian cereals, maize was the most contaminated commodity, among others (wheat and barley), with *Aspergilli*, particularly *A. flavus* and *A. fumigatus* [28]. In the present study, *Aspergillus* was the most prominent fungi in wheat, rice, maize and oats (Table 1).

Table 1. Frequency (Fr) and relative density (RD) of *Aspergillus* and *Penicillium* in the cereal samples.

Item	Fungi Isolated	No. of Isolates	Isolation Frequency (%)	Relative Density (%)
Wheat (n = 17)	*Aspergillus*	25	88.23 ± 13.33 [a]	67.56 ± 11.73
	Penicillium	3	11.76 ± 5.21	8.57 ± 2.27 [b]
Rice (n = 14)	*Aspergillus*	17	78.57 ± 14.81 [a]	73.91 ± 16.20
	Penicillium	2	14.28 ± 6.67	8.69 ± 0.94 [b]
Maize (n = 13)	*Aspergillus*	31	100 ± 0.00 [a]	65.95 ± 8.59
	Penicillium	7	30.76 ± 6.53	14.89 ± 3.36 [a]
Oats (n = 14)	*Aspergillus*	5	14.28 ± 3.25 [b]	55.56 ± 7.31
	Penicillium	0	0	0
Breakfast Cereals (n = 17)	*Aspergillus*	0	0	0
	Penicillium	0	0	0

The frequency (Fr) and relative density of the isolated *Aspergillus* and *Penicillium* species were determined in cereal samples by using the formulas given in methodology sections. *Aspergillus* spp. are found in abundance in cereal samples except breakfast cereals. *Penicillium* was the least present in all cereals, with no contamination in oats and breakfast cereals. Values in columns with different superscripts are significantly different from each other at $p \leq 0.05$.

All maize samples (100%) were contaminated with *Aspergilli*, followed by wheat (88.23%), rice (78.57%) and oats (14.28%). Oats were the cereals least contaminated with fungal communities. Among all the isolated fungi, *Aspergillus* constituted 73.91% in rice, 67.65% in wheat, 65.95% in maize and 55.56% in oats. In line with the present study, in Tunisia, [29] detected the highest prevalence of *Aspergillus* spp. in maize, particularly *A. flavus* and *A. parasiticus*.

2.2. Morphological Identification of the Isolated Fungi

In the present study, fungal species were initially identified on the basis of their morphological characteristics, including colony size, shape, color (observe and reverse), spore size and shape and microscopic appearance [30]. Two fungi, *A. flavus* and *A. parasiticus*, showed closely related morphological characteristics, such as greenish-yellow, olive green or deep green conidia on CYA. Similarly, the size of the colonies on CYA was 60–70 mm, and on MEA it was 50–70 mm. The differentiation between these two species was made by microscopic examination. *A. flavus* produced conidia having different sizes and shapes, with relatively thin walls, and ranging from smooth to rough. On the other hand, *A. parasiticus* conidia were spherical with thick and rough walls. Furthermore, *A. flavus* vesicles were larger, reaching 50 μm in diameter and bearing metulae, while *A. parasiticus* vesicle diameters rarely exceeded 30 μm and metulae were not seen. These findings were in line with the key characteristics described by Pitt and Hocking [30]. In accordance with the present study, *A. flavus* from four districts in Kenya was identified in maize and soil samples using the morphological approach [31]. In 2014, Iheanacho et al. [32] also used the same morphological and molecular approaches to identify and distinguish *A. flavus* and *A. parasiticus* from the compound feed samples in South Africa. Two other species, *A. niger* and *A. carbonarius*, both producing black sporulation, were differentiated on the basis of conidia color. In *A. niger*, the conidia color was brownish black, while in the case of *A. carbonarius*, it was jet black.

2.3. Molecular Identification of Mycotoxigenic Fungi Isolated from Cereal Samples

Universal primers (ITS1/ITS4) amplified all DNA samples with a single band of 600 bp confirming that the quality of DNA was appropriate for PCR reactions [33]. Species-specific PCR primers were used for the amplification of DNA samples from the selected fungal isolates. Six isolates (coded as 3MZd, 9MRc, 9MRg) showed specific amplification using ITS1/NIG primer pair (product size 420 bp), confirming the species to be *A. niger* (Figure 1). Three isolates (3RCb, 5OTe, 4WTa) produced a single amplification band of 500 bp, using FLA1/FLA2 primer pair, confirming the species as *A. flavus*. One isolate (1WTb) was confirmed as *A. parasiticus* as it showed a single amplification band (430 bp) with the primer pair PAR1/PAR2. Two isolates (4MZd, 5 MZa) showed an amplification of a 439 bp sequence using primer pair OCRAF/OCRAR and confirming them as *A. ochraceus*. In line with the present study, *A. flavus* from wheat flour was identified by using FLA1/FLA2 primer pair with a PCR amplification product of 500 bp [34]. In another study, ITS-based primers were used for the selective identification of black *Aspergilli*. ITS1/NIG primer pair was used for *A. niger* and CAR1/CAR2 primer pair for *A. carbonarius*, and selected primer pairs allowed for the successful differentiation of two species [35]. In line with the present study, *A. flavus*, *A. parasiticus*, *A. niger* and *A. carbonarius* isolated from the animal feed samples marketed in Qatar were identified using species-specific primers [10]. These findings confirm the accuracy and suitability of PCR-based identification for toxigenic fungi.

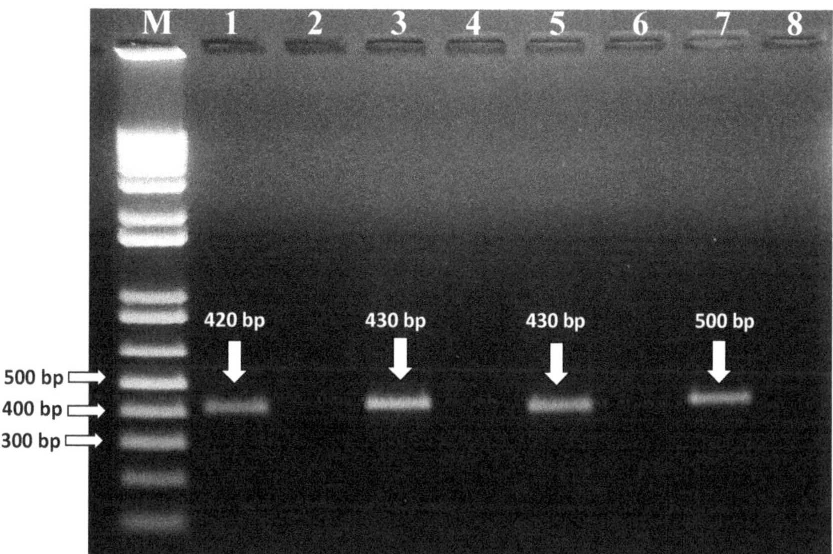

Figure 1. PCR amplification using species-specific primers for fungal isolates. Lanes: 1, primer ITS1/NIG with DNA from *A. niger* (amplicon size 420 bp); 3, primer PAR1/PAR2 with template DNA from *A. parasiticus* (amplicon size 430 bp); 5, primer OCRAF/OCRAR with DNA from *A. ochraceus* (amplicon size 430 bp); 7, primer FLA1/FLA2 and DNA from *A. flavus* (amplicon size 500). Lanes 2, 4, 6 and 8 represent the non-template control of their previous lane. Lane M, 1kb plus DNA marker with fragment sizes 12,000, 11,000, 10,000, 9000, 8000, 7000, 6000, 5000, 4000, 3000, 2000, 1000, 850, 650, 500, 400, 300, 200 and 100 bp.

2.4. Levels of AFs and OTA in Cereals

Human exposure to mycotoxins is mainly due to the ingestion of grains and grain-based products [5]. There are several reports on mycotoxin occurrence in cereal grains and legumes. The levels of mycotoxins depend on the types of cereals and climatic conditions, in addition to many other factors. In Nigeria, Makun et al. [36] detected AFs at 34.1 µg/kg and OTA at 188.2 µg/kg in rice. In the present study, among the tested cereal samples, 7 samples (41%) of wheat and 10 (71%) of rice were positive for AF contamination. However, all the samples (100%) of maize, oats and breakfast cereals were found to be contaminated with AFs (Table 2).

Table 2. Mycotoxin levels in different cereal samples.

Item	Mycotoxin Concentration (µg/kg)			
	Aflatoxins (AFs)		Ochratoxin A (OTA)	
	Range (Min–Max)	Mean ± SD	Range (Min–Max)	Mean ± SD
Wheat (n = 17)	ND *–3.14	2.81 ± 1.21 [b]	1.91–2.79	2.31 ± 0.43
Rice (n = 14)	ND–3.50	2.23 ± 1.07 [b]	1.84–2.91	2.41 ± 1.43
Maize (n = 13)	1.93–6.7	3.89 ± 1.32 [a]	1.57–2.58	2.05 ± 0.89
Oats (n = 14)	1.8–2.58	1.87 ± 0.86 [b]	0	0
Breakfast Cereals (n = 17)	2.23–3.62	2.59 ± 1.60 [b]	0	0

The values in the columns with different superscript letters are significantly different from each other at $p \leq 0.05$; * ND = not detected.

The aflatoxin contamination for different cereals was in the range of nd–3.14 µg/kg for wheat, nd–3.50 µg/kg for rice, 1.93–6.70 µg/kg for maize, 1.8–2.58 µg/kg for oats and 2.23–3.62 µg/kg for breakfast cereals. None of the cereal samples showed aflatoxin levels higher than the EU maximum limit of 4 µg/kg.

Since maize is intended to be further processed before human consumption, its set limit is 10 µg/kg [37,38]. None of the tested samples showed levels higher than this limit. The European Union maximum levels for cereals excluding maize are 4 µg/kg, while maize that is intended to be further processed before human consumption has levels of 10 µg/kg [37,38]. Although all the tested samples of maize, oats and breakfast cereal samples were positive for AF contamination, the levels were within the EU's permissible limits. This high percentage of samples showing AF contamination is predominantly related to the occurrence of aflatoxigenic fungi in the maize, oats and mixed-grain cereals (breakfast cereals). *A. flavus* and *A. parasiticus* are commonly detected in maize and oats as compared to rice and wheat [30,39]. The differentiation witnessed in the fungal ecology might be responsible for the high incidence of AFs in maize and its products. Moreover, the origin of the samples is an important factor, as in the present work, the majority of the maize and oat samples originated from South East Asia, where the occurrence of aflatoxigenic strains is reported more frequently due to favorable weather conditions and relatively poor cereals storage [30].

In the case of OTA contents, all samples (100%) of wheat, rice and maize were found contaminated with mycotoxins, while none of the oats and breakfast cereals showed the presence of OTA (Table 2). Like AFs, none of the samples among the wheat, rice and maize had OTA contents higher than the EU maximum limit of 4 µg/kg. The contamination of cereal within the EU permissible limits indicates the good quality of the cereals marketed in the Qatari market. These findings are in line with those of AbdulKardar et al. [9], who reported similar levels (1.65–1.95 µg/kg) of OTA in rice samples marketed in Qatar. However, in the same study, they did not detect OTA in any wheat, wheat flour or corn flack samples. The findings are in accordance with the incidence of ochratoxigenic fungi, which were likely less commonly found in the rice and oats samples.

2.5. Effects of Cyberlindnera Jadinii 273 Volatiles on Fungal Growth

In the presence of yeast volatile organic compounds (VOCs), the growth of *A. parasiticus* was significantly inhibited compared to the unexposed control. The effect of yeast volatiles was significant at days 3, 5 and 7 of co-incubation. The diameter of *A. parasiticus* colony exposed to *Cyberlindnera jadinii* 273 VOCs at days 3, 5 and 7 was 22.56 ± 2.51 mm, 26.63 ± 2.67 mm and 29.50 ± 4.21 mm, respectively. These values were significantly lower than the control, whereas at days 3, 5 and 7 the diameters were 32.5 ± 0.9 mm, 35.0 ± 1.0 mm and 40.0 ± 1.2 mm, respectively (Figures 2 and 3A). In line with these findings, diameters of *A. flavus* [23,25] and *A. carbonarius* [40] colonies were significantly reduced on exposure to yeast VOC. The biocontrol activity of the yeast VOCs is attributed to the release of different antifungal compounds, mainly 2-phenylethanol, which is known to inhibit the growth of fungi even at very low concentrations [41]. Another mechanism of inhibition was noted in *Pichia anomala* volatiles, in which the growth of *A. flavus* was inhibited by downregulating the expression of the genes associated with fungal vegetative growth [42].

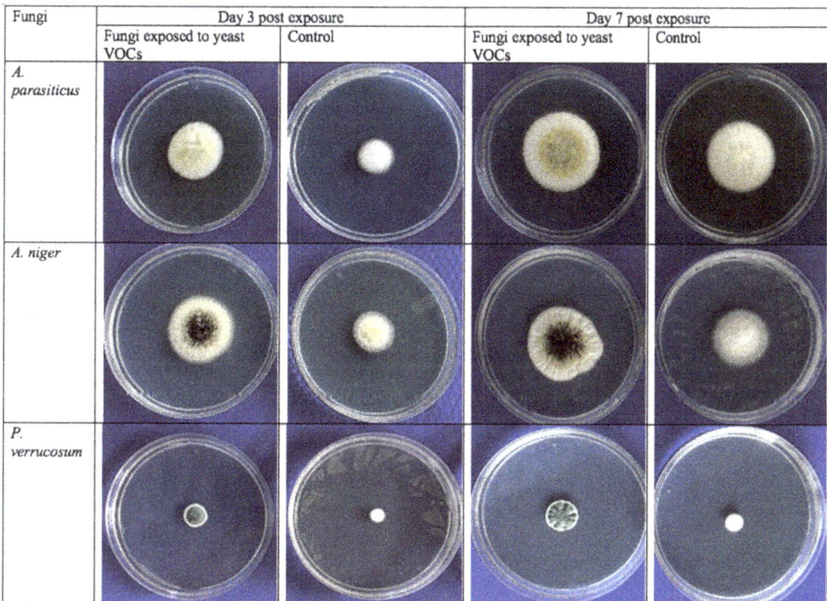

Figure 2. Effect of *Cyberlindnera jadinii* 273 VOCs on the colony sizes and sporulation of the fungal strains isolated from the cereal samples. *A. parasiticus*, *A. niger* and *P. verrucosum* were incubated in the environment of yeast volatiles for 3, 5 and 7 days. Yeast volatiles significantly inhibited fungal growth and sporulation.

Similarly, the colony size of *A. niger* was measured at days 3, 5 and 7 of exposure to yeast volatiles and was compared with the control (fungi not exposed to yeast VOCs). As shown in Figures 2 and 3B, yeast volatiles significantly inhibited the growth and sporulation of *A. niger*. At days 3, 5 and 7, the colony size of fungi exposed to yeast VOCs were 25.54 ± 4.32 mm, 28.60 ± 3.2 mm and 29.80 ± 2.5 mm compared to 34.50 ± 1.3 mm, 35.0 mm ± 1.9 and 36.0 ± 2.0 mm for the control, respectively. In compliance with the present study, *B. simplex* volatiles, composed mainly of quinoline and benzenemethanamine, significantly inhibited *A. flavus* and *A. carbonarius* in coffee beans [43]. These findings suggest potential application of *Cyberlindnera jadinii* 273 during storage of food items in commercial as well as domestic settings.

Finally, *P. verrucosum*, an important ochratoxigenic species for fruits and cereals, was exposed to *Cyberlindnera jadinii* 273 volatile in co-incubation assays. The growth of *P. verrucosum* was monitored at days 3, 5 and 7 in the treated fungi, as shown in Figures 2 and 3C. The colony diameters of the fungi were 6.5 ± 0.5 mm, 6.5 ± 0.75 mm and 7.0 ± 0.81 mm as compared to 13.0 ± 1.5 mm, 20.0 ± 2. 6 mm and 20.0 ± 2.9 mm in the untreated control fungi, respectively. This growth retardation might be associated with the downregulation of genes associated with fungal growth [41,44]. In a similar approach [23], there were significant reductions in the growth and OTA synthesis ability of *P. verrucosum* upon exposure to the VOCs of a low-fermenting yeast, *Lachancea thermotolerans*. At days 3, 5 and 7, the growth of *P. verrucosum* was inhibited at rates of 32%, 37.91% and 43.72%, respectively [23]. These activities are mainly associated with 2-phenylethanol (2-PE), which was a major component of the head-space volatiles of this yeast strain [41]. This assumption can be confirmed from the findings of Tilocca and others [44], where an artificial mixture of commercial 2-PE showed a significant inhibition of toxigenic *A. carbonarius*.

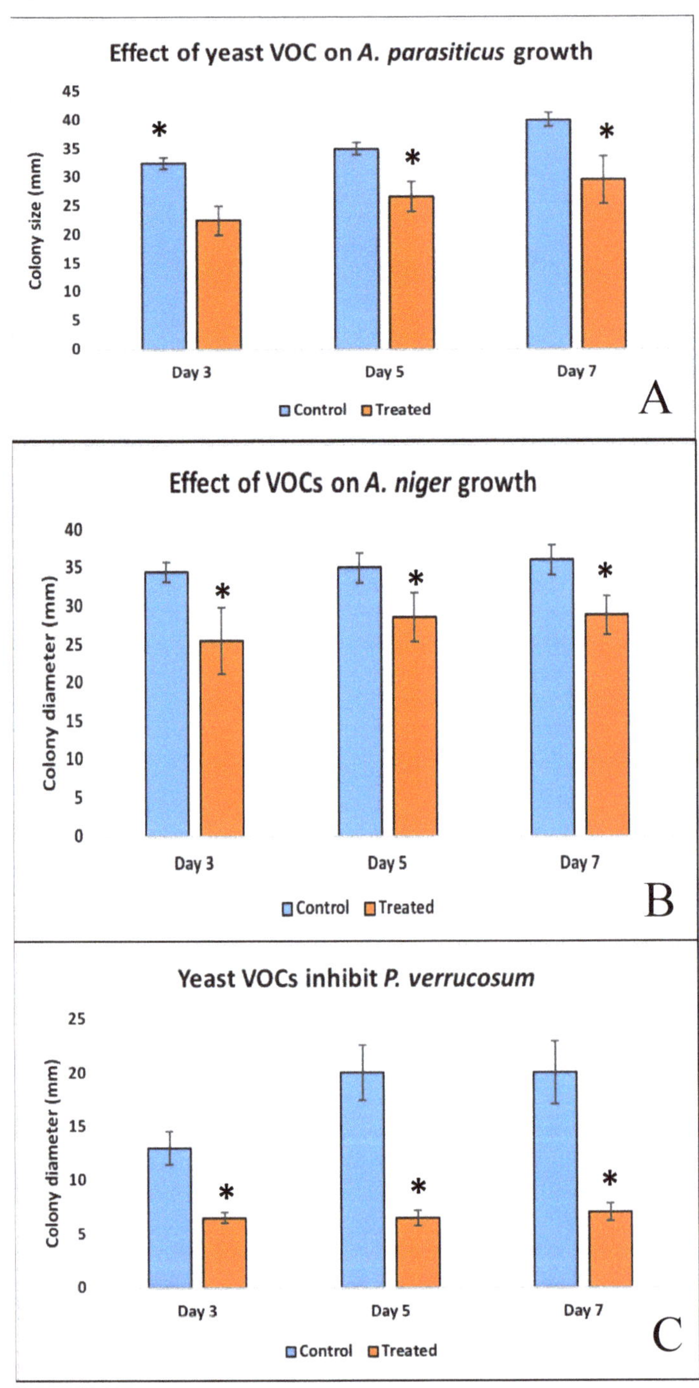

Figure 3. Effect of yeast volatiles on colony diameter of three toxigenic fungi. All three fungal strains *A. parasiticus* (**A**), *A. niger* (**B**) and *P. verricusum* (**C**) were significantly inhibited by *Cyberlindnera jadinii* 273 at all three time points i.e., 3, 5 and 7 days. (*), treated.

3. Conclusions

In this study, 75 cereal samples (including wheat, rice, maize, oats and breakfast cereals) were tested for the presence of mycotoxins and toxigenic fungi. *Aspergillus* spp. were the most frequent contaminating fungi of the cereals, with the highest isolation frequency 100% in maize samples. The occurrence of key toxigenic fungi, such as *A. flavus*, *A. parasiticus*, *A. ochraceus*, *A. carbonarius* and *A. niger,* was confirmed by their morphological characteristics and molecular profiles. All types of cereals were contaminated with AFs, with significantly higher levels in maize samples. Levels of OTA were negligible in all cereals and were non-significant in wheat, rice and maize samples. Overall, none of the toxins in any of the cereals was higher than the EU's permissible limits, suggesting no expected threat of mycotoxin exposure to humans. The volatiles of *Cyberlindnera jadinii* (an antagonistic yeast) significantly inhibited the growth and spread of the *A. parasiticus*, *A. niger* and *P. verrucosum* strains. These findings suggest the presence of toxigenic fungi and mycotoxins in marketed cereals and the biocontrol activities of *Cyberlindnera jadinii* against toxigenic fungi, which potentially can replace the application of synthetic fungicides in the agriculture and food industry.

4. Materials and Methods

4.1. Material and Supplies

Cyberlindnera jadinii 273 was kindly provided by Prof. Quirico Migheli, UNISS, Italy. RIDASCREEN® Ochratoxin A 30/15 and Aflatoxin Total ELISA kits and data reduction software R9996 RIDA® were obtained from R-Biopharm, Dermstadt, Germany. Tecan Sunrise® microplate absorbance reader (Tecan, Grödig, Austria). The fungal isolation media, potato dextrose agar (PDA), Czapek Dox Yeast Extract Agar (CYA), Malt extract agar (MEA), Dichloran Rose Bengal Chloramphenicol Agar (DRBC) and Glycerol 25% nitrate (G25N) were all purchased from Sigma, Dermstadt, Germany. All the buffer solutions were prepared in the lab.

4.2. Sampling

In total, 75 samples of cereals, including wheat grain (n = 17), rice (n = 14), maize (n = 13), oat (n = 14) and breakfast cereals (n = 17) were collected from the local markets in Doha (Qatar). Since toxigenic fungi and mycotoxins have a heterogeneous distribution in food, the contents of bags were well mixed before collecting a representative sample. Each sample of food item was collected from each different lot at three different points and pooled together to generate at least 1 kg of final sample. All the samples were aseptically transported to the research lab at Qatar University and kept at 4 °C prior to analysis.

4.3. Isolation and Morphological Identification of Fungi

After surface disinfection of the grains with 1.5% bleach, they were directly plated on DRBC agar plates. Based on the size of grains, up to 20 grains were aseptically placed on the media. In case of powdered samples, dilution plating method of Pitt and Hocking was adopted [30]. Samples were diluted in sterile distilled water and 100 μL was plated on DRBC media plates. All the plates were incubated at 28 °C for five days. In the case of grains, the fungal infection was calculated in percentage (%) by using following the formula:

$$\text{Grains infection } (\%) = \frac{No.\ of\ grains\ showing\ at\ least\ a\ single\ fungal\ sp.}{Total\ No.\ of\ grains} \times 100$$

For the ground samples (powder), the fungal infection was calculated as CFU/g of the cereal samples [10].

To calculate the isolation frequency (*Fr*) of *Aspergillus* and *Penicillium* in the cereal samples, the following equation was used:

$$Fr\ (\%)\ of\ Aspergillus\ or\ Penicillium = \frac{No.\ of\ samples\ with\ a\ specie\ of\ genus}{Total\ No.\ of\ samples} \times 100$$

However, relative density (*RD*) of genus *Aspergillus* and *Penicillium* was calculated by the following formula:

$$RD\ (\%) = \frac{No.\ of\ isolates\ of\ Aspergillus\ or\ Penicillum}{Total\ No.\ of\ Isolates} \times 100$$

All the fungal strains were purified by monosporic isolation using the method described by Balmas et al. [45]. Pure strains were transferred to identification media CYA, MEA and G25N and were incubated at 30 °C, 37 °C and 5 °C for 7 days. Colony morphological characteristics, such as size, color, extent of sporulation, and microscopic features, were all pooled to match the *Aspergillus* and *Penicillium* key [30].

4.4. Molecular Identification of Fungi Isolated from the Cereal Samples

Fungal DNA was extracted using QIAGEN Plant DNeasy kit as described by Hassan et al. [10]. Briefly, fungal spores from five-day-old pure colonies were suspended in 100 mL of potato dextrose broth (PDB) and incubated in shaking incubator (180 RPM) at 28 °C for 24 hrs. Freshly growing mycelia were obtained by filtration using Whatman No. 1 filter papers. Colonies were ground to powder in liquid nitrogen and DNA extraction protocol was followed as described in QIAGEN manual. Extracted DNA was tested for suitability in PCR reaction using ITS1 (TCC GTA GGT GAA CCT GCG G) ITS4 (TCC TCC GCT TAT TGA TAT GC) primers. For the specific identification of mycotoxigenic species, primer pair FLA1 (GTAGGGTTCCTAGCGAGCC) FLA2 (GGAAAAAGATTGATTTGCGTC) for *A. flavus*, PAR1 (GTCATGGCCGCCGGGGGCGTC), PAR2 (CCTGGAAAAAATGGTTGTTT-TGCG) for *A. parasiticus*, ITS1 (TCCGTAGGTGAACCTGCGG) NIG (CCGGAGAGAGGGG-ACGGC) for *A. niger*, CAR1 (GCATCTCTGCCCCTCGG) CAR2 (GGTTGGAGTTGTCG-GCAG) for *A. carbonarius*, OCRAF (CTTTTTCTTTTAGGGGGCACAG), OCRAR (CAAC-CTGGAAAAATAGTTGGTTG) for *A. ochraceus* and WESTF (CTTCCTTAGGGGTGGCACAG), WESTR (CAACCTGATGAAATAGATTGGTTG) for *A. westerdijkiae* were used. All the PCR mixes and conditions were adopted [10]. Amplified PCR products were analyzed by 1% agarose gel DNA electrophoresis and visualized by UV.

4.5. Monitoring of Aflatoxins (AFs) and Ochratoxin A (OTA) in Cereal Samples

Mycotoxin analysis was carried out using ELISA kit procured from R-Biopharm, Germany. Manufacturer's guidelines were followed for the extraction of mycotoxins from the cereal matrices. For the extraction of AFs, cereals samples were suspended in methanol and filtered [10]. After dilution, 50 μL of the analyte was used in ELISA wells. On the other hand, cereal samples were first acidified with 1 N HCl and then suspended in dichloromethane (DCM) for the extraction of OTA. Samples were diluted in dihydrogen carbonate buffer before loading in ELISA wells. All the absorbance values were measured at 450 nm and values of the unknown samples were calculated by generating a calibration curve based on the absorbance values of known standard solution provided with the kits. The software RIDA®SOFT Win (Art. No. Z9999) was used for these calculations.

4.6. Biocontrol of Toxigenic Fungi Using Antagonistic Yeast (Cyberlindnera jadinii 273) Volatiles

The antagonistic yeast (*Cyberlindnera jadinii* 273) was used to produce antifungal volatile organic compounds (VOCs) to inhibit the growth of selected toxigenic fungi isolated from the cereal samples. For this purpose, yeast cells were pre-cultured on yeast extract peptone dextrose broth (YPDB) for 24 hrs. A suspension of 100 μL of yeast cells was plated on YPDA plates and incubated at 28 °C for 24 hrs [40]. For co-incubation assay, the lid of yeast plates was replaced by bottom PDA plates freshly inoculated in the center with 10 µL spores of *A. flavus*, *A. niger* and *P. verrucosum*. After tight sealing to inhibit the evaporation of VOCs and allow the direct exposure of fungal spores to yeast compounds, the plates were incubated at 28 °C. Fungi in the control plates were incubated with YPDA media without inoculated yeast cells. After days 3, 5 and 7 of incubation, colony diameters of treated fungi were measured and compared to the control. Moreover, fungal growth inhibition ration (FGI %) was calculated as given below.

$$Fungal\ growth\ inhibition\ (\%) = \frac{C - T}{C} \times 100$$

where, C is the diameter (mm) of control fungi and T is the diameter of treated fungi at different time points. In total, 12 replicates were prepared for each fungus and its respective controls.

4.7. Statistical Analysis

The analysis of variance test (ANOVA) was performed using SPSS software (ver. 23). Different group means were compared either by Duncan's multiple range (DMR) test or by student t-test [46]. Colony-forming units (CFU/g), relative density (RD), isolation frequencies and fungal growth inhibition ratios were calculated by using specific equations given in their respective sections.

Author Contributions: Conceptualization, S.J., Z.U.H. and R.A.-T.; methodology, A.A., S.J., Z.U.H. and R.Z.; validation, A.A. and Z.U.H.; formal analysis, Z.U.H.; investigation, S.J.; resources, S.J.; data curation, Z.U.H.; writing—original draft preparation, A.A. and Z.U.H.; writing—review and editing, S.J. and R.A.-T.; supervision, S.J.; project administration, S.J.; funding acquisition, S.J. All authors have read and agreed to the published version of the manuscript.

Funding: This research was funded by the Qatar Foundation, grant number NPRP8-392-4-003, and Qatar University, grant number QUST-1-CAS-2018-45.

Institutional Review Board Statement: Not applicable.

Informed Consent Statement: Not applicable.

Data Availability Statement: Data can be provided by the corresponding author on email request.

Acknowledgments: We greatly acknowledge the Qatar Foundation and Qatar University for funding this study. The statements made herein are solely the responsibility of the authors.

Conflicts of Interest: The authors declare no conflict of interest.

References

1. Tacer-Caba, Z.; Nilufer-Erdil, D.; Ai, Y. Chemical Composition of Cereals and Their Products. In *Handbook of Food Chemistry*; Cheung, P.C.K., Mehta, B.M., Eds.; Springer: Berlin/Heidelberg, Germany, 2015; pp. 1–23, ISBN 978-3-642-41609-5.
2. Laskowski, W.; Górska-Warsewicz, H.; Rejman, K.; Czeczotko, M.; Zwolińska, J. How Important Are Cereals and Cereal Products in the Average Polish Diet? *Nutrients* **2019**, *11*, 679. [CrossRef] [PubMed]
3. Food and Agriculture Organization (FAO). *World Agriculture: Towards 2015/2030: Summary Report*; FAO. Rome, Italy, 2002.
4. FAO. *Crop Prospects and Food Situation*; Quarterly Global Report No. 4, December 2021; FAO: Rome, Italy, 2021. [CrossRef]
5. Oliveira, G.; da Silva, D.M.; Alvarenga Pereira, R.G.F.; Paiva, L.C.; Prado, G.; Batista, L.R. Effect of Different Roasting Levels and Particle Sizes on Ochratoxin A Concentration in Coffee Beans. *Food Control* **2013**, *34*, 651–656. [CrossRef]
6. Palma-Guerrero, J.; Chancellor, T.; Spong, J.; Canning, G.; Hammond, J.; McMillan, V.E.; Hammond-Kosack, K.E. Take-All Disease: New Insights into an Important Wheat Root Pathogen. *Trends Plant Sci.* **2021**, *26*, 836–848. [CrossRef]
7. Różewicz, M.; Wyzińska, M.; Grabiński, J. The Most Important Fungal Diseases of Cereals—Problems and Possible Solutions. *Agronomy* **2021**, *11*, 714. [CrossRef]
8. Cimbalo, A.; Alonso-Garrido, M.; Font, G.; Manyes, L. Toxicity of Mycotoxins in Vivo on Vertebrate Organisms: A Review. *Food Chem. Toxicol.* **2020**, *137*, 111161. [CrossRef]
9. Abdulkadar, A.H.W.; Al-Ali, A.A.; Al-Kildi, A.M.; Al-Jedah, J.H. Mycotoxins in Food Products Available in Qatar. *Food Control* **2004**, *15*, 543–548. [CrossRef]
10. Hassan, Z.U.; Al-Thani, R.F.; Migheli, Q.; Jaoua, S. Detection of Toxigenic Mycobiota and Mycotoxins in Cereal Feed Market. *Food Control* **2018**, *84*, 389–394. [CrossRef]

11. Hassan, Z.U.; Al Thani, R.; Balmas, V.; Migheli, Q.; Jaoua, S. Prevalence of Fusarium Fungi and Their Toxins in Marketed Feed. *Food Control* **2019**, *104*, 224–230. [CrossRef]
12. Alshannaq, A.; Yu, J.-H. Occurrence, Toxicity, and Analysis of Major Mycotoxins in Food. *Int. J. Environ. Res. Public Health* **2017**, *14*, 632. [CrossRef]
13. Ul Hassan, Z.; Al Thani, R.; Atia, F.A.; Al Meer, S.; Migheli, Q.; Jaoua, S. Co-Occurrence of Mycotoxins in Commercial Formula Milk and Cereal-Based Baby Food on the Qatar Market. *Food Addit. Contam. Part B* **2018**, *11*, 191–197. [CrossRef]
14. Neme, K.; Mohammed, A. Mycotoxin Occurrence in Grains and the Role of Postharvest Management as a Mitigation Strategies. A Review. *Food Control* **2017**, *78*, 412–425. [CrossRef]
15. Ul-Hassan, Z.; Zargham Khan, M.; Khan, A.; Javed, I. Immunological Status of the Progeny of Breeder Hens Kept on Ochratoxin A (OTA)- and Aflatoxin B_1 (AFB_1)-Contaminated Feeds. *J. Immunotoxicol.* **2012**, *9*, 381–391. [CrossRef] [PubMed]
16. Kifer, D.; Jakšić, D.; Šegvić Klarić, M. Assessing the Effect of Mycotoxin Combinations: Which Mathematical Model Is (the Most) Appropriate? *Toxins* **2020**, *12*, 153. [CrossRef] [PubMed]
17. Smith, M.-C.; Madec, S.; Coton, E.; Hymery, N. Natural Co-Occurrence of Mycotoxins in Foods and Feeds and Their in vitro Combined Toxicological Effects. *Toxins* **2016**, *8*, 94. [CrossRef]
18. Pitt, J.I.; Hocking, A.D. *Fungi and Food Spoilage*, 3rd ed.; Springer: New York, NY, USA, 2009. [CrossRef]
19. Susca, A.; Villani, A.; Moretti, A.; Stea, G.; Logrieco, A. Identification of Toxigenic Fungal Species Associated with Maize Ear Rot: Calmodulin as Single Informative Gene. *Int. J. Food Microbiol.* **2020**, *319*, 108491. [CrossRef]
20. Krulj, J.; Ćurčić, N.; Stančić, A.B.; Kojić, J.; Pezo, L.; Tukuljac, L.P.; Solarov, M.B. Molecular Identification and Characterisation of *Aspergillus flavus* Isolates Originating from Serbian Wheat Grains. *Acta Aliment.* **2020**, *49*, 382–389. [CrossRef]
21. Yu, J.J.; Whitelaw, C.A.; Nierman, W.C.; Bhatnagar, D.; Cleveland, T.E. *Aspergillus flavus* expressed sequence tags for identification of genes with putative roles in aflatoxin contamination of crops. *FEMS Microbiol. Lett.* **2004**, *237*, 333–340. [CrossRef]
22. Sipos, P.; Peles, F.; Brassó, D.L.; Béri, B.; Pusztahelyi, T.; Pócsi, I.; Győri, Z. Physical and Chemical Methods for Reduction in Aflatoxin Content of Feed and Food. *Toxins* **2021**, *13*, 204. [CrossRef]
23. Zeidan, R.; Ul-Hassan, Z.; Al-Thani, R.; Balmas, V.; Jaoua, S. Application of Low-Fermenting Yeast *Lachancea thermotolerans* for the Control of Toxigenic Fungi *Aspergillus parasiticus*, *Penicillium verrucosum* and *Fusarium graminearum* and Their Mycotoxins. *Toxins* **2018**, *10*, 242. [CrossRef]
24. Ul Hassan, Z.; Al Thani, R.; Alnaimi, H.; Migheli, Q.; Jaoua, S. Investigation and Application of *Bacillus licheniformis* Volatile Compounds for the Biological Control of Toxigenic *Aspergillus* and *Penicillium* spp. *ACS Omega* **2019**, *4*, 17186–17193. [CrossRef]
25. Alasmar, R.; Ul-Hassan, Z.; Zeidan, R.; Al-Thani, R.; Al-Shamary, N.; Alnaimi, H.; Migheli, Q.; Jaoua, S. Isolation of a Novel *Kluyveromyces Marxianus* Strain QKM-4 and Evidence of Its Volatilome Production and Binding Potentialities in the Biocontrol of Toxigenic Fungi and Their Mycotoxins. *ACS Omega* **2020**, *5*, 17637–17645. [CrossRef] [PubMed]
26. Wilson, J.P.; Jurjevic, Z.; Hannah, W.; Wilson, D.M.; Potter, T.L.; Coy, A.E. Host-specific variation in infection by toxigenic fungi and contamination by mycotoxins in pearl millet and corn. *Mycopathologia* **2006**, *161*, 101–107. [CrossRef] [PubMed]
27. Ranjbar, R.; Roayaei Ardakani, M.; Mehrabi Kushki, M.; Kazeminezhad, I. Identification of Toxigenic Aspergillus Species from Rice of Khuzestan and Mycotoxins in Imported Cereals. *Iran J. Med. Microbiol.* **2019**, *13*, 355–373. [CrossRef]
28. Tabuc, C.; Marin, D.; Guerre, P.; Sesan, T.; Bailly, J.D. Molds and Mycotoxin Content of Cereals in Southeastern Romania. *J. Food Prot.* **2009**, *72*, 662–665. [CrossRef]
29. Jedidi, I.; Cruz, A.; Gonzalez-Jaen, M.T.; Said, S. Aflatoxins and ochratoxin a and their aspergillus causal species in Tunisian cereals. *Food Addit. Contam. Part B* **2017**, *10*, 51–58. [CrossRef]
30. Pitt, J.I.; Hocking, A.D. *Fungi and Food Spoilage*, 2nd ed.; Blackie Academic and Professional: London, UK, 1997.
31. Thathana, M.; Murage, H.; Abia, A.; Pillay, M. Morphological Characterization and Determination of Aflatoxin-Production Potentials of Aspergillus Flavus Isolated from Maize and Soil in Kenya. *Agriculture* **2017**, *7*, 80. [CrossRef]
32. Iheanacho, H.E.; Njobeh, P.B.; Dutton, F.M.; Steenkamp, P.A.; Steenkamp, L.; Mthombeni, J.Q.; Daru, B.H.; Makun, A.H. Morphological and Molecular Identification of Filamentous *Aspergillus flavus* and *Aspergillus parasiticus* Isolated from Compound Feeds in South Africa. *Food Microbiol.* **2014**, *44*, 180–184. [CrossRef]
33. White, T.J.; Bruns, T.; Lee, S.; Taylor, J.W. Amplification and direct sequencing of fungal ribosomal RNA genes for phylogenetics. In *PCR Protocols: A Guide to Methods and Applications*; Innis, M.A., Gelfand, D.H., Sninsky, J.J., White, T.J., Eds.; Academic Press Inc.: New York, NY, USA, 1990; pp. 315–322.
34. González-Salgado, A.; Patiño, B.; Vázquez, C.; González-Jaén, M.T. Discrimination of *Aspergillus niger* and other *Aspergillus* species belonging to section *Nigri* by PCR assays. *FEMS Microbiol. Lett.* **2005**, *245*, 353–361. [CrossRef]
35. Oliveri, C.; Torta, L.; Catara, V. A *polyphasic* approach to the identification of ochratoxin A-producing black *Aspergillus* isolates from vineyards in Sicily. *Int. J. Food Microbiol.* **2008**, *127*, 147–154. [CrossRef]
36. Makun, H.A.; Dutton, M.F.; Njobeh, P.B.; Mwanza, M.; Kabiru, A.Y. Natural multi-occurrence of mycotoxins in rice from Niger State, Nigeria. *Mycotoxin Res.* **2011**, *27*, 97–104. [CrossRef]
37. European Commission. Commission Recommandation of 17 August 2006 on the Presence of Deoxynivalenol, Zearalenone, Ochratoxin A, T-2 and HT-2 and Fumonisins in Products Intended for Animal Feeding. 2006. Available online: http://eur-lex.europa.eu/legal-content/EN/TXT/PDF/?uri1/4CELEX:32006H0576&from1/4EN (accessed on 18 May 2022).
38. European Union. Commission Regulation (EU) No 1058/2012 of 12 November 2012. Amending Regulation (EC) No 1881/2006 as regards maximum levels for aflatoxins in dried figs. *Off. J. Eur. Union* **2010**, *L313*, 14–15.

39. Giorni, P.; Magan, N.; Pietri, A.; Bertuzzi, T.; Battilani, P. Studies on Aspergillus section Flavi isolated from maize in northern Italy. *Int. J. Food Microbiol.* **2007**, *113*, 330–338. [CrossRef] [PubMed]
40. Fiori, S.; Urgeghe, P.P.; Hammami, W.; Razzu, S.; Jaoua, S.; Migheli, Q. Biocontrol Activity of Four Non- and Low-Fermenting Yeast Strains against Aspergillus Carbonarius and Their Ability to Remove Ochratoxin A from Grape Juice. *Int. J. Food Microbiol.* **2014**, *189*, 45–50. [CrossRef] [PubMed]
41. Farbo, M.G.; Urgeghe, P.P.; Fiori, S.; Marcello, A.; Oggiano, S.; Balmas, V.; Hassan, Z.U.; Jaoua, S.; Migheli, Q. Effect of Yeast Volatile Organic Compounds on Ochratoxin A-Producing *Aspergillus carbonarius* and *A. Ochraceus. Int. J. Food Microbiol.* **2018**, *284*, 1–10. [CrossRef]
42. Hua, S.S.T.; Beck, J.J.; Sarreal, S.B.L.; Gee, W. The major volatile compound 2-phenylethanol from the biocontrol yeast, *Pichia anomala*, inhibits growth and expression of aflatoxin biosynthetic genes of *Aspergillus flavus. Mycotoxin Res.* **2014**, *30*, 71–78. [CrossRef]
43. Al Attiya, W.; Hassan, Z.U.; Al-Thani, R.; Jaoua, S. Prevalence of Toxigenic Fungi and Mycotoxins in Arabic Coffee (*Coffea Arabica*): Protective Role of Traditional Coffee Roasting, Brewing and Bacterial Volatiles. *PLoS ONE* **2021**, *16*, e0259302. [CrossRef]
44. Tilocca, B.; Balmas, V.; Hassan, Z.U.; Jaoua, S.; Migheli, Q. A Proteomic Investigation of *Aspergillus carbonarius* Exposed to Yeast Volatilome or to Its Major Component 2-Phenylethanol Reveals Major Shifts in Fungal Metabolism. *Int. J. Food Microbiol.* **2019**, *306*, 108265. [CrossRef]
45. Balmas, V.; Migheli, Q.; Scherm, B.; Garau, P.; O'Donnell, K.; Ceccherelli, G.; Kang, S.; Geiser, D.M. Multilocus phylogenetics show high levels of endemic fusaria inhibiting Sardinian soils (Tyrrhenian island). *Mycologia* **2010**, *4*, 803–812. [CrossRef]
46. Steel, R.G.D.; Torrie, J.H.; Dicky, D.A. *Principles and Procedures of Statistics, A Biometrical Approach*, 3rd ed.; McGraw Hill, Inc.: New York, NY, USA; Book Co.: New York, NY, USA, 1997; pp. 352–358.

Article

Exploration of Mycotoxin Accumulation and Transcriptomes of Different Wheat Cultivars during *Fusarium graminearum* Infection

Kailin Li [1], Dianzhen Yu [1], Zheng Yan [1], Na Liu [1], Yingying Fan [2], Cheng Wang [2] and Aibo Wu [1,*]

[1] SIBS-UGENT-SJTU Joint Laboratory of Mycotoxin Research, CAS Key Laboratory of Nutrition, Metabolism and Food Safety, Shanghai Institute of Nutrition and Health, University of Chinese Academy of Sciences, Chinese Academy of Sciences, Shanghai 200031, China; likailin2018@sibs.ac.cn (K.L.); dzyu@sibs.ac.cn (D.Y.); zyan@sibs.ac.cn (Z.Y.); liuna@sibs.ac.cn (N.L.)

[2] Institute of Quality Standards & Testing Technology for Agro-Products, Xinjiang Academy of Agricultural Sciences, Key Laboratory of Agro-Products Quality and Safety of Xinjiang, Laboratory of Quality and Safety Risk Assessment for Agro-Products (Urumqi), Ministry of Agriculture and Rural Affairs, Urumqi 830091, China; fyyxaas@sina.com (Y.F.); wangchengxj321@sina.com (C.W.)

* Correspondence: abwu@sibs.ac.cn; Tel.: +86-21-54920716

Citation: Li, K.; Yu, D.; Yan, Z.; Liu, N.; Fan, Y.; Wang, C.; Wu, A. Exploration of Mycotoxin Accumulation and Transcriptomes of Different Wheat Cultivars during *Fusarium graminearum* Infection. *Toxins* **2022**, *14*, 482. https://doi.org/10.3390/toxins14070482

Received: 8 June 2022
Accepted: 12 July 2022
Published: 13 July 2022

Publisher's Note: MDPI stays neutral with regard to jurisdictional claims in published maps and institutional affiliations.

Abstract: *Fusarium graminearum* is one of the most devastating diseases of wheat worldwide, and can cause Fusarium head blight (FHB). *F. graminearum* infection and mycotoxin production mainly present in wheat and can be influenced by environmental factors and wheat cultivars. The objectives of this study were to examine the effect of wheat cultivars and interacting conditions of temperature and water activity (a_w) on mycotoxin production by two strains of *F. graminearum* and investigate the response mechanisms of different wheat cultivars to *F. graminearum* infection. In this regard, six cultivars of wheat spikes under field conditions and three cultivars of post-harvest wheat grains under three different temperature conditions combined with five water activity (a_w) conditions were used for *F. graminearum* infection in our studies. Liquid chromatography tandem mass spectrometry (LC–MS/MS) analysis showed significant differences in the concentration of Fusarium mycotoxins deoxynivalenol (DON) and its derivative deoxynivalenol-3-glucoside (D3G) resulting from wheat cultivars and environmental factors. Transcriptome profiles of wheat infected with *F. graminearum* revealed the lower expression of disease defense-factor-related genes, such as mitogen-activated protein kinases (MAPK)-encoding genes and hypersensitivity response (HR)-related genes of infected Annong 0711 grains compared with infected Sumai 3 grains. These findings demonstrated the optimal temperature and air humidity resulting in mycotoxin accumulation, which will be beneficial in determining the conditions of the relative level of risk of contamination with FHB and mycotoxins. More importantly, our transcriptome profiling illustrated differences at the molecular level between wheat cultivars with different FHB resistances, which will lay the foundation for further research on mycotoxin biosynthesis of *F. graminearum* and regulatory mechanisms of wheat to *F. graminearum*.

Keywords: wheat; *Fusarium graminearum*; mycotoxin; water activity; temperature; transcriptome

Key Contribution: Mycotoxin production by *F. graminearum* in different FHB-resistant wheat spikes under field conditions and in post-harvest wheat grains under different laboratory conditions and transcriptomes of different FHB-resistant wheat cultivars during *F. graminearum* infection were investigated, which provides references for mycotoxin control and mechanism research governing the response of wheat to *F. graminearum*.

1. Introduction

Wheat is an essential food source for humans. Plant diseases and insect pests such as head blight, rusts, powdery mildew, leaf blotch, and wheat curl mite negatively affect the quality and yield of wheat [1,2]. Fusarium head blight (FHB) is a devastating disease that occurs widely in wheat crops in humid and semihumid regions of the world [3]. The average yearly occurrence of FHB has caused severe yield losses [4,5]. During recent decades, many efforts have been deployed to dissect FHB resistance, investigating both the wheat responses to infection and the fungal determinants of pathogenicity [6]. From this, different cultivars of wheat with FHB resistance have been widely conducted as research objects [7–9].

FHB can be caused by a variety of *Fusarium graminearum* species complexes (FGSC), and among them, *F. graminearum* is the most prevalent and aggressive pathogen of FHB in wheat [5,10,11]. During infection of wheat, *F. graminearum* can synthesize a large amount of deoxynivalenol (DON) and its derivates, and different *F. graminearum* strains show differences in the capacity of infection and toxin biosynthesis [12–15]. DON can cause acute physiological effects in humans and animals including vomiting, diarrhea, intestinal inflammation, and gastrointestinal hemorrhage [16]. In some cases, DON can be degraded into masked forms by phase I metabolism or phase II metabolism [17]. Owing to the low toxicity, deoxynivalenol-3-glucoside (D3G) is generally regarded as a detoxification product of DON in plants, and its production is usually related to wheat resistance [18–21]. Biotransformation of DON in Fusarium-resistant and -susceptible wheat lines shows differences [18,22]. More importantly, it has been reported that D3G can be converted into DON in some food-processing processes, such as dough extrusion, fermentation, and steaming [23,24]. Studies have also shown that some microbiotas in intestines of animals and even humans can rapidly hydrolyze D3G into DON, which provides reasons for much more attention on D3G [25–31].

In previous studies, the environmental effects on fungal growth and potential mycotoxin contamination were demonstrated by inoculating *F. graminearum*, *F. verticillioides*, *F. langsethiae*, and *F. meridionale* in different cereal matrixes such as maize, oat, and soybean [15,32–34]. It has been reported that biosynthesis of DON and D3G is usually affected by environmental temperature, humidity, and hosts [22,34–37]. However, studies on the effect of *F. graminearum* strains and abiotic factors on mycotoxin production and response mechanisms in wheat-based matrixes are still not comprehensive.

In order to demonstrate that some plant functions and the expression of specific genes are needed to promote FHB, an increasing list of effectors, genes, and mechanisms in the development of FHB have been found using omics approach [20,38]. In particular, the increasing application of transcriptomes has successfully helped researchers map the regulatory responses, which provides an efficient tool for mechanism investigation [39–41]. In our study, six cultivars of wheat spikes and three cultivars of post-harvest wheat grains were used for two strains of *F. graminearum* infection to demonstrate differences between wheat cultivars. Further, three different temperature conditions combined with five water activity (a_w) conditions were applied to investigate the interacting effect of wheat cultivars and environmental factors on mycotoxin production. Furthermore, we analyzed the transcriptomic profiles of wheat grains infected with *F. graminearum* F1. Using mycotoxin production analysis combined with transcriptomic analysis, we revealed the differences in toxin concentration and gene expression caused by different *F. graminearum* strains, environmental factors, and wheat cultivars. The results of our study may provide a reference for wheat breeding and wheat storage to reduce the FHB incidence and mycotoxin accumulation in wheat and wheat products, preventing harm to humans. Furthermore, our transcriptome profiling will lay the foundation for further research on mycotoxin biosynthesis and regulatory mechanisms of wheat to *F. graminearum*.

2. Results

2.1. Evaluation of Toxin Accumulation of Six Wheat Cultivars under Field Conditions

Field experiments showed that there were significant differences in resistance to FHB among the wheat cultivars. Both *F. graminearum* PH−1 and *F. graminearum* F1 produced spikelets with blight symptoms on these wheat cultivars (Figure S1). The average symptomatic spikelet numbers of Sumai 3 and Wangshuibai were significantly lower than those of other cultivars of wheat (Figure 1A and Figure S1). In these *F. graminearum* PH$-$1-infected wheat groups, Sumai 3 was the most resistant, with the lowest symptomatic spikelet rate of 5.21%, while the highest symptomatic spikelet rate was 98.00% of ZK001 (Figure 1A). In the *F. graminearum* F1$-$infected groups, Wangshuibai was the most resistant cultivar with the lowest symptomatic spikelet rate of 1.54%, while the highest symptomatic spikelet rates were in Nanda 2419 and ZK001, which were up to 90.0% (Figure 1A). Then, wheat spikelets were collected and subjected to mycotoxin determination. The results showed that there were differences in the accumulation of DON between the six varieties of wheat spikelets (Figure 1B). Among them, the concentration of DON produced by *F. graminearum* PH$-$1 was highest in ZK001 and Aikang 58 followed by Zhongmai 66B (Figure 1B). When inoculated with *F. graminearum* F1, the concentration of DON was highest in Zhongmai 66B followed by ZK001, with values of 8211 µg/kg and 5218 µg/kg, respectively (Figure 1B). The content of DON was lowest in Sumai 3 and Wangshuibai and had no significant difference between these two cultivars when they were infected by *F. graminearum* PH$-$1 and F1 (Figure 1B). However, compared with DON, the accumulation of D3G was significantly lower (Figure 1C). Furthermore, D3G content was highest in ZK001 and lowest in Nanda 2491 when spikes were infected with *F. graminearum* PH$-$1 and was highest in Zhongmai 66B and lowest in Aikang 58 and Nanda 2491 when spikes were infected by *F. graminearum* F1 (Figure 1C). By using Nonlinfit analysis, we found that there was a certain negative correlation showing an exponential model change between the ratio of D3G and total DON with symptomatic spikelet rate (Figure 1D). Obviously, the strength of the correlation between D3G/total DON and the symptomatic spikelet rate varies among *F. graminearum* strains (Figure 1D). The R^2 value was as high as 0.9405 between D3G/total DON and the symptomatic spikelet rate when they were infected by *F. graminearum* PH$-$1 and 0.8794 between D3G/total DON and the symptomatic spikelet rate when they were infected by *F. graminearum* F1 (Figure 1D). The ratios of D3G/total DON of the FHB-resistant cultivars Sumai 3 and Wangshuibai were higher than other cultivars (Figure 1D).

2.2. Accumulation of DON and D3G in F. graminearum PH$-$1-Infected Wheat Grains under Different a_w and Temperature Conditions

To explore the influence of abiotic factors (temperature and a_w) on the accumulations of toxins in wheat grains, we selected Sumai 3 (highly resistant), Annong 0711 (moderately resistant), and Zhongmai 66B (susceptive) for *F. graminearum* PH$-$1 infection. When the infection time reached one week, DON and D3G had significantly accumulated in the matrixes (Figure 2A,B). Toxin accumulation was most significant at a_w 0.99, but varied with temperature (Figure 2A). The production trends of DON and D3G were consistent at each temperature, which were higher in susceptive cultivars than resistant cultivars (Figure 2A,B). The concentration of DON reached a maximum at 25 °C (Figure 2A). For D3G, the maximum content was at 25 °C followed by 20 °C (Figure 2B). At 25 °C, the accumulation of DON was approximately 10 times higher in Annong 0711 grains and 5 times higher in Zhongmai 66B grains than in Sumai 3 grains, and the content of D3G was much lower in proportion, reflecting the difference in fungal resistance of wheat cultivars (Figure 2A,B). However, when the a_w was below 0.99, the accumulation of DON and D3G was not obviously detected in any of the three cultivars, which indicated the importance of a_w to the accumulation of DONs. Furthermore, the production of DONs at different temperatures did not show significant differences between the three cultivars of wheat grains after infection with *F. graminearum* PH$-$1 for 7 days (Figure 2A,B). In Figure 2C, the ratio of D3G and total DON did not show a significant difference between these groups.

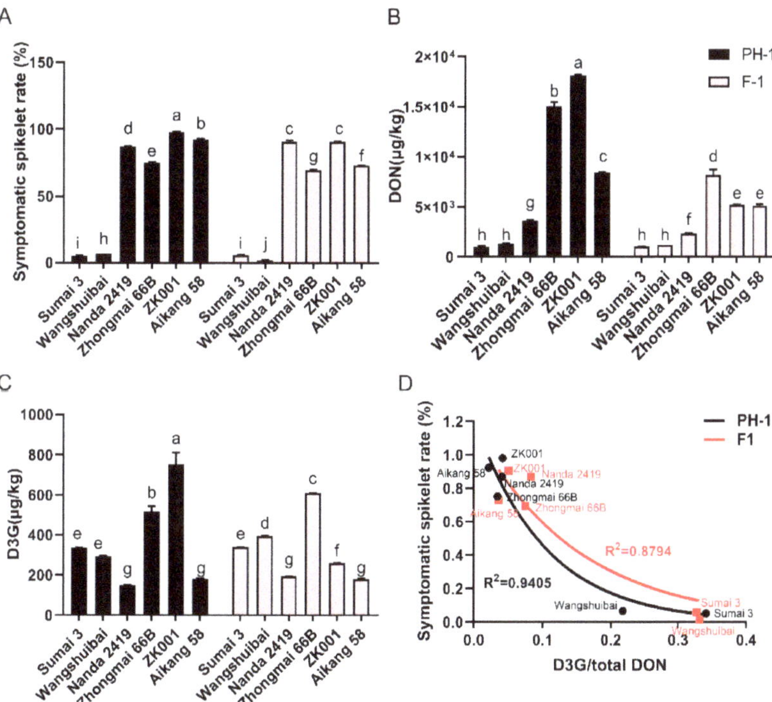

Figure 1. Differences in toxin production levels and symptomatic spikelet numbers between different wheat cultivars. (**A**) Symptomatic spikelet rate of 6 different cultivars of wheat spikes after inoculation with *Fusarium graminearum* PH−1 and *F. graminearum F1*. (**B**) Deoxynivalenol (DON) concentration in 6 varieties of wheat spikes infected with *F. graminearum* PH−1 and *F. graminearum* F1. (**C**) Deoxynivalenol-3-glucoside (D3G) concentration in 6 varieties of wheat spikes infected with *F. graminearum* PH−1 and *F. graminearum* F1. (**D**) Correspondence between symptomatic spikelet rates and D3G/total DON of different wheat cultivars infected with *F. graminearum* PH−1 and F1. Bars with different letters represent significant differences ($p < 0.05$) according to two-way ANOVA; the correspondence was analyzed by Nonlin fit.

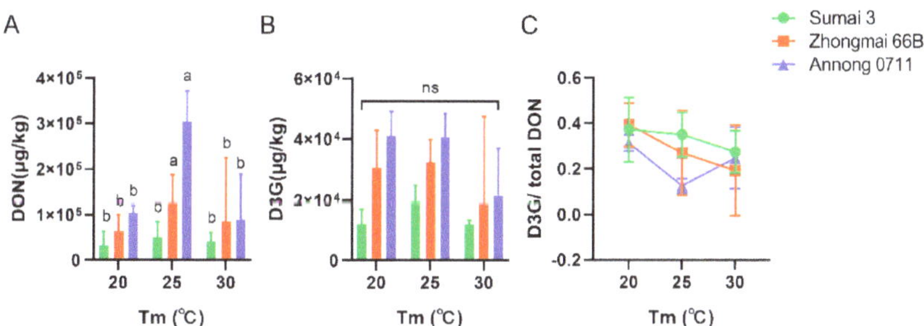

Figure 2. DON (**A**) and D3G (**B**) produced by *F. graminearum* PH−1 and the ratios of D3G/total DON (**C**) of Sumai 3, Zhongmai 66B, and Annong 0711 groups at 20, 25, and 30 °C with water activity (a_w) of 0.99. Bars with different letters represent significant differences ($p < 0.05$) according to two-way ANOVA. The ratio of D3G/total DON was presented as the mean ± SD and analyzed by two-way ANOVA.

2.3. Accumulation of DON and D3G in F. graminearum F1-Infected Wheat Grains under Different a_w and Temperature Conditions

Since wheat contamination in nature is not limited to a single *F. graminearum* strain, we also used *F. graminearum* F1 to infect these cultivars of wheat grains to better illustrate the difference between different FHB-resistance wheat cultivars under different a_w and temperature conditions. After infection with *F. graminearum* F1 for 7 days, DON and D3G were not obviously detected in Sumai 3 and Zhongmai 66B wheat grains. However, the accumulation of toxins in Annong 0711 was very significant but was lower than that when grains were infected by *F. graminearum* PH−1 (Figure S2).

To clearly measure DON and D3G content differences among the three wheat cultivars, we extended infection time to 14 days. The accumulation of DON and D3G in all three cultivars was much lower under a_w below 0.99 than above; among these cultivars, DON was highest in Annong 0711 wheat grains but not more than 1500 μg/kg (Table S1). Under the condition of a_w 0.99, DON in Sumai 3 was highest at 25 °C, followed by 20 °C, while in Zhongmai 66B and Annong 0711, DON was highest at 25 °C, followed by 30 °C (Figure 3A). For D3G, the accumulation in Sumai 3 and Zhongmai 66B reached the highest level at 20 °C, followed by 25 °C. The content of D3G in Annong 0711 was highest at 30 °C, followed by 20 °C (Figure 3B).

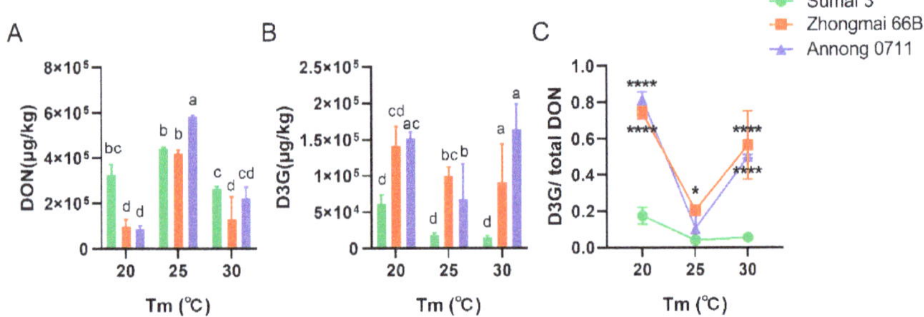

Figure 3. DON (**A**) and D3G (**B**) produced by *F. graminearum* F1 and the ratios of D3G/total DON (**C**) of Sumai 3, Zhongmai 66B, and Annong 0711 groups at 20, 25, and 30 °C with an a_w of 0.99. Bars with different letters represent significant differences ($p < 0.05$) according to two-way ANOVA. The ratio of D3G/total DON is presented as the mean ± SD and was analyzed by two-way ANOVA. * $p < 0.05$, **** $p < 0.0001$: Zhongmai 66B group and Annong 0711 group, respectively, compared with the Sumai 3 group at the three temperatures.

The concentration of DON in Zhongmai 66B showed significant differences between 20 °C and 25 °C, 25 °C and 30 °C, and the DON contents in Sumai 3 and Annong 0711 were significantly different under the three temperature conditions (Figure 3A). For D3G, the contents in Sumai 3 showed no differences at the three temperature conditions, but in Zhongmai 66B and Annong 0711 the contents showed a significant difference under the three temperature conditions (Figure 3B).

At 20 °C, the DON and D3G contents showed significant differences between Sumai 3 and Zhongmai 66B, Sumai 3 and Annong 0711, and the D3G content significantly differed in all three cultivars (Figure 3A,B). At 25 °C, the DON content showed a significant difference between Sumai 3 and Annong 0711, Zhongmai 66B and Annong 0711, and the D3G content showed a significant difference between the three cultivars (Figure 3A,B). At 30 °C, the DON content showed a significant difference among the three groups, and the D3G content showed a significant difference between Sumai 3 and Zhongmai 66B, Sumai 3 and Annong 0711(Figure 3A,B). Compared with *F. graminearum* F1-infected Sumai 3, the ratios of D3G and total DON of *F. graminearum* F1-infected Zhongmai 66B and Annong 0711 groups were significantly different at 20 °C and 30 °C (Figure 3C). At 25 °C, only the ratios of Zhongmai 66B and Sumai 3 were statistically different (Figure 3C).

2.4. Overview of Differentially Expressed Genes (DEG) between F. graminearum F1-Infected Sumai 3 and Annong 0711 Wheat Grains

To understand the molecular mechanisms underlying the phenotypic differences between different wheat cultivars, we sequenced the interaction transcriptome of wheat grains and *F. graminearum* F1 under the infection condition of 20 °C and a_w 0.99 (GEO: GSE188959). Based on the sequencing data, the rates of total mapped clean reads of Sumai 3 and Annong 0711 averaged 64.72% and 66.57%, respectively, which indicated similar infection degrees of the two groups, while the low mapping rate of Zhongmai 66B wheat reads indicated that susceptible varieties were very weak in resistance to *F. graminearum* infection (Figure 4A). Then, Venn analysis was performed for the indicated two groups (Figure 4B). The transcriptome profile of wheat grains showed a significant difference between the infected Sumai 3 and Annong 0711. The results obtained from a differential expression analysis of *F. graminearum* F1-infected Annong 0711 compared with F1-infected Sumai 3 wheat grains showed that 1583 out of 4131 genes were upregulated, while 2548 genes were downregulated, and these genes were annotated based on the clusters orthologous groups (COG) database (Figure 4C).

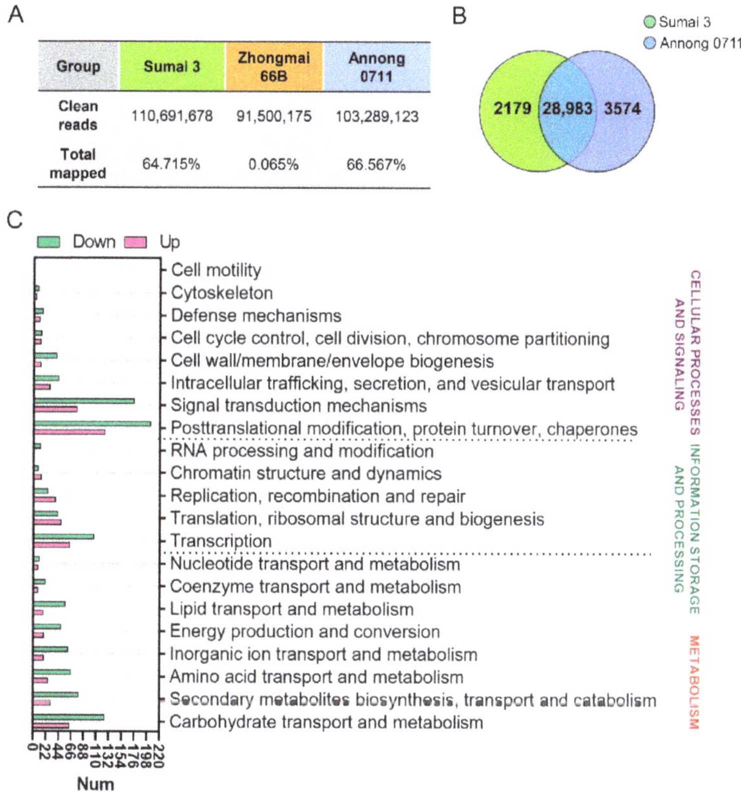

Figure 4. RNA sequencing analysis of Sumai 3 and Annong 0711 wheat gene responses to *F. graminearum* F1. (**A**) Ratio of mapped RNA sequencing reads of the indicated wheat to all sequencing reads after inoculation with *F. graminearum* F1. (**B**) Venn diagram illustration of the genes between the *F. graminearum* F1-treated Sumai 3 group and Annong 0711 group. (**C**) Clusters orthologous groups (COG) annotation of differentially expressed genes (DEGs) between the *F. graminearum* F1-infected Sumai 3 group and Annong 0711 group.

2.5. Gene Ontology (GO) and Kyoto Encyclopedia of Genes and Genomes (KEGG) Enrichment Analyses of DEGs in Wheat

The GO analysis placed the DEGs into three categories based on their functions: biological process (BP), cellular component (CC), and molecular function (MF). Upregulated DEGs were significantly enriched in eight GO terms; among these terms, functions associated with CC followed by MF accounted for a large proportion (Table S2). KEGG pathway maps included seven categories: metabolism, genetic information processing, environmental information processing, cellular processes, organismal systems, human diseases, and drug development. Upregulated genes were significantly enriched in "glutathione metabolism" and "ribosome biogenesis in eukaryotes" (Table S3). For downregulated genes, the top 20 significantly enriched terms mainly belonged to BP and MF, among which the gene number in "defense response to fungi" was the largest (Figure 5A). According to the KEGG analysis, the top 20 maps of downregulated genes were distributed in metabolism, environmental information processing, and organismal systems categories (Figure 5B).

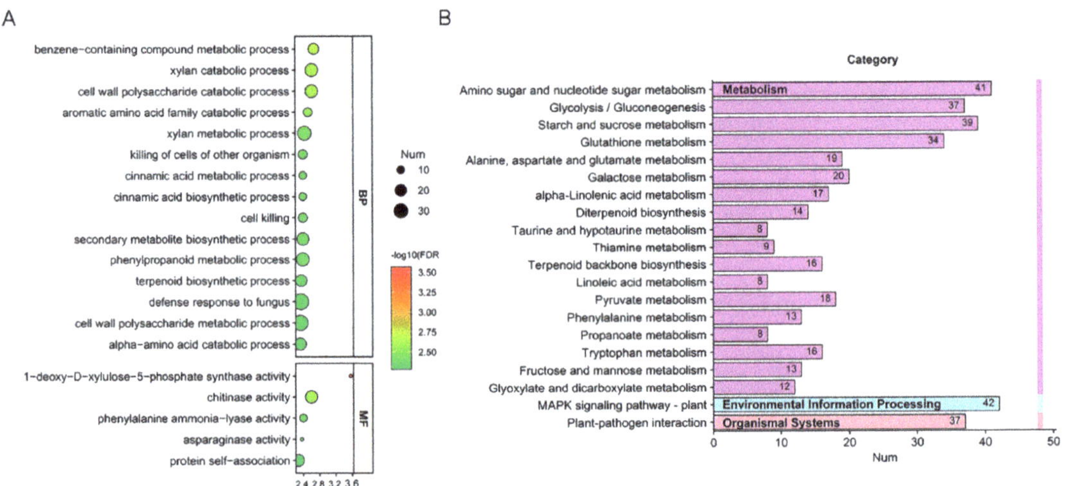

Figure 5. Enrichment analysis of downregulated genes based on gene ontology (GO) and Kyoto encyclopedia of genes and genomes (KEGG) analyses. The top 20 GO categories (**A**) and KEGG pathways (**B**) of the downregulated genes in infected Annong 0711 grains vs. infected Sumai 3 grains were separately grouped and arranged based on their *p* value ranks. The size of the circle represents the gene number.

In these maps, "mitogen-activated protein kinases (MAPK) signaling pathway" and "plant–pathogen interaction" accounted for the largest proportions (Figure 5B). Accordingly, the GO function and KEGG pathway enrichment analyses revealed that compared with the noninfected wheat grains, genes associated with plant defense in the infected Sumai 3 and Annong 0711 grains were both upregulated, while the expression in Annong 0711 wheat grains was significantly lower than that in Sumai 3 (Figure 5 and Table S2). In addition, the expression of genes related to "glutathione metabolism", "phenylalanine metabolism", "glycolysis/gluconeogenesis", "alpha-linolenic acid metabolism", "galactose metabolism", "amino sugar and nucleotide sugar metabolism", and "taurine and hypotaurine metabolism" was also upregulated in the two cultivars of infected wheat grains but was significantly lower in Annong 0711 grains than in Sumai 3 grains, which may illustrate the difference in toxin accumulation between these two wheat cultivars (Figure 5B and Table S3).

Investigating the pathways related to plant defense against pathogen infection, we found that the expression of genes related to MAPK and the hypersensitive response (HR) was lower in *F. graminearum* F1-infected Annong 0711 wheat grains than in Sumai 3 wheat grains (Figure 6). The low-level expression of these genes directly or indirectly negatively affected the fungal resistance response, including the defense response for pathogens, the HR, and defense-related downstream gene induction of Annong 0711 wheat.

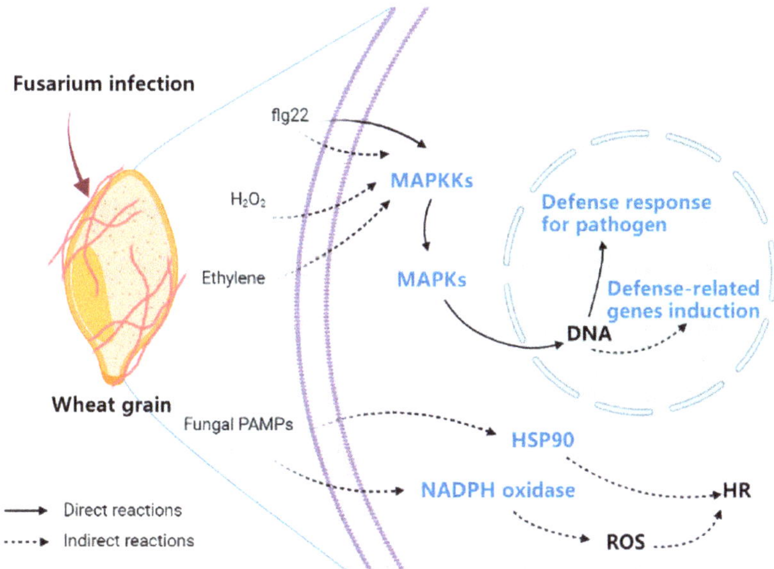

Figure 6. Schematic diagram of Annong 0711 wheat grain genes enriched in the "mitogen activated protein kinases (MAPK) signaling pathway" (map04016) and "plant–pathogen interaction pathway" (map04626). The blue font indicated that the genes that encode the protein were relatively lowly expressed in Annong 0711 compared to Sumai 3 wheat grains.

3. Discussion

Fusarium graminearum is the major agent of FHB that causes wheat diseases and reduces seed yield worldwide, and different *F. graminearum* strains show differences in infection and toxin production ability [14,42–44]. Furthermore, the production of DON and its derivatives was usually affected by environmental conditions such as temperature, humidity and host species [15,22,34,36,37]. In previous studies, the circumstantial effects from temperature and a_w were demonstrated for potential contamination by inoculating *F. graminearum*, *F. verticillioides*, *F. langsethiae*, and *F. meridionale* in a cereal matrix such as maize, oat, or soybean [15,32–34]. However, studies on the effect of *F. graminearum* strains and abiotic factors on mycotoxin production in wheat-based matrixes are still not comprehensive. In our study, we compared the differences in the accumulation of DON and D3G and rates of symptomatic spikelets between six cultivars of wheat spikes that were infected by *F. graminearum* PH−1 and *F. graminearum* F1 under field conditions. We found that there were significant differences in mycotoxin concentration and symptomatic spikelet rates between these wheat cultivars and that the DON and D3G contents were exponentially related to the rates of symptomatic spikelets (Figure 1).

To investigate the effect of abiotic factors on DON and D3G accumulation, we analyzed the mycotoxin assay using three wheat cultivars of harvested wheat grains with known resistance to FHB under indicated temperature and a_w conditions in a laboratory. Then, we found that the concentration of DON and D3G was significantly increased in Sumai 3, Zhongmai 66B, and Annong 0711 wheat grains that were infected by *F. graminearum* PH−1 for 7 days, while the compounds did not show a significant difference under different temperatures between these cultivars due to the high pathogenicity of PH−1 (Figure 2A,B).

Additionally, when the wheat grains were infected by *F. graminearum* F1 for 7 days, the accumulation of DON and D3G only occurred in Annong 0711 wheat (Figure S2). Accordingly, the accumulation of DON and D3G strongly related to the infectivity, genotype and chemotype of *F. graminearum* strains. When the *F. graminearum* F1 infection time was extended to 14 days, the accumulation of DON and D3G and the ratio of D3G to DON showed significant differences between these wheat cultivars but had no relation with the FHB resistance of cultivars (Figure 2C,D and Figure 3 and Table S1). We speculated that under optimal inoculation conditions, the toxin concentration was more affected by the nutrient content in wheat grains.

In order to explore the changes at the molecular level during interaction, we employed RNA-Seq to perform a transcriptomic study and analyzed the changes in gene expression in *F. graminearum* F1-Sumai 3, *F. graminearum* F1-Zhongmai 66B, and *F. graminearum* F1-Annong 0711 wheat grains. Our results had demonstrated significantly DEGs between the *F. graminearum* F1-Sumai 3 and *F. graminearum* F1-Annong 0711 libraries. In previous studies, MAPK genes have been investigated in the plant response to fungal pathogens [45]. It has been reported that DON exerts its effects at the cellular level by activating MAPK through a process known as the ribotoxic stress response, and the outcome of DON-associated MAPK activation is dose- and duration-dependent [16]. In the transcriptome assay, we found that 42 MAPK genes were upregulated in the infected wheat grains; however, compared with infected Sumai 3, the expression of these genes was lower in infected Annong 0711 grains, which is consistent with the DON concentration result (Figures 3 and 5 and Table S3). The hypersensitivity response (HR) is found in all higher plants and is an extremely effective component of the plant immune system [46]. In our study, the expression of genes related to the defense response, especially HR, such as heat shock protein 90 (HSP90) and NADPH oxidase, was also lower in Annong 0711 wheat grains (Figures 5A and 6). These results indicated that the difference in FHB resistance between Sumai 3 and Annong 0711 wheat cultivars was associated with the wheat HR to *F. graminearum*. However, the numbers of DEGs between *F. graminearum* F1 on Sumai 3 and Annong 0711 and enriched pathways were very small (Table S4).

In conclusion, our results illustrated the effect of wheat cultivars, temperature and water activity on mycotoxin production by combining field and laboratory treatments, which will be beneficial in determining the conditions of the relative level of risk of contamination with mycotoxins and providing control strategies to reduce the risk of the occurrence of mycotoxins in pre- and post-harvest wheat. Furthermore, our transcriptome results demonstrated molecular changes in wheat with different FHB resistance and *F. graminearum*, which will lay the foundation for further research on mycotoxin biosynthesis of *F. graminearum* and the regulatory mechanisms of wheat to *F. graminearum*.

4. Materials and Methods

4.1. Wheat Sample and F. graminearum Strains

Wheat cultivars Sumai 3, Wangshuibai, Zhongmai 66B, ZK001, Aikang 58, and Nanda 2419 were provided by the Hefei Institute of Physical Science, Chinese Academy of Sciences (Table 1). *F. graminearum* F1 and *F. graminearum* PH−1 species were donated by Huazhong Agricultural University. All *F. graminearum* strains were stored as spore suspensions in 20% glycerol at −80 °C.

4.2. Chemicals and Reagents

DON and D3G standard solution were purchased from Sigma-Aldrich (St. Louis, MO, USA). Ultrapure water (18.2 MΩ cm) used in our experiments was supplied by Millipore (Bedford, MA, USA). Acetonitrile and methanol (HPLC-grade) were purchased from Honeywell (Shanghai, China). Formic acid (HPLC-grade) was obtained from Anpel (Shanghai, China). Potato dextrose agar medium (PDA) was purchased from BD Difco (San Diego, CA, USA).

Table 1. List of the *F. graminearum* strains, wheat cultivars, and experimental conditions used for infection.

Strain	Wheat Cultivar	FHB Resistance	Conditions	Subject
F. graminearum PH−1 *F. graminearum* F1	Sumai 3 Wangshuibai ZK001 Nanda 2419 Aikang 58 Zhongmai 66B	Resistant Resistant Moderately Resistant Moderately Resistant Susceptible Susceptible	Field conditions	Wheat spikes * [9]
F. graminearum PH−1 *F. graminearum* F1	Sumai 3 Annong 0711 Zhongmai 66B	Resistant Moderately Resistant Susceptible	a_w: 0.80, 0.85, 0.9, 0.95, 0.99 T (°C): 20, 25, 30 [15]	Post-harvest wheat grains

* *F. graminearum* spores were inoculated on wheat florets.

4.3. Inoculation and Incubation Conditions

Florets of 6 cultivars of wheat including Sumai 3, Wangshuibai, Zhongmai 66B, ZK001, Aikang 58, and Nanda 2419 were used for *F. graminearum* PH−1 and *F. graminearum* F1 infection. Then, 20 μL of *F. graminearum* spore suspension (5×10^5 per mL) was inoculated on the florets (Table 1). Symptomatic spikelets on wheat spikes were measured at 21 days post inoculation, and the rate of symptomatic spikelets was calculated based on the following formula:

$$\text{Symptomatic spikelets rate (\%)} = (\text{number of symptomatic spikelets/total wheat spikes number}) \times 100 \tag{1}$$

For laboratory conditions, one portion of every 25 g of post-harvest grains was irradiated at 8 kGy using a cobalt radiation source and then stored aseptically at 4 °C before utilization. One portion for every 25 g of Sumai 3, Zhongmai 66B, and Annong 0711 wheat grain was plated into a 100 mL sterile conical flask, and the initial a_w of the wheat grain was 0.572, which was confirmed by using a Novasina Labmaster-Neo water activity meter (Novasina, Inc., WA, Swit). Then, sterile distilled water was added to rehydrate to the required a_w (a_w: 0.80, 0.85, 0.9, 0.95, and 0.99). Flasks were subsequently refrigerated at 4 °C for 72 h with periodic shaking to ensure uniform absorption and equilibration of water. After three days of equilibration, the wheat grains were inoculated centrally with 4 mm agar plugs taken from the margin of 7-day-old colonies of *F. graminearum* grown on PDA at 25 °C. For all temperatures (20, 25, 30 °C) and a_w treatments, three replicates per strain were used. The total number of treatments was 3 wheat cultivars × 2 *F. graminearum* strains × 3 temperatures conditions × 5 a_w conditions × 3 replicates (Table 1). All treatment groups were dried at 60 °C after 7 or 14 days post inoculation and stored at −20 °C until mycotoxin extraction was carried out.

4.4. Mycotoxin Extraction

Mycotoxin extractions were determined according to a published method with minor modifications [47]. All samples were ground into a homogenized powder, weighed into 50 mL centrifuge tubes, and then mixed with 10 mL of acetonitrile/water (84:16, *v/v*). The tubes were shaken at 2500 rpm/min at 25 °C in an orbital shaker for 60 min and then ultrasonicated for 40 min. Then, tubes were centrifuged at 4000 rpm/min for 30 min. Then, 2 mL of supernatant was transferred to a 15 mL centrifuge tube, 150 mg of anhydrous magnesium sulphate was added, and the tube was vortexed. Then, the supernatant was transferred to a new 15 mL centrifuge tube and 1 mL of n-hexane was added to a biosafety cabinet for degreasing, shaken vigorously, centrifuged to remove n-hexane, and completely dried in a stream of nitrogen. All dried extracts were dissolved in 1 mL of acetonitrile:water (20:80 *v/v*). The purified supernatant was filtered through a 0.22 μm nylon filter and stored in sampler vials at −20 °C until LC–MS/MS analysis.

4.5. Mycotoxin Determination by LC–MS/MS

LC–MS/MS analysis was performed as described by Yu et al. [15]. Mycotoxins were quantified by an Accela 1250 UPLC system (Thermo Fisher Scientific, San Jose, CA, USA) coupled to a TSQ Vantage™ (Thermo Fisher Scientific, San Jose, CA, USA) triple-stage quadruple mass spectrometer. An Agilent Extend C18 chromatographic column (100 mm × 4.6 mm, 3.5 µm) was used at a flow rate of 0.35 mL/min at 30 °C and with a 10 µL injection volume. The mobile phase consisted of 5 mM ammonium acetate (A) and 100% methanol (B). The gradient was as follows: 0 min 15% B, 1 min 15% B, 6.5 min 90% B, 8.5 min 90% B, 9 min 15% B, and 12 min 15% B. Mass spectrometry analysis was carried out in both positive (ESI + 3.5 kV) and negative (ESI − 2.5 kV) ionization modes using selected reaction monitoring (SRM). For the MS/MS analysis, both the vaporizer and capillary temperatures were 300 °C, the sheath gas pressure was 50 psi, and the aux gas pressure was 5 psi. Raw data were analyzed using Xcalibur™ software (Thermo Fisher Scientific, San Jose, CA, USA, 2011).

The ratio of D3G and total DON was calculated based on the following formula:

$$\text{D3G/total DON} = m\text{D3G}/(m\text{DON} + n\text{D3G} * \text{MDON}) \tag{2}$$

4.6. Total RNA Extraction

The total RNA of the mixture samples of Sumai 3, Zhongmai 66B, and Annong 0711 wheat and *F. graminearum* F1 for 14 days was extracted using Plant RNA Purification Reagent for plant tissue (Invitrogen, Carlsbad, CA, USA) according the manufacturer's instructions, respectively. Genomic DNA was removed by DNaseI (Takara, Beijing, China). The RNA quality and concentration were determined using a NanoDrop 2000 (Agilent Technologies, Santa Clara, CA, USA).

4.7. Library Preparation and Sequencing

The RNA-seq transcriptome library was prepared using a TruSeq™ RNA sample preparation kit from Illumina (San Diego, CA, USA) and 1 µg of total RNA. Second, double-stranded cDNA was synthesized using a SuperScript double-stranded cDNA synthesis kit (Invitrogen, CA, USA), 300 bp fragmented mRNA, and random hexamer primers. The synthesized cDNA was then subjected to end repair and adaptor ligation according to Illumina's library construction protocol. Then, cDNA target fragments of 300 bp were amplified using Phusion DNA polymerase (NEB) for 15 PCR cycles. After quantification by TBS380, the paired-end sequencing library was sequenced using the Illumina HiSeq Xten sequencer (2 × 150 bp read length) [48].

4.8. Read Mapping and DEG Analysis

The raw paired-end reads were optimized by SeqPrep (https://github.com/jstjohn/SeqPrep (accessed on 7 June 2022)) and Sickle (https://github.com/najoshi/sickle, accessed on 7 June 2022) using default parameters to obtain clean reads. Then, clean reads were separately aligned to the reference genome with orientation mode using HISAT2 (http://ccb.jhu.edu/software/hisat2/index.shtml, accessed on 7 June 2022) software [49]. Gene abundances were quantified using RSEM (http://deweylab.biostat.wisc.edu/rsem/, accessed on 7 June 2022) [50]. A DEG analysis was performed using DESeq2 with a $|\text{log2FC}| > 1$ and BH-corrected p value ≤ 0.05 [51].

4.9. Functional Annotation and Enrichment

A Clusters of Orthologous Groups of proteins (COG) annotation analysis was performed using HMMER [52]. Gene Ontology (GO) functional enrichment and Kyoto Encyclopedia of Genes and Genome (KEGG) pathway analysis were carried out by Goatools (https://github.com/tanghaibao/Goatools, accessed on 7 June 2022) and KOBAS (http://kobas.cbi.pku.edu.cn/home.do, accessed on 7 June 2022) at a p value ≤ 0.05 or corrected p value ≤ 0.05 [53,54].

4.10. Quantitative Real-Time PCR Analysis

Total RNA of the mixed culture samples of Sumai 3, Zhongmai 66B, and Annong 0711 wheat and *F. graminearum* F1 for 14 days were prepared using TRIzol reagent (Invitrogen, CA, USA). cDNA was synthesized using PrimeScript™ RT reagent Kit with gDNA Eraser (Takara, Beijing, China). Each sample was quantified using TB Green Premix Ex Taq II (Takara, Beijing, China) following the instructions of the manufacturer and applied to the QuantStudio™ Real-Time PCR System (Applied Biosystems™). The primers employed in this experiment were listed in Table S5. Actin was used as an internal control in this experiment. The $2^{-\Delta\Delta Ct}$ method was used to calculate the expression level and three replicates were employed for every gene [55].

4.11. Statistical Analysis

The mycotoxin concentration analysis was constructed using GraphPad Prism 8 (GraphPad Software Inc., San Diego, CA, USA) by two-way ANOVA. Post hoc Tukey's testing was used to evaluate changes between groups. A *p* value < 0.05 was considered statistically significant.

Supplementary Materials: The following supporting information can be downloaded at: https://www.mdpi.com/article/10.3390/toxins14070482/s1, Figure S1: Representative images of 6 different cultivars of wheat spikelets at 21 days after inoculation with *F. graminearum* PH−1 and *F. graminearum* F1; Figure S2: The accumulation of DON and D3G in Annong 0711 grains that were infected by *F. graminearum* F1 at different temperatures and a_w for 7 days; Table S1: DON and D3G accumulation in Sumai 3, Zhongmai 66B, and Annong 0711 wheat grains infected by *F. graminearum* F1 under different conditions; Table S2: GO and KEGG enrichment analysis of upregulated genes of infected Annong 0711 wheat grains; Table S3: Significantly enriched (corrected *p* value < 0.05) KEGG pathways of upregulated genes in infected wheat grains; Table S4: GO and KEGG enrichment analysis of DEGs of *F. graminearum* F1 in Sumai 3 and Annong 0711 wheat grains; Table S5: Primer sequences used for RT-qPCR amplification of the differentially expressed genes selected for validation.

Author Contributions: Conceptualization, K.L. and A.W.; Methodology, K.L.; Software, K.L.; Validation, D.Y., Z.Y., N.L. and A.W.; Formal Analysis, K.L.; Investigation, K.L.; Resources, Y.F.; C.W. and A.W.; Data Curation, K.L.; Writing—Original Draft Preparation, K.L.; Writing—Review and Editing, D.Y., N.L. and A.W.; Visualization, K.L.; Supervision, D.Y., N.L. and A.W.; Project Administration, Y.F., C.W. and A.W.; Funding Acquisition, Y.F., C.W. and A.W. All authors have read and agreed to the published version of the manuscript.

Funding: This research was funded by the National Key Research and Development Program of China (2019YFC1604502), National Science Fund for Distinguished Young Scholars (32025030).

Institutional Review Board Statement: Not applicable.

Informed Consent Statement: Not applicable.

Data Availability Statement: The data that support the findings of this study are available from the corresponding author upon reasonable request.

Conflicts of Interest: The authors declare no conflict of interest.

Abbreviations

FHB	Fusarium head blight
DON	deoxynivalenol
D3G	deoxynivalenol-3-glucoside
HPLC	high performance liquid chromatography
LC-MS/MS	liquid chromatography tandem mass spectrometry
a_w	water activity

DEG	differentially expressed gene
COG	Clusters Orthologous Groups
GO	Gene Ontology
KEGG	Kyoto Encyclopedia of Genes and Genome
BP	biological process
CC	cellular component
MF	molecular function
MAPK	mitogen activated protein kinases
HR	hypersensitivity response

References

1. Scala, V.; Pietricola, C.; Farina, V.; Beccaccioli, M.; Zjalic, S.; Quaranta, F.; Fornara, M.; Zaccaria, M.; Momeni, B.; Reverberi, M.; et al. Tramesan Elicits Durum Wheat Defense against the Septoria Disease Complex. *Biomolecules* **2020**, *10*, 608. [CrossRef] [PubMed]
2. Kiani, M.; Bryan, B.; Rush, C.; Szczepaniec, A. Transcriptional Responses of Resistant and Susceptible Wheat Exposed to Wheat Curl Mite. *Int. J. Mol. Sci.* **2021**, *22*, 2703. [CrossRef]
3. Kikot, G.E.; Moschini, R.; Consolo, V.F.; Rojo, R.; Salerno, G.; Hours, R.A.; Gasoni, L.; Arambarri, A.M.; Alconada, T.M. Occurrence of Different Species of Fusarium from Wheat in Relation to Disease Levels Predicted by a Weather-Based Model in Argentina Pampas Region. *Mycopathologia* **2011**, *171*, 139–149. [CrossRef]
4. Zhao, Y.; Guan, X.; Zong, Y.; Hua, X.; Xing, F.; Wang, Y.; Wang, F.; Liu, Y. Deoxynivalenol in wheat from the Northwestern region in China. *Food Addit. Contam. Part B* **2018**, *11*, 281–285. [CrossRef] [PubMed]
5. Chen, Y.; Kistler, H.C.; Ma, Z. *Fusarium graminearum* Trichothecene Mycotoxins: Biosynthesis, Regulation, and Management. *Annu. Rev. Phytopathol.* **2019**, *57*, 15–39. [CrossRef] [PubMed]
6. Fabre, F.; Rocher, F.; Alouane, T.; Langin, T.; Bonhomme, L. Searching for FHB Resistances in Bread Wheat: Susceptibility at the Crossroad. *Front. Plant Sci.* **2020**, *11*, 731. [CrossRef] [PubMed]
7. Yan, H.; Li, G.; Shi, J.; Tian, S.; Zhang, X.; Cheng, R.; Wang, X.; Yuan, Y.; Cao, S.; Zhou, J.; et al. Genetic control of Fusarium head blight resistance in two Yangmai 158-derived recombinant inbred line populations. *Theor. Appl. Genet.* **2021**, *134*, 3037–3049. [CrossRef]
8. Zhang, Y.; Yang, Z.; Ma, H.; Huang, L.; Ding, F.; Du, Y.; Jia, H.; Li, G.; Kong, Z.; Ran, C.; et al. Pyramiding of Fusarium Head Blight Resistance Quantitative Trait Loci, Fhb1, Fhb4, and Fhb5, in Modern Chinese Wheat Cultivars. *Front. Plant Sci.* **2021**, *12*, 694023. [CrossRef]
9. Jia, L.-J.; Tang, H.-Y.; Wang, W.-Q.; Yuan, T.-L.; Wei, W.-Q.; Pang, B.; Gong, X.-M.; Wang, S.-F.; Li, Y.-J.; Zhang, D.; et al. A linear nonribosomal octapeptide from *Fusarium graminearum* facilitates cell-to-cell invasion of wheat. *Nat. Commun.* **2019**, *10*, 922. [CrossRef]
10. Chiotta, M.L.; Alaniz Zanon, M.S.; Palazzini, J.M.; Alberione, E.; Barros, G.G.; Chulze, S.N. *Fusarium graminearum* species complex occurrence on soybean and *F. graminearum* sensu stricto inoculum maintenance on residues in soybean-wheat rotation under field conditions. *J. Appl. Microbiol.* **2021**, *130*, 208–216. [CrossRef]
11. van der Lee, T.; Zhang, H.; van Diepeningen, A.; Waalwijk, C. Biogeography of *Fusarium graminearum* species complex and chemotypes: A review. *Food Addit. Contam. Part A Chem. Anal. Control Expo Risk Assess* **2015**, *32*, 453–460. [CrossRef] [PubMed]
12. Palacios, S.A.; Erazo, J.G.; Ciasca, B.; Lattanzio, V.M.; Reynoso, M.M.; Farnochi, M.C.; Torres, A.M. Occurrence of deoxynivalenol and deoxynivalenol-3-glucoside in durum wheat from Argentina. *Food Chem.* **2017**, *230*, 728–734. [CrossRef]
13. Yan, P.; Liu, Z.; Liu, S.; Yao, L.; Liu, Y.; Wu, Y.; Gong, Z. Natural Occurrence of Deoxynivalenol and Its Acetylated Derivatives in Chinese Maize and Wheat Collected in 2017. *Toxins* **2020**, *12*, 200. [CrossRef]
14. Crippin, T.; Renaud, J.B.; Sumarah, M.W.; Miller, J.D. Comparing genotype and chemotype of *Fusarium graminearum* from cereals in Ontario, Canada. *PLoS ONE* **2019**, *14*, e0216735. [CrossRef]
15. Yu, S.; Jia, B.; Li, K.; Zhou, H.; Lai, W.; Tang, Y.; Yan, Z.; Sun, W.; Liu, N.; Yu, D.; et al. Pre-warning of abiotic factors in maize required for potential contamination of fusarium mycotoxins via response surface analysis. *Food Control* **2021**, *121*, 107570. [CrossRef]
16. Hooft, J.M.; Bureau, D.P. Deoxynivalenol: Mechanisms of action and its effects on various terrestrial and aquatic species. *Food Chem. Toxicol.* **2021**, *157*, 112616. [CrossRef]
17. Guo, H.; Ji, J.; Wang, J.; Sun, X. Deoxynivalenol: Masked forms, fate during food processing, and potential biological remedies. *Compr. Rev. Food Sci. Food Saf.* **2020**, *19*, 895–926. [CrossRef] [PubMed]
18. Cirlini, M.; Generotti, S.; Dall'Erta, A.; Lancioni, P.; Ferrazzano, G.; Massi, A.; Galaverna, G.; Dall'Asta, C. Durum Wheat (*Triticum Durum* Desf.) Lines Show Different Abilities to Form Masked Mycotoxins under Greenhouse Conditions. *Toxins* **2013**, *6*, 81–95. [CrossRef]
19. He, Y.; Ahmad, D.; Zhang, X.; Zhang, Y.; Wu, L.; Jiang, P.; Ma, H. Genome-wide analysis of family-1 UDP glycosyltransferases (UGT) and identification of UGT genes for FHB resistance in wheat (*Triticum aestivum* L.). *BMC Plant Biol.* **2018**, *18*, 67. [CrossRef]
20. He, Y.; Wu, L.; Liu, X.; Jiang, P.; Yu, L.; Qiu, J.; Wang, G.; Zhang, X.; Ma, H. TaUGT6, a Novel UDP-Glycosyltransferase Gene Enhances the Resistance to FHB and DON Accumulation in Wheat. *Front. Plant Sci.* **2020**, *11*, 574775. [CrossRef]

21. Sharma, P.; Gangola, M.P.; Huang, C.; Kutcher, H.R.; Ganeshan, S.; Chibbar, R.N. Single Nucleotide Polymorphisms in B-Genome Specific UDP-Glucosyl Transferases Associated with Fusarium Head Blight Resistance and Reduced Deoxynivalenol Accumulation in Wheat Grain. *Phytopathology* **2018**, *108*, 124–132. [CrossRef] [PubMed]
22. Kluger, B.; Bueschl, C.; Lemmens, M.; Michlmayr, H.; Malachova, A.; Koutnik, A.; Maloku, I.; Berthiller, F.; Adam, G.; Krska, R.; et al. Biotransformation of the Mycotoxin Deoxynivalenol in Fusarium Resistant and Susceptible Near Isogenic Wheat Lines. *PLoS ONE* **2015**, *10*, e0119656. [CrossRef] [PubMed]
23. Zhang, H.; Wang, B. Fate of deoxynivalenol and deoxynivalenol-3-glucoside during wheat milling and Chinese steamed bread processing. *Food Control* **2014**, *44*, 86–91. [CrossRef]
24. Zhang, H.; Wang, B. Fates of deoxynivalenol and deoxynivalenol-3-glucoside during bread and noodle processing. *Food Control* **2015**, *50*, 754–757. [CrossRef]
25. Broekaert, N.; Devreese, M.; van Bergen, T.; Schauvliege, S.; De Boevre, M.; De Saeger, S.; Vanhaecke, L.; Berthiller, F.; Michlmayr, H.; Malachova, A.; et al. In vivo contribution of deoxynivalenol-3-β-D-glucoside to deoxynivalenol exposure in broiler chickens and pigs: Oral bioavailability, hydrolysis and toxicokinetics. *Arch. Toxicol.* **2017**, *91*, 699–712. [CrossRef]
26. Mengelers, M.; Zeilmaker, M.; Vidal, A.; De Boevre, M.; De Saeger, S.; Hoogenveen, R. Biomonitoring of Deoxynivalenol and Deoxynivalenol-3-glucoside in Human Volunteers: Renal Excretion Profiles. *Toxins* **2019**, *11*, 466. [CrossRef]
27. Gratz, S.W.; Duncan, G.; Richardson, A.J. The Human Fecal Microbiota Metabolizes Deoxynivalenol and Deoxynivalenol-3-Glucoside and May Be Responsible for Urinary Deepoxy-Deoxynivalenol. *Appl. Environ. Microbiol.* **2013**, *79*, 1821–1825. [CrossRef]
28. Gratz, S.W.; Currie, V.; Richardson, A.J.; Duncan, G.; Holtrop, G.; Farquharson, F.; Louis, P.; Pinton, P.; Oswald, I.P. Porcine Small and Large Intestinal Microbiota Rapidly Hydrolyze the Masked Mycotoxin Deoxynivalenol-3-Glucoside and Release Deoxynivalenol in Spiked Batch Cultures In Vitro. *Appl. Environ. Microbiol.* **2018**, *84*, e02106-17. [CrossRef]
29. Nagl, V.; Schwartz, H.; Krska, R.; Moll, W.-D.; Knasmüller, S.; Ritzmann, M.; Adam, G.; Berthiller, F. Metabolism of the masked mycotoxin deoxynivalenol-3-glucoside in rats. *Toxicol. Lett.* **2012**, *213*, 367–373. [CrossRef]
30. Nagl, V.; Woechtl, B.; Schwartz-Zimmermann, H.E.; Hennig-Pauka, I.; Moll, W.-D.; Adam, G.; Berthiller, F. Metabolism of the masked mycotoxin deoxynivalenol-3-glucoside in pigs. *Toxicol. Lett.* **2014**, *229*, 190–197. [CrossRef]
31. Li, K.; Wang, L.; Yu, D.; Yan, Z.; Liu, N.; Wu, A. Cellobiose inhibits the release of deoxynivalenol from transformed deoxynivalenol-3-glucoside from *Lactiplantibacillus plantarum*. *Food Chem. Mol. Sci.* **2022**, *4*, 100077. [CrossRef] [PubMed]
32. Verheecke-Vaessen, C.; Garcia-Cela, E.; Lopez-Prieto, A.; Jonsdottir, I.O.; Medina, A.; Magan, N. Water and temperature relations of Fusarium langsethiae strains and modelling of growth and T-2 and HT-2 mycotoxin production on oat-based matrices. *Int. J. Food Microbiol.* **2021**, *348*, 109203. [CrossRef] [PubMed]
33. Rybecky, A.; Chulze, S.; Chiotta, M. Effect of water activity and temperature on growth and trichothecene production by Fusarium meridionale. *Int. J. Food Microbiol.* **2018**, *285*, 69–73. [CrossRef]
34. Peter Mshelia, L.; Selamat, J.; Iskandar Putra Samsudin, N.; Rafii, M.Y.; Abdul Mutalib, N.A.; Nordin, N.; Berthiller, F. Effect of Temperature, Water Activity and Carbon Dioxide on Fungal Growth and Mycotoxin Production of Acclimatised Isolates of *Fusarium verticillioides* and *F. graminearum*. *Toxins* **2020**, *12*, 478. [CrossRef] [PubMed]
35. Schiro, G.; Müller, T.; Verch, G.; Sommerfeld, T.; Mauch, T.; Koch, M.; Grimm, V.; Müller, M.E. The distribution of mycotoxins in a heterogeneous wheat field in relation to microclimate, fungal and bacterial abundance. *J. Appl. Microbiol.* **2018**, *126*, 177–190. [CrossRef]
36. Fan, P.; Gu, K.; Wu, J.; Zhou, M.; Chen, C. Effect of wheat (*Triticum aestivum* L.) resistance, *Fusarium graminearum* DNA content, strain potential toxin production, and disease severity on deoxynivalenol content. *J. Basic Microbiol.* **2019**, *59*, 1105–1111. [CrossRef]
37. Belizán, M.M.; Gomez, A.D.L.A.; Baptista, Z.P.T.; Jimenez, C.M.; Matías, M.D.H.S.; Catalán, C.A.; Sampietro, D.A. Influence of water activity and temperature on growth and production of trichothecenes by *Fusarium graminearum* sensu stricto and related species in maize grains. *Int. J. Food Microbiol.* **2019**, *305*, 108242. [CrossRef]
38. Wang, H.; Sun, S.; Ge, W.; Zhao, L.; Hou, B.; Wang, K.; Lyu, Z.; Chen, L.; Xu, S.; Guo, J.; et al. Horizontal gene transfer of *Fhb7* from fungus underlies *Fusarium* head blight resistance in wheat. *Science* **2020**, *368*, 822. [CrossRef]
39. Kazan, K.; Gardiner, D.M. Transcriptomics of cereal-*Fusarium graminearum* interactions: What we have learned so far. *Mol. Plant Pathol.* **2018**, *19*, 764–778. [CrossRef]
40. Chen, L.; Wang, H.; Yang, J.; Yang, X.; Zhang, M.; Zhao, Z.; Fan, Y.; Wang, C.; Wang, J. Bioinformatics and Transcriptome Analysis of CFEM Proteins in *Fusarium graminearum*. *J. Fungi* **2021**, *7*, 871. [CrossRef]
41. Buhrow, L.M.; Liu, Z.; Cram, D.; Sharma, T.; Foroud, N.A.; Pan, Y.; Loewen, M.C. Wheat transcriptome profiling reveals abscisic and gibberellic acid treatments regulate early-stage phytohormone defense signaling, cell wall fortification, and metabolic switches following *Fusarium graminearum*-challenge. *BMC Genom.* **2021**, *22*, 798. [CrossRef] [PubMed]
42. Haile, J.K.; N'Diaye, A.; Walkowiak, S.; Nilsen, K.T.; Clarke, J.M.; Kutcher, H.R.; Steiner, B.; Buerstmayr, H.; Pozniak, C.J. Fusarium Head Blight in Durum Wheat: Recent Status, Breeding Directions, and Future Research Prospects. *Phytopathology* **2019**, *109*, 1664–1675. [CrossRef] [PubMed]
43. Goswami, R.S.; Kistler, H.C. Pathogenicity and In Planta Mycotoxin Accumulation among Members of the *Fusarium graminearum* Species Complex on Wheat and Rice. *Phytopathology* **2005**, *95*, 1397–1404. [CrossRef] [PubMed]

44. Gautam, P.; Dill-Macky, R. Impact of moisture, host genetics and *Fusarium graminearum* isolates on Fusarium head blight development and trichothecene accumulation in spring wheat. *Mycotoxin Res.* **2011**, *28*, 45–58. [CrossRef] [PubMed]
45. Meng, X.; Zhang, S. MAPK Cascades in Plant Disease Resistance Signaling. *Annu. Rev. Phytopathol.* **2013**, *51*, 245–266. [CrossRef]
46. Balint-Kurti, P. The plant hypersensitive response: Concepts, control and consequences. *Mol. Plant Pathol.* **2019**, *20*, 1163–1178. [CrossRef]
47. Tian, Y.; Tan, Y.; Liu, N.; Yan, Z.; Liao, Y.; Chen, J.; De Saeger, S.; Yang, H.; Zhang, Q.; Wu, A. Detoxification of Deoxynivalenol via Glycosylation Represents Novel Insights on Antagonistic Activities of Trichoderma when Confronted with *Fusarium graminearum*. *Toxins* **2016**, *8*, 335. [CrossRef]
48. Yu, X.; Wang, T.; Zhu, M.; Zhang, L.; Zhang, F.; Jing, E.; Ren, Y.; Wang, Z.; Xin, Z.; Lin, T. Transcriptome and physiological analyses for revealing genes involved in wheat response to endoplasmic reticulum stress. *BMC Plant Biol.* **2019**, *19*, 193. [CrossRef]
49. Kim, D.; Langmead, B.; Salzberg, S.L. HISAT: A fast spliced aligner with low memory requirements. *Nat. Methods* **2015**, *12*, 357–360. [CrossRef]
50. Li, B.; Dewey, C.N. RSEM: Accurate transcript quantification from RNA-Seq data with or without a reference genome. *BMC Bioinform.* **2011**, *12*, 323. [CrossRef]
51. Love, M.I.; Huber, W.; Anders, S. Moderated estimation of fold change and dispersion for RNA-seq data with DESeq2. *Genome Biol.* **2014**, *15*, 550. [CrossRef] [PubMed]
52. Feldbauer, R.; Gosch, L.; Lüftinger, L.; Hyden, P.; Flexer, A.; Rattei, T. DeepNOG: Fast and accurate protein orthologous group assignment. *Bioinformatics* **2020**, *36*, 5304–5312. [CrossRef]
53. Klopfenstein, D.V.; Zhang, L.; Pedersen, B.S.; Ramírez, F.; Vesztrocy, A.W.; Naldi, A.; Mungall, C.J.; Yunes, J.M.; Botvinnik, O.; Weigel, M.; et al. GOATOOLS: A Python library for Gene Ontology analyses. *Sci. Rep.* **2018**, *8*, 10872. [CrossRef]
54. Shen, S.; Kong, J.; Qiu, Y.; Yang, X.; Wang, W.; Yan, L. Identification of core genes and outcomes in hepatocellular carcinoma by bioinformatics analysis. *J. Cell. Biochem.* **2019**, *120*, 10069–10081. [CrossRef] [PubMed]
55. Pfaffl, M.W. A new mathematical model for relative quantification in real-time RT-PCR. *Nucleic Acids Res.* **2001**, *29*, e45. [CrossRef] [PubMed]

 *toxins*

Article

Lethal and Sublethal Toxicity Assessment of Cyclosporin C (a Fungal Toxin) against *Plutella xylostella* (L.)

Jianhui Wu [1,2,†], Xiaochen Zhang [1,2,†], Muhammad Hamid Bashir [3] and Shaukat Ali [1,2,*]

[1] Key Laboratory of Bio-Pesticide Innovation and Application, College of Plant Protection, South China Agricultural University, Guangzhou 510642, China; jhw@scau.edu.cn (J.W.); zhangxiaoying@stu.scau.edu.cn (X.Z.)
[2] Engineering Research Center of Biological Control, Ministry of Education and Guangdong Province, South China Agricultural University, Guangzhou 510642, China
[3] Department of Entomology, Faculty of Agriculture, University of Agriculture, Faisalabad 38000, Pakistan; hamid_uaf@yahoo.com
[*] Correspondence: aliscau@scau.edu.cn
[†] These authors contributed equally to this work.

Abstract: Secondary metabolites/toxins produced by *Purpeocillium lilacinum* (Hypocreales; Phiocordycipitaceae), a well-known insect pathogen, can be used for the management of different insect pests. We report the lethal and sublethal effects of cyclosporin C (a toxin produced by *P. lilacinum*) against a major vegetable pest, *Plutella xylostella*, at specific organismal (feeding rate, larval growth, adult emergence, fecundity, and adult longevity) and sub-organismal levels (changes in antioxidant and neurophysiological enzyme activities). The toxicity of cyclosporin C against different larval instars of *P. xylostella* increased with increasing concentrations of the toxin and the maximum percent mortality rates for different *P. xylostella* larval instars at different times were observed for the 300 µg/mL cyclosporin C treatment, with an average mortality rate of 100% for all larval instars. The median lethal concentrations (LC_{50}) of cyclosporin C against the first, second, third, and fourth larval instars of *P. xylostella* 72 h post-treatment were 78.05, 60.42, 50.83, and 83.05 µg/mL, respectively. Different concentrations of cyclosporin C caused a reduction in the average leaf consumption and average larval weight. Different life history parameters, such as the pupation rate (%), adult emergence (%), female fecundity, and female longevity were also inhibited when different concentrations of cyclosporin C were applied topically. The cyclosporin C concentrations inhibited the activities of different detoxifying (glutathione S-transferase, carboxylesterase, and acetylcholinesterase) and antioxidant enzyme (superoxide dismutase, catalase, and peroxidase) activities of *P. xylostella* when compared to the control. These findings can serve as baseline information for the development of cyclosporin C as an insect control agent, although further work on mass production, formulation, and field application is still required.

Keywords: biopesticides; cyclosporin C; diamondback moth; toxicity; antifeedant activity; fecundity; enzyme activities

Key Contribution: This study reports the lethal and sublethal effects of cyclosporin C (a toxin produced by *P. lilacinum*) against a major vegetable pest, *P. xylostella*, at specific organismal (feeding rate, body weight, fecundity, and adult longevity) and sub-organismal levels (changes in antioxidant and neurophysiological enzyme activities).

Citation: Wu, J.; Zhang, X.; Bashir, M.H.; Ali, S. Lethal and Sublethal Toxicity Assessment of Cyclosporin C (a Fungal Toxin) against *Plutella xylostella* (L.). *Toxins* **2022**, *14*, 514. https://doi.org/10.3390/toxins14080514

Received: 25 June 2022
Accepted: 25 July 2022
Published: 28 July 2022

Publisher's Note: MDPI stays neutral with regard to jurisdictional claims in published maps and institutional affiliations.

1. Introduction

Different species of insect pathogenic fungi are possible alternatives to chemical insecticides for insect pest management. Their diversity (among taxa), production of secondary metabolites, and environmental safety make them more practically applicable biological control agents [1]. Members of fungal taxa, *Purpureocillium* Luangsa-ard, are pathogens of plant-feeding insects and nematodes [2,3]. *Purpureocillium lilacinum* is a well-known insect pathogen that has been used to control different crop pests, such as aphids, whiteflies, thrips, fruit flies, and plant parasitic nematodes [2–4]. *Purpureocillium lilacinum* isolates produce different secondary compounds/toxins known as paecilotoxins [5]. These toxins demonstrate oral toxicity against insects and nematodes [6].

Cyclosporins are nonpolar cyclic oligopeptides that are produced by different fungal species belonging to genera such as *Beauveria*, *Lecanicillum*, *Purpeocillium*, and *Tolypocladium* [5,7]. During insect infection, cyclosporins are secreted at later stages of host infection and disrupt the immune response of the insect host, ultimately leading to insect mortality [8–10]. Cyclosporins act through the blocked glycoprotein-related pumps involved in the removal of xenobiotics out of haemolymph [11]. Weiser and Matha [12] reported the toxicity mechanism of cyclosporin A against *Culex pipiens*. Cyclosporin A toxicity induced different histopathological changes to *C. pipiens* tissues (formation of swollen mitochondria and vacuolization of the endoplasmic reticulum) [13]. Vilcinskas et al. [5] reported that an artificial diet supplemented with a sublethal concentration of cyclosporin A reduced the phagocytosis of plasmatocytes in *Galleria mellonella*. Mosquito (*Aedes aegypti*) larvae treated with dried methanolic extract of *Tolypocladium* (with cyclosporin A as the predominant metabolite) caused high mortality because of the disruption of midgut mitochondrial cells [14]. Cyclosporins have been reported to show a narrow spectrum of antibiotic and immuno-suppressive activities in humans, recently being used for the cure of different diseases [15–19], but the use of cyclosporins at high doses (above 25 mg/kg) can cause adverse effects, including nephrotoxicity, diarrhoea, headache, anxiety, and metabolic toxicity [20,21]. However, the use of cyclosporins at lower doses (below 3 mg/kg) was reported to have no side effects [22]. Recent studies have shown that the immuno-suppressive effects of cyclosporin C are lower than that of other cyclosporins [15], and therefore, it would be of interest to study methods of mass production and the insecticidal efficacy of cyclosporin C produced by insect pathogenic fungi *P. lilacinum*, keeping in mind the ecological implications of its application. Since cyclosporin C has been reported to be produced by different plant pathogenic filamentous fungi (belonging to the genera *Aspergillus* and *Fusarium*) [23], cyclosporin C should be used at low doses (below 3 mg/kg) for insect pest management to avoid the risks of mycotoxin contamination and indirectly to avoid public health risks from the consumption of mycotoxin-contaminated foods [24].

Plutella xylostella (L.) (Lepidoptera: Plutellidae) is a major threat to different vegetable crops across the globe. The annual yield losses caused by *P. xylostella* worldwide are estimated to be USD 4–5 billion [25]. This insect pest has a strong ability to develop insecticide resistance [26]. Therefore, the use of insect pathogenic fungi or their secondary metabolites can serve as an alternate strategy for *P. xylostella* management. Amiri et al. [27] showed a reduction in the feeding rate of *P. xylostella* in response to destruxin (a fungal toxin produced by *Metarhizium anisopliae*) application. The observed antifeedant index (AI) was dose-related, and there were significant differences between the treated and untreated leaves. Huang et al. [28] observed increased mortality of *P. xylostella* larvae with increasing concentrations of destruxin A.

So far, very minimal work has been reported on the lethal and sublethal effects of cyclosporin C on target insect hosts [29]. The sublethal effects have been observed on a life history basis (fecundity/life table parameters) [28] and an immune response basis (variations in activities of antioxidant and neurophysiological enzymes) [30]. The present work was performed to assess the lethal and sublethal toxicity of cyclosporin C against *P. xylostella*. The lethal toxicity was studied by testing the efficacy of different cyclosporin C concentrations against different larval instars of *P. xylostella*. The sublethal toxicity was observed at specific organismal (feeding rate, change in larval weight, adult emergence, fecundity, and adult longevity) and sub-organismal levels (changes in antioxidant and neurophysiological enzyme activities). We hope that this study will provide the basis for the future development of cyclosporin C as a biological control agent for the management of insect pests such as *P. xylostella*.

2. Results

2.1. Toxicity of Cyclosporin C against P. xylostella

2.1.1. Concentration–Mortality Response of P. xylostella to Cyclosporin C

After 3 days of treatment, the percent mortality rates of different *P. xylostella* larval instars were significantly different among the different treatments and the control ($F_{18,56} = 6\,6.45$; $p < 0.01$). The mortality of *P. xylostella* increased with an increasing concentration of cyclosporin C (Figure 1). The maximum percent mortality rates for different *P. xylostella* larval instars at different times were observed for the 300 µg/mL cyclosporin C treatment, with an average mortality rate of 100% for all larval instars.

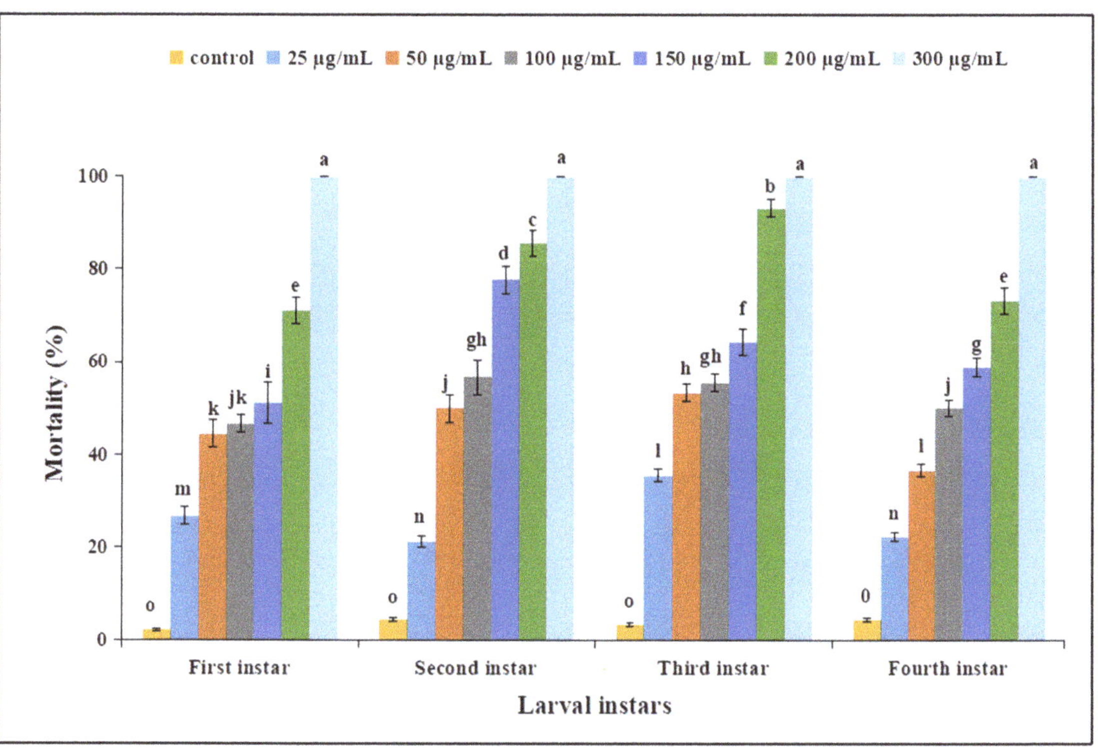

Figure 1. Mortality (%) of different larval instars of *Plutella xylostella* treated with different concentrations of cyclosporin C. Bars represent standard error of means (based on three independent replicates). Bars with different letters represent significantly different means (Tukey's, $p < 0.05$).

2.1.2. Transmission Electron Microscopic Examination of P. xylostella Midgut Following Cyclosporin C Treatment

As shown in Figure 2, the nuclei of untreated cells had smooth nuclear membranes, but after ingesting cyclosporin C, the nuclei of midgut cells were enlarged, and the nuclear membranes appeared to be folded. The degree of nuclei enlargement increased, and the nuclear membrane folds became more obvious with an increase in time post-treatment. However, the nuclei and nuclear membranes of the untreated midgut cells did not change much.

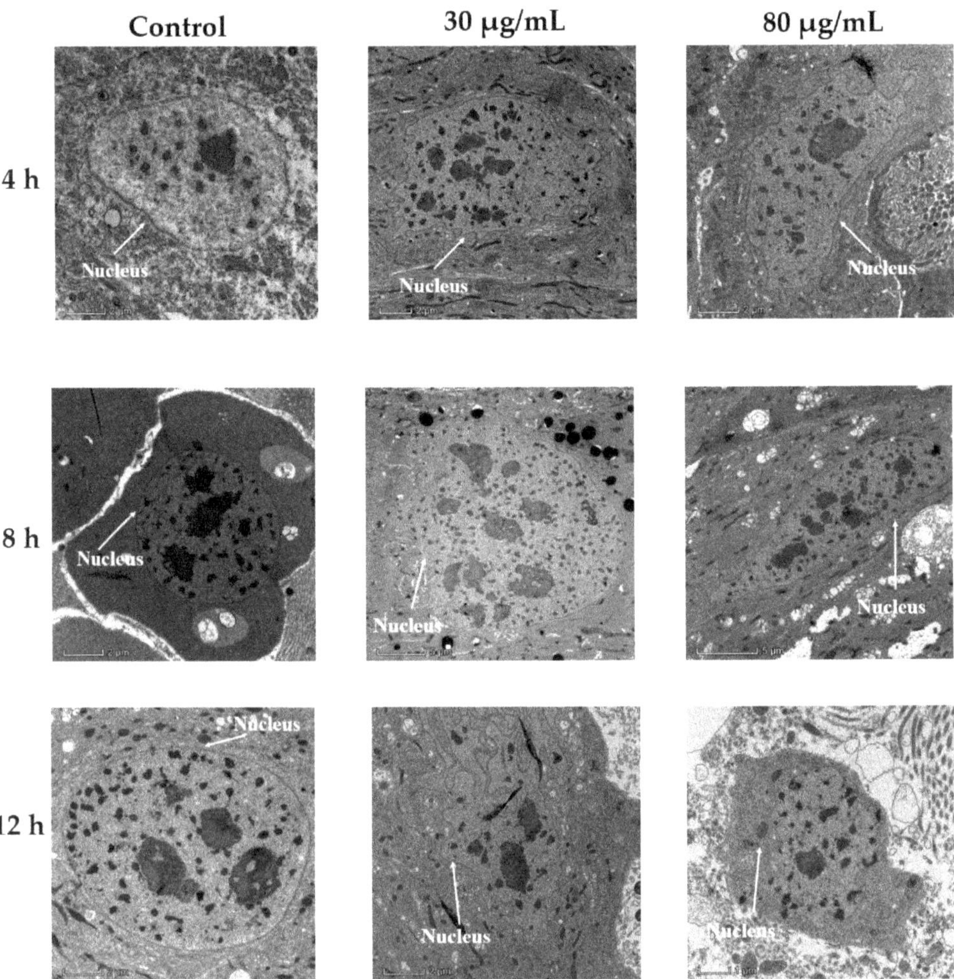

Figure 2. Changes in nucleus region of fourth *Plutella xylostella* (fourth instar larvae) midgut cells following cyclosporin C treatment.

The microvilli of treated cells became shorter and more disorganized with the passage of time (Figure 3). The microvilli of the midgut cells from *P. xylostella* larvae treated with an 80 μg/mL concentration became completely detached and minute in size after 12 h of treatment when compared to the control cells (Figure 3).

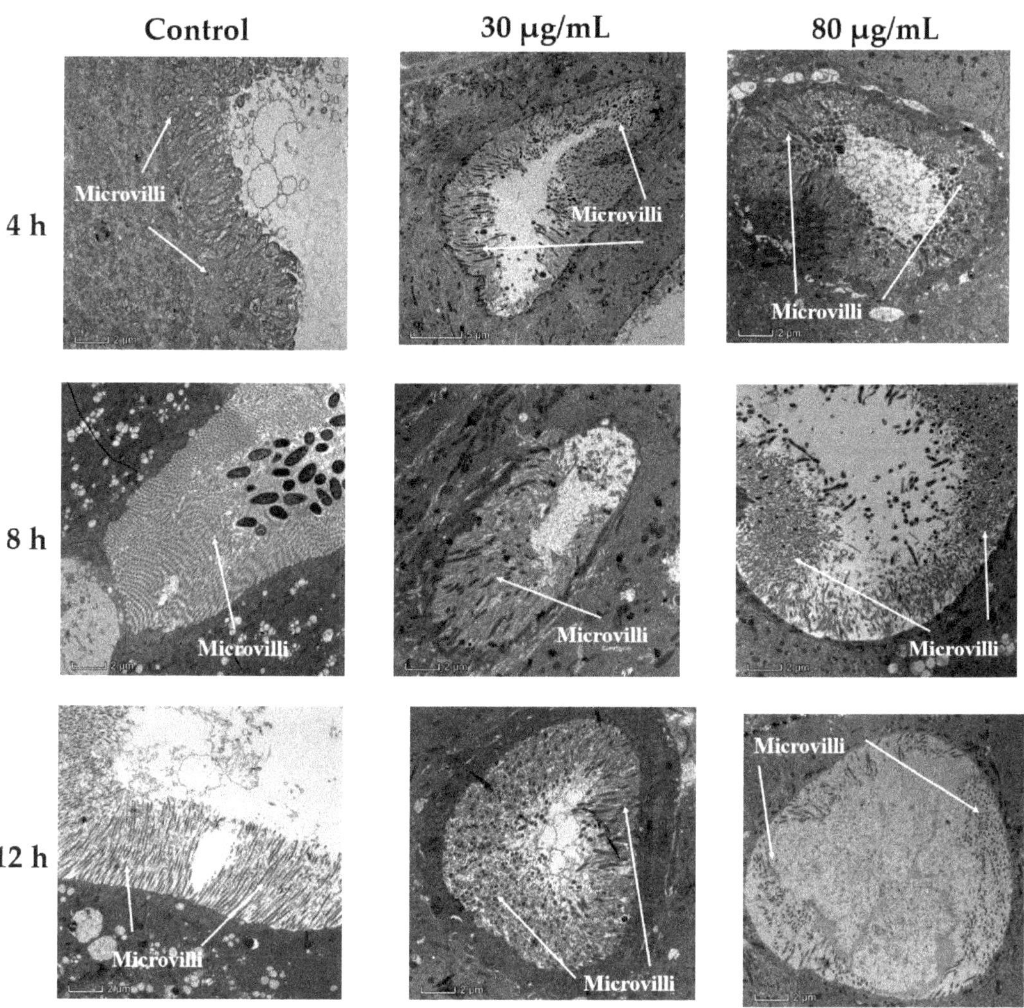

Figure 3. Changes in microvilli of *Plutella xylostella* (fourth instar larvae) midgut cells following cyclosporin C treatment.

2.2. Sublethal Effects of Cyclosporin C against P. xylostella

2.2.1. Effect of Cyclosporin C on Feeding and Larval Weight of *P. xylostella*

The average feeding rates (milligrams of leaves consumed per larvae) of *P. xylostella* larvae at different time intervals were significantly different among the cyclosporin C treatments and the control ($F_{14,68}$ = 26.92, $p < 0.0001$). A decrease in the amount of leaf consumed (mg) by *P. xylostella* larvae was observed with increases in the cyclosporin C concentration. The amount of leaf consumed (mg) by *P. xylostella* larvae treated with the 30 μg/mL cyclosporin C treatment across the experimental period was significantly higher than that of those treated with the 80 μg/mL cyclosporin C treatment, but it was still significantly lower than that of the control (Figure 4A).

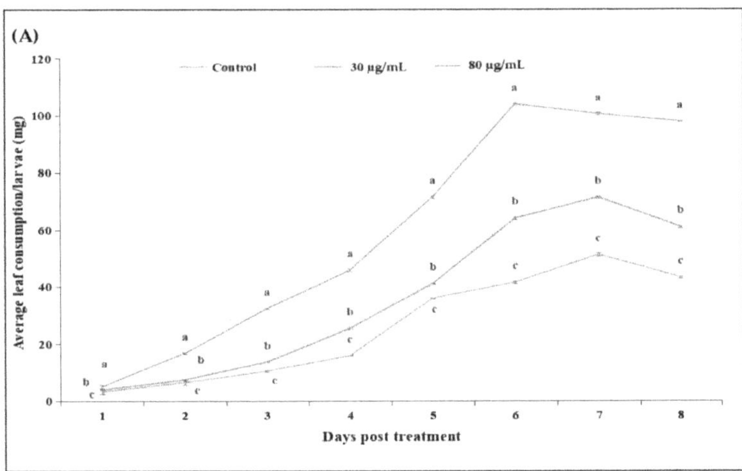

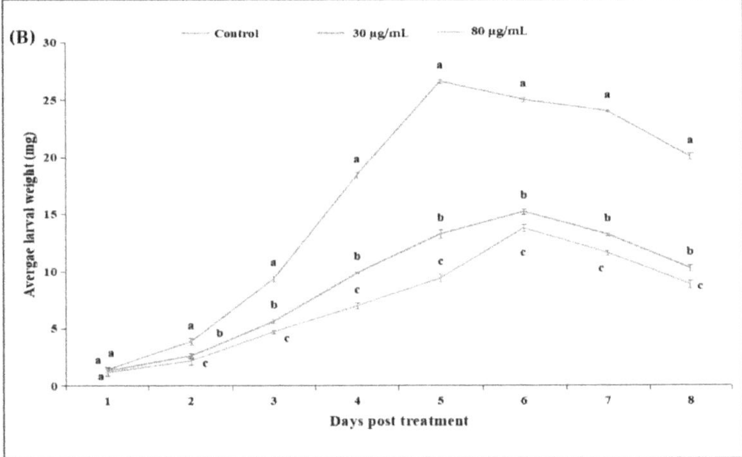

Figure 4. Amount of cabbage leaf consumed (**A**) and change in larval body weight (**B**) of *Plutella xylostella* larvae treated with cyclosporin C at different time intervals. Lines represent standard error of means (based on three independent replicates). Lines with different letters represent significantly different means (Tukey's, $p < 0.05$).

The average weight (mg) of *P. xylostella* larvae at different time intervals were significantly different among the cyclosporin C treatments and the control ($F_{14,68} = 26.92$, $p < 0.0001$). A decrease in the average weight (mg) of *P. xylostella* larvae was observed with increases in the cyclosporin C concentration. The average weight (mg) of *P. xylostella* larvae treated with 80 µg/mL cyclosporin C was significantly lower than the larval weights observed for those under the 30 µg/mL cyclosporin C treatment (except one day post-treatment) (Figure 4B).

2.2.2. Sublethal Toxicity of Cyclosporin C

The pupation rate (%) of *P. xylostella* after 10 days of treatment was significantly different among the different cyclosporin C treatments and the control ($F_{2,5} = 21.69$, $p < 0.001$). Maximum pupation (96.66 ± 1.66%) was observed for the control, whereas the rates of pupation observed for the 30 µg/mL and 80 µg/mL cyclosporin C treatments were 58.33 ± 1.40 and 14.44 ± 0.93%, respectively (Table 1).

Table 1. Sub-lethal effects of cyclosporin C on life history parameters of *Plutella xylostella*.

Life History Parameters	Control	30 µg/mL	80 µg/mL
Pupation (%)	96.66 ± 1.66 a	58.33 ± 1.40 b	14.44 ± 0.93 c
Adult emergence (%)	66.66 ± 2.33 a	23.33 ± 1.66 b	7.78 ± 1.11 c
Pre-oviposition period (h)	7.25 ± 0.66 c	9.33 ± 0.50 b	11.50 ± 0.33 a
Fecundity (eggs/female)	197 ± 4.48 a	137 ± 2.93 b	82 ± 2.34
Female longevity (d)	12 ± 0.33 a	9 ± 0.66 b	7 ± 0.66 c

Means (±standard error) in same row followed by different letters (a, b, c) are significantly different from each other (Tukey's, $p < 0.05$).

The adult emergence (%) of *P. xylostella* after 10 days of treatment was significantly different among the different cyclosporin C treatments and the control ($F_{2,5}$ = 34.84, $p < 0.001$). The maximum adult emergence (66.66 ± 2.33%) was observed for the control, whereas the rates of adult emergence observed for the 30 µg/mL and 80 µg/mL cyclosporin C treatments were 23.33± 1.66 and 7.78 ± 1.11%, respectively (Table 1).

The pre-oviposition periods of *P. xylostella* after 10 days of treatment were significantly different among the different cyclosporin C treatments and the control ($F_{2,5}$ = 36.97, $p < 0.001$). The longest pre-oviposition period (11.50 ± 0.33 hrs) was observed for 80 µg/mL cyclosporin C, whereas the pre-oviposition periods of *P. xylostella* observed for the 30 µg/mL treatment and the control were 9.33 ± 0.50 and 7.25 ± 0.66 hrs, respectively (Table 1).

The fecundity of *P. xylostella* adults that emerged from fourth instar larvae treated with different cyclosporin C concentrations was significantly different among different treatments and the control ($F_{2,5}$ = 43.26, $p < 0.001$). The maximum fecundity (197 ± 4.48 eggs/female) was observed for the control, whereas the *P. xylostella* female fecundity observed for the 30 µg/mL and 80 µg/mL cyclosporin C treatments were 137 ± 2.93 and 82 ± 2.34 eggs/female, respectively (Table 1).

The longevity of *P. xylostella* adults that emerged from fourth instar larvae treated with different cyclosporin C concentrations was significantly different among different treatments and the control ($F_{2,5}$ = 32.97, $p < 0.001$). The maximum *P. xylostella* female longevity (12 ± 0.33 days) was observed for the control, whereas the *P. xylostella* female longevity observed for the 30 µg/mL and 80 µg/mL cyclosporin C treatments were 9 ± 0.66 and 7 ± 0.66 days, respectively (Table 1).

2.2.3. Effect of Cyclosporin C Treatment on Neurophysiological and Antioxidant Enzyme Activities of *P. xylostella*

The acetylcholinesterase (AChE) activities of *P. xylostella* were significantly different among the cyclosporin C treatments and the control at different time intervals. The AChE activities of *P. xylostella* larvae treated with 30 µg/mL or 80 µg/mL cyclosporin C were significantly lower than those observed for the control. The lowest AChE activities were observed for the 80 µg/mL cyclosporin C treatment throughout the experimental period, whereas the highest (AChE) activities were observed for the control (Figure 5A).

The glutathione S-transferase (GST) activities of *P. xylostella* showed significant differences among the different cyclosporin C treatments and the control at different time intervals. The GST activities in response to different treatments showed an increasing trend until 36 h post-treatment, whereas a sharp decline in GST activities was observed at 36 h post-treatment until the end of the experimental periods. The lowest GST activities at different time intervals were observed for the 80 µg/mL cyclosporin C treatment, whereas the highest (AChE) activities at different time intervals were observed for the control treatments (Figure 5B).

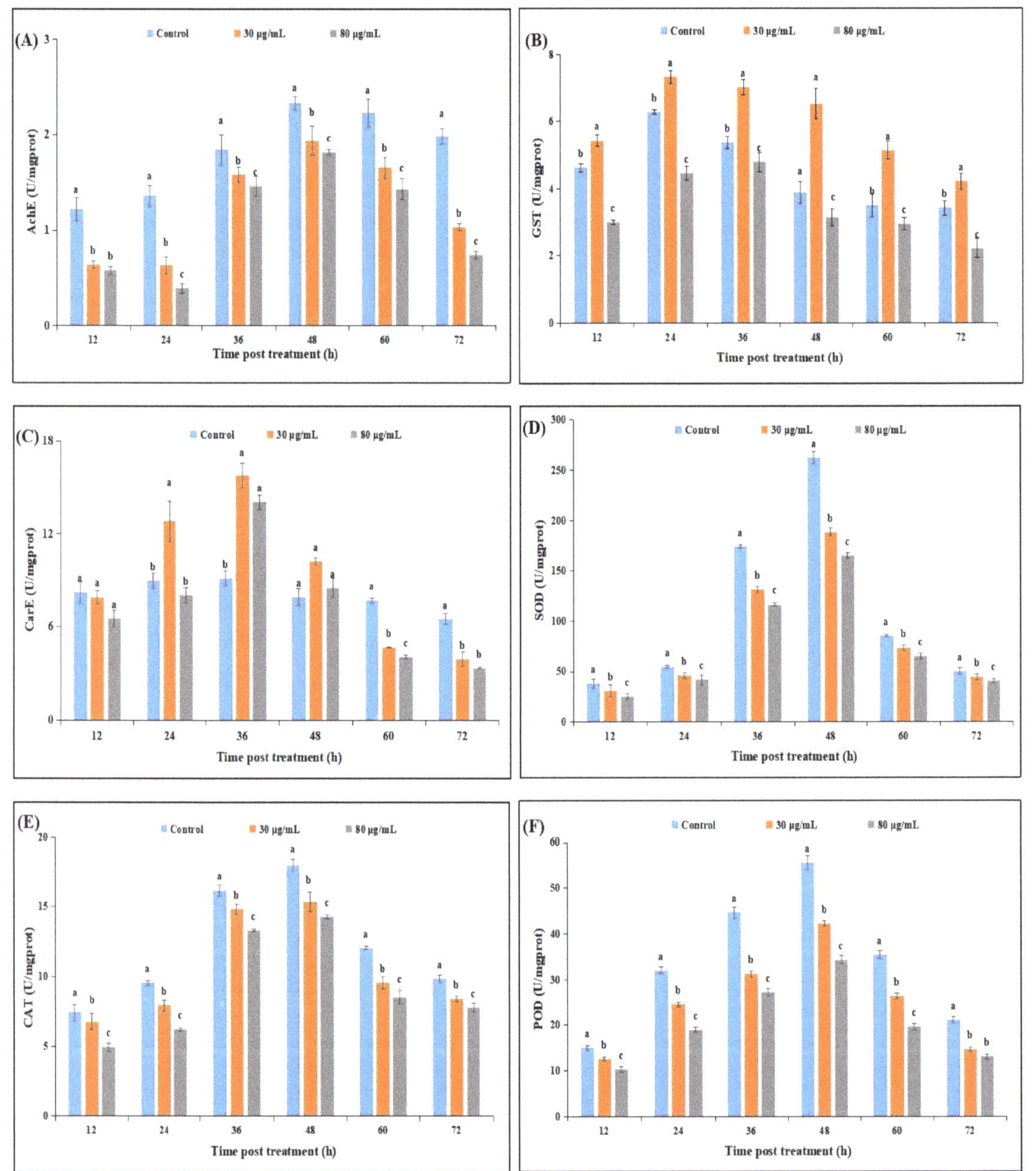

Figure 5. Effect of cyclosporin C on neurophysiological and antioxidant enzyme activities. (**A**) Acetylcholinesterase activity (AchE); (**B**) glutathione S-transferase (GST); (**C**) carboxylesterase (CarE); (**D**) superoxide dismutase (SOD); (**E**) catalase (CAT), and (**F**) peroxidase (POD). Bars represent standard error of means (based on three independent replicates). Bars with different letters represent significantly different means at different time intervals (Tukey's, $p < 0.05$).

The carboxylesterase (CarE) activities of *P. xylostella* were significantly different among the cyclosporin C treatments and the control at different time intervals. The CarE activities in response to different treatments showed an increasing trend until 36 h post-treatment, whereas a sharp decline in CarE activities was observed at 36 h post-treatment until the end of the experimental periods. The CarE activities of *P. xylostella* larvae treated with 30 μg/mL or 80 μg/mL cyclosporin C after 24, 36, and 48 h of treatment were significantly higher than the CarE activities observed for the control treatment. The lowest CarE activities throughout the experimental period (except 48 h post-treatment) were observed for the 80 μg/mL cyclosporin C treatment (Figure 5C).

The superoxide dismutase (SOD), catalase (CAT), and peroxidase (POD) activities of *P. xylostella* were significantly different among the cyclosporin C treatments and the control at different days post-treatment. The antioxidant enzyme activities in response to different treatments showed an increasing trend until 48 h post-treatment, followed by a sharp decline until the end of the experimental periods. The lowest antioxidant enzyme activities throughout the experimental period were observed for the 80 μg/mL cyclosporin C treatment, whereas the highest activities of antioxidant enzymes were observed for the control (Figure 5D–F).

3. Discussion

Data on the toxicity of microbial toxins (such as cyclosporins) against different insect pests are very limited to date [31]. This study demonstrated that cyclosporin C produced by entomopathogenic fungus *P. lilacinum* can induce lethal and sublethal effects against *P. xylostella*, which provides initial information about the potential of cyclosporin C for *P. xylostella* management.

Bioassay studies have revealed that different larval instars of *P. xylostella* were sensitive to different concentrations of cyclosporin C, showing a dose-dependent response. The median lethal concentrations (LC_{50}) of cyclosporin C against first, second, third, and fourth instar larvae 72 h after application were 78.05, 60.42, 50.83, and 83.05 μg/mL, respectively. The transmission electron microscopy of the midgut of fourth instar *P. xylostella* larvae showed nuclei enlargement, folding of the nuclear membrane, and detachment of the microvilli from the midgut cells. Our results are similar to those of Weiser and Matha [12], who reported the toxicity of cyclosporin A against *Culex pipiens*, accompanied by histopathological changes to *C. pipiens* tissues (the formation of swollen mitochondria and vacuolization of the endoplasmic reticulum). Our results are also consistent with those of Vilcinskas et al. [5], who reported that an artificial diet supplemented with a sublethal concentration of cyclosporin A reduced the phagocytosis of plasmatocytes in *Galleria mellonella*.

To date, not many studies on microbial toxins have focused on their effects on the feeding rate, fecundity, and adult longevity of target insect pests, as these parameters can influence the ultimate population dynamics of insect hosts. Observations on antifeedant activity, larval growth rate, and adult emergence are very vital, as these variables are directly related to growth and reproduction [31]. Our results show a significant reduction in the amount of leaf consumed by *P. xylostella* per day, as well as an increase in the larval weight per day compared to the control after treatment with cyclosporin C. Different life history parameters, such as pupation rate (%), adult emergence (%), female fecundity, and female longevity, were also inhibited by different concentrations of cyclosporin C. These toxic properties of cyclosporin C can be related to the mode of action of cyclosporins. This group of fungal toxins has an ionophore-like mode of action that targets the Ca^{+2} level of invertebrate cells [32,33].

The biochemical responses by insect hosts to fungal are the least explained, and further elaboration of these responses is required to improve our knowledge of the toxicity of fungal cyclosporin C. Significant fluctuations in detoxifying and antioxidant enzyme activities of *P. xylostella* were observed post-cyclosporin C treatment. Insect immune systems respond to outer threats (pesticides or microorganisms) by regulating the enzyme activities of detoxifying enzymes, such as esterases and glutathione-S-transferase [34]. This study has revealed increased CarE and GSTs activities in *P. xylostella* larvae treated with cyclosporin

C, confirming their role in cyclosporin C detoxification by *P. xylostella*. Our results are consistent with those of Luo and Zhang [35], who observed similar changes in the CarE and GSTs activities of *P. xylostella* treated with matrine. Acetylcholinesterase (AChE) terminates insect nerve impulses through acetylcholine breakdown at insects' synapses [36]. Our results show an increase in AchE activities until 48 h post-cyclosporin C treatment, followed by a sharp decline. Antioxidant enzymes (SOD, CAT, and POD) are responsible for reducing damage caused by reactive oxygen species (ROS) [29]. We observed an increase in the SOD, CAT, and POD activities during the initial 48 h, followed by a reduction in enzyme activities post-cyclosporin C treatment. The reduced antioxidant enzyme activities might have resulted in the lower removal of the ROS and the denaturation of different bio-molecules, which led to higher insect mortality [37].

4. Conclusions

Our results show promising findings about the lethal and sublethal effects of cyclosporin C against *P. xylostella*, showing its potential for future development as a biocontrol agent of insects, such as the diamondback moth. These findings can serve as baseline information for the development of cyclosporin C as an insect control agent. However, further studies on the mass production, formulation, dose selection, and field application of cyclosporin C, as well as the ecological implications of its application (including potential biosafety studies on natural enemies and other living organism side effects) are still required for the successful development of cyclosporin C as a biopesticide.

5. Materials and Methods

5.1. Plutella xylostella and Cyclosporin C

The *Plutella xylostella* larvae used in this study were obtained from the stock cultures reared on *Brassica campestris* L. (Datou Aijiao cultivar; Guangzhou Changhe Seeds, Guangzhou China) at the Guangdong Key Laboratory of Biopesticide Innovation and Application, South China Agricultural University, Guangzhou [38]. Freshly hatched first instar larvae were used for anti-feedant assays, while fourth instar larvae were used for life history studies and enzyme assays.

Cyclosporin C was extracted and purified from strain XI-5 of *Purpureocillium lilacinum* (NCBI accession No. MW386435) following the method described by Burhan et al. [39]. The cyclosporin C was diluted to 500 μg/mL with dimethyl sulfoxide (DMSO) and stored at 4 °C until further use.

5.2. Toxicity of Cyclosporin C against P. xylostella

5.2.1. Toxicity Assays

Different concentrations of cyclosporin C (25, 50, 100, 150, 200, and 300 μg/mL) were prepared from a stock solution through serial dilution using a 15% sucrose solution. The toxicity assays of cyclosporin C against different larval instars of *P. xylostella* were performed by oral route following the method by Ali et al. [40]. Briefly, cabbage leaf discs (10 cm^2) were dipped for 30 s in different concentrations of cyclosporin C and air-dried (for 2 h) on paper. The leaf discs similarly treated with ddH$_2$O served as the control. Freshly hatched first, second, third, and fourth instar *P. xylostella* larvae (30 individuals each) were transferred to *B. campestris* leaves from individual treatments (different concentrations and the control) via a camel hairbrush. *P. xylostella* larvae were fed on treated leaf discs for 48 h, followed by feeding on normal *B. campestris* leaf discs. The leaf discs were replaced every 48 h. The Petri dishes were incubated at 26 ± 1 °C and 55 ± 10% R.H. and 14:10 h (light/dark photoperiod). The leaf discs (treated and control) were replaced on a daily basis. The number of dead insects was counted on a daily basis. All the treatments and the control were repeated thrice with freshly hatched *P. xylostella* larvae.

5.2.2. Transmission Electron Microscopic Examination of *P. xylostella* Midgut Following Cyclosporin C Treatment

Fourth instar *P. xylostella* larvae were treated with two concentrations of cyclosporin C (30 and 80 µg/mL) and the control treatment (ddH$_2$O) following the method described in Section 5.2.1. Changes in the appearance of the infected *P. xylostella* midgut were directly monitored at 4, 8, and 12 h post-treatment under a JEM1011 transmission electron microscope (Nikon Co. Ltd., Tokyo, Japan) following the method by Du et al. [36]. The treated larvae were sampled at 4, 8, and 12 h post-treatment and dissected under a stereo microscope (Stemi 508, ZEISS, Germany) to obtain midgut samples. The samples were fixed overnight in 2.5% glutaraldehyde +2% paraformaldehyde solution at 4 °C, followed by rinsing with PBS buffer (0.1 M). The samples were then stained overnight with 1% uranyl acetate at 4 °C, followed by dehydration with gradient concentrations of ethanol. The dehydrated tissues were embedded in silica gel blocks, and sections were cut using an automatic microwave tissue-processing instrument (EM AMW, Leica, Germany), and a cryo-ultramicrotome (EM UC7/FC7, Leica, Germany).

5.3. Antifeedant and Sublethal Toxicity of Cyclosporin C

5.3.1. Antifeedant Activity Assays

For antifeedant assays, *B. campestris* leaf discs (10 cm^2) were dipped in different concentrations of cyclosporin C (30 and 80 µg/mL) for 30 s, followed by air-drying (for 1 h) on Whatman Grade 1 filter paper. Leaf discs dipped in distilled water served as the control. Freshly hatched first instar *P. xylostella* larvae were transferred to the treatment and control leaf disks. The leaf disks were placed in Petri dishes lined with Whatman Grade 1 filter paper. The Petri dishes were incubated at 26 ± 1 °C and 55 ± 10% R.H. and 14:10 h (light/dark photoperiod). The *P. xylostella* larvae were fed on the treated leaf discs for 48 h, followed by feeding on normal *B. campestris* leaf discs. The leaf discs were replaced every 24 h. The feeding rate was defined as the change in weight of the *B. campestris* leaf discs after 24 h. The *P. xylostella* larvae from each treatment were weighed every 24 h on a weighing balance (Sartorius BCE Entriss II Analytical Balance) and the difference in the larval weight every 24 h was used as the change in larval weight per day. The effect of feeding on larval development was measured by changes in body weight during the observation period. All the treatments and the control were repeated thrice with freshly hatched *P. xylostella* larvae (four leaf discs with 20 larvae each).

5.3.2. Sublethal Toxicity of Cyclosporin C

To determine the sublethal effect of cyclosporin C on pupation, adult emergence, adult longevity, and fecundity, a partial life cycle test was performed by feeding fourth instar *P. xylostella* larvae on *B. campestris* leaf discs (10 cm^2) dipped in two concentrations of cyclosporin C (30 and 80 µg/mL) and the control treatment (ddH$_2$O), as described in Section 5.3.1. The leaf disks were placed in Petri dishes lined with Whatman Grade 1 filter paper. The Petri dishes were incubated at 26 ± 1 °C and 55 ± 10% R.H. and 14:10 h (light/dark photoperiod). The *P. xylostella* larvae were fed on the treated leaf discs for 48 h followed by feeding on normal *B. campestris* leaf discs. The leaf discs were replaced every 48 h. The *P. xylostella* larvae were observed on a daily basis until adult emergence. All of the treatments and the control were repeated thrice. The pupation (%) and adult emergence (%) were calculated as follows:

$$\text{Pupation (\%)} = (\text{Number of successfully pupating larvae/Total number of larvae}) \times 100$$

$$\text{Adult emergence (\%)} = (\text{Number of adults emerged/Total number of pupae}) \times 100$$

The emerged adults (from each treatment group) were reared following the method of Ali et al. [40] to calculate the average number of eggs laid/day, and adult longevity.

5.4. Effect of Cyclosporin C on Neurophysiological and Antioxidant Enzyme Activities

Fourth instar *P. xylostella* larvae (50 individuals each) were treated with two concentrations of cyclosporin C (30 and 80 µg/mL) and the control treatment (ddH$_2$O). Larvae from each treatment (5 individuals) were collected every 12 h post-treatment for enzyme assays. The larvae were homogenized in specific buffers for each enzyme at 4 °C following the method by Wu et al. [41]. Detailed information on the enzymatic assays can be seen in the Supplementary Materials. A total protein assay was performed with a total protein assay kit (Beyotime Biotech Inc., Shanghai, China), using bovine albumin serum as the standard [42].

Glutathione S-transferase (GSTs) activity assays were carried out according to Habig et al. [36]. Changes in absorbance were recorded spectrophotometrically (iMark microplate absorbance reader, Biorad, Feldkirchen, Germany) at 340 nm. A reduction in glutathione per minute per mg protein was used as a unit of enzyme activity.

Carboxylesterase (CarE) was quantified using 1-naphthyl acetate as a substrate [43]. The reaction mixture (0.45 mL 0.05 mol/L sodium phosphate buffer, 1.8 mL 0.00003 mol/L 1-naphthyl acetate, and 0.05 mL sample) was incubated in water at 30 °C for 15 min, followed by the addition of 0.9 mL 1% fast blue RR salt to terminate the reaction. The change in absorbance was measured at 600 nm. The enzyme activity was determined by the protein changes per unit over time.

The acetylcholinesterase (AChE) contents of *M. usitatus* from the different treatments were observed following the methods of Wu et al. [41] and Ellman et al. [44]. Changes in the absorbance were recorded spectrophotometrically (iMark microplate absorbance reader, Biorad, Feldkirchen, Germany) at 405 nm for 40 min. A unit of enzyme activity was defined as changes in the absorbance per minute per milligram of total protein.

The antioxidant enzyme (superoxide dismutase (SOD), catalase (CAT), and peroxidase (POD)) assays were performed with the respective enzyme assay kits provided by Nanjing Jiancheng Bioengineering Institute, Nanjing China. One unit of SOD was defined as the amount of enzyme required to suppress 50% of nitroblue tetrazolium at 560 nm [45]. One unit of CAT activity was defined as the amount of enzyme required to decompose 1 mmol H$_2$O$_2$/min [46]. One unit of POD activity was defined as changes in the absorbance per minute per milligram of protein [47].

5.5. Data Analysis

The data regarding mortality were arcsine-transformed, the means were compared using analysis of variance (ANOVA-2), and the significance of the means was compared using Tukey's HSD test at a 5% level of significance. The data regarding the feeding rate, larval weight, percent of pupation, and percent of adult emergence were subjected to analysis of variance (ANOVA-1), and the significance of the means was compared using Tukey's HSD test at a 5% level of significance. SAS 9.2 was used for all statistical analyses [48].

Supplementary Materials: The following are available online at https://www.mdpi.com/article/10.3390/toxins14080514/s1.

Author Contributions: Conceptualization, J.W. and S.A.; methodology, S.A.; formal analysis, X.Z.; investigation, X.Z.; resources, J.W.; data curation, X.Z.; writing—original draft, J.W. and S.A.; writing—review and editing, M.H.B. and S.A.; supervision, S.A.; funding acquisition, J.W. and S.A. All authors have read and agreed to the published version of the manuscript.

Funding: This research was funded by grants from The Key Area Research and Development Program of Guangdong Province (2018B020205003).

Institutional Review Board Statement: Not applicable.

Informed Consent Statement: Not applicable.

Data Availability Statement: The raw data supporting the conclusion will be made available by the corresponding author on request.

Acknowledgments: The authors want to thank the handling editor and anonymous reviewers for their constructive comments and suggestions.

Conflicts of Interest: The authors declare no conflict of interest.

References

1. Wang, X.S.; Xu, J.; Wang, X.M.; Qiu, B.L.; Cuthbertson, A.G.S.; Du, C.L.; Wu, J.H.; Ali, S. *Isaria fumosorosea*-based-zero-valent iron nanoparticles affect the growth and survival of sweet potato whitefly, *Bemisia tabaci* (Gennadius). *Pest Manag. Sci.* **2019**, *75*, 2174–2181. [CrossRef] [PubMed]
2. Goffré, D.; Folgarait, P.J. *Purpureocillium lilacinum*, potential agent for biological control of the leaf-cutting ant *Acromyrmex lundii*. *J. Invertebr. Pathol.* **2015**, *130*, 107–115. [CrossRef] [PubMed]
3. Sun, T.F.; Wu, J.H.; Ali, S. Morphological and molecular identification of four *Purpureocillium* isolates and evaluating their efficacy against the sweet potato whitefly, *Bemisia tabaci* (Genn.) (Hemiptera: Aleyrodidae). *Egypt. J. Biol. Pest. Cont.* **2021**, *31*, 27. [CrossRef]
4. Amala, U.; Jiji, T.; Naseema, A. Laboratory evaluation of local isolate of entomopathogenic fungus, *Paecilomyces lilacinus* Thom Samson (ITCC 6064) against adults of melon fruit fly, *Bactrocera cucurbitae* Coquillett (Diptera; Tephritidae). *J. Trop. Agri.* **2013**, *51*, 132–134.
5. Vilcinskas, A.; Jegorov, A.; Landa, Z.; Götz, P.; Matha, V. Effects of beauverolide L and cyclosporin A on humoral and cellular immune response of the greater wax moth, *Galleria mellonella*. *Comp. Biochem. Physiol.* **1999**, *122*, 83–92. [CrossRef]
6. Khan, A.; Williams, K.; Nevalaine, H. Testing the nematophagous biological control strain *Paecilomyces lilacinus* 251 for paecilotoxin production. *FEMS Microbiol. Lett.* **2003**, *227*, 107–111. [CrossRef]
7. Fiolka, M.J. Immuno-suppressive effect of cyclosporin A on insect humoral immune response. *J. Invert. Pathol.* **2008**, *98*, 287–292. [CrossRef]
8. Jegorov, A.; Matha, V.; Weiser, J. Production of cyclosporins by entomogenous fungi. *Microbiol. Lett.* **1990**, *45*, 65–69.
9. Dumas, C.; Ravallec, M.; Matha, V.; Vey, A. Comparative study of the cytological aspects of the mode of action of destruxins and other peptidic fungal metabolites on target epithelial cells. *J. Invertebr. Pathol.* **1996**, *67*, 137–146. [CrossRef]
10. Gillespie, J.P.; Bailey, A.M.; Cobb, B.; Vilcinskas, A. Fungi as elicitors of insect immune responses. *Arch. Insect Biochem. Physiol.* **2000**, *44*, 49–68. [CrossRef]
11. Podsiadlowski, L.; Matha, V.; Vilcinskas, A. Detection of a P-glycoprotein related pump in chironomus larvae and its inhibition by verapamil and cyclosporin A. *Comp. Biochem. Physiol.* **1998**, *121*, 443–450. [CrossRef]
12. Weiser, J.; Matha, V. The insecticidal activity of cyclosporines on mosquito larvae. *J. Invertebr. Pathol.* **1988**, *51*, 92–93. [CrossRef]
13. Weiser, J.; Matha, V.; Zizka, Z.; Jegorov, A. Pathology of cyclosporin A in mosquito larvae. *Cytobios* **1989**, *59*, 143–150. [PubMed]
14. Matha, V.; Jegorov, A.; Weiser, J.; Pillai, J.S. The mosquitocidal activity of conidia of *Tolypocladium tundrense* and *Tolypocladium terricola*. *Cytobios* **1992**, *69*, 163–170.
15. Moussaif, M.; Jacques, P.; Schaarwachter, P.; Budzikiewicz, H.; Thonart, P. Cyclosporin C is the main antifungal compound produced by *Acremonium luzulae*. *Appl. Environ. Mirobiol.* **1997**, *63*, 1739–1743. [CrossRef] [PubMed]
16. Kanakala, S.; Ghanim, M. Implication of the Whitefly *Bemisia tabaci* cyclophilin B protein in the transmission of tomato yellow leaf curl virus. *Front. Plant Sci.* **2016**, *7*, 1702. [CrossRef]
17. Soleymani, T.; Vassantachart, J.M.; Wu, J.J. Comparison of guidelines for the use of cyclosporine for psoriasis: A critical appraisal and comprehensive review. *J. Drugs Dermatol.* **2016**, *15*, 293–301.
18. Buscher, A.K.; Beck, B.B.; Melk, A.; Hoefele, J.; Kranz, B.; Bamborschke, D.; Baig, S.; Lang-Sperandio, B.; Jungraithmayr, T.; Weber, L.T.; et al. Rapid response to cyclosporin a and favorable renal outcome in nongenetic versus genetic steroid-resistant nephrotic syndrome. *Clin. J. Am. Soc. Nephrol.* **2016**, *11*, 245–253. [CrossRef] [PubMed]
19. Chighizola, C.B.; Ong, V.H.; Meroni, P.L. The Use of Cyclosporine a in rheumatology: A 2016 comprehensive review. *Clin. Rev. Allergy Immunol.* **2017**, *52*, 401–423. [CrossRef]
20. Kahan, B.D. Cyclosporine. *N. Engl. J. Med.* **1989**, *321*, 1725–1738.
21. Calne, R.Y.; Thiru, S.; McMaster, P.; Craddock, G.N.; White, D.J.G.; Evans, D.B. Cyclosporin A in patients receiving renal allografts from cadaver donors. *Lancet* **1978**, *312*, 1323–1327. [CrossRef]
22. Flores, C.; Fouquet, G.; Moura, I.C.; Maciel, T.T.; Hermine, O. Lessons to learn from low-dose cyclosporin-A: A new Approach for unexpected clinical applications. *Front. Immunol.* **2019**, *10*, 588. [CrossRef] [PubMed]
23. Ezekiel, C.N.; Kraak, B.; Sandoval-Denis, M.; Sulyok, M.; Oyedele, O.A.; Ayeni, K.I.; Makinde, O.M.; Akinyemi, O.M.; Krska, R.; Crous, P.W.; et al. Diversity and toxigenicity of fungi and description of *Fusarium madaense* sp. nov. from cereals, legumes and soils in north-central Nigeria. *MycoKeys* **2020**, *67*, 95–124. [CrossRef] [PubMed]
24. Avery, S.V.; Singleton, I.; Magan, N.; Goldman, G.H. The fungal threat to global food security. *Fungal Biol.* **2019**, *123*, 555–557. [CrossRef]
25. Zalucki, M.P.; Shabbir, A.; Silva, R.; Adamson, D.; Shu-Sheng, L.; Furlong, M.J. Estimating the economic cost of one of the world's major insect pests, *Plutella xylostella* (Lepidoptera: Plutellidae): Just how long is a piece of string? *J. Econ. Entomol.* **2012**, *105*, 1115–1129. [CrossRef]

26. Shakeel, M.; Farooq, M.; Nasim, W.; Akram, W.; Khan, F.Z.A.; Jaleel, W. Environment polluting conventional chemical control compared to an environmentally friendly IPM approach for control of diamondback moth, *Plutella xylostella* (L.), in China: A review. *Environ. Sci. Pollut. Res.* **2017**, *24*, 14537–14550. [CrossRef] [PubMed]

27. Amiri, B.; Ibrahim, L.; Butt, T.M. Antifeedant properties of destruxins and their potential use with the entomogenous fungus *Metarhizium anisopliae* for improved control of crucifer pests. *Biocont. Sci. Technol.* **1999**, *9*, 487–498. [CrossRef]

28. Huang, S.; Ali, S.; Ren, S.X. The combination of destruxin A with insecticide chlorantraniliprole increases virulence against *Plutella xylostella*. *Pak. J. Zool.* **2016**, *48*, 1849–1855.

29. Arthurs, S.; Dara, S.K. Microbial biopesticides for invertebrate pests and their markets in the United States. *J. Invertebr. Pathol.* **2018**, *165*, 13–21. [CrossRef]

30. Bordalo, M.D.; Gravato, C.; Beleza, S.; Campos, D.; Lopes, I.; Pestana, J.L.T. Lethal and sublethal toxicity assessment of *Bacillus thuringiensis* var. *israelensis* and *Beauveria bassiana* based bioinsecticides to the aquatic insect *Chironomus riparius*. *Sci. Total Environ.* **2020**, *698*, 134155. [CrossRef]

31. Duchet, C.; Franquet, E.; Lagadic, L.; Lagneau, C. Effects of *Bacillus thuringiensis israelensis* and spinosad on adult emergence of the non-biting midges *Polypedilum nubifer* (Skuse) and *Tanytarsus curticornis* Kieffer (Diptera: Chironomidae) in coastal wetlands. *Ecotoxicol. Environ. Saf.* **2015**, *115*, 272–278. [CrossRef] [PubMed]

32. Vereb, G.; Panyi, J.G.; Balazs, M.; Matyus, L.; Matko, J.; Damjanovich, S. Effcet of cyclosporin A on the membrane potential and Ca^{++} level of human lymphoid cell lines and mouse thymocytes. *Biochim. Biophys. Acta* **1990**, *109*, 159–165. [CrossRef]

33. Griffiths, E.J.; Halestrap, A. Further evidence that cyclosporin A protects mitochondria from calcium overload by inhibiting a matrix peptidyl-prolyl cis-trans isomerase. *Biochem. J.* **1991**, *274*, 611–614. [CrossRef] [PubMed]

34. Claudianoc, C.; Ranson, H.; Jhonson, R.M.; Biswas, S.; Schuler, M.A.; Berenbaum, M.R.; Feyereisen, R.; Oakeshott, J.G. A deficit of detoxification enzymes: Pesticide sensitivity and environmental response in the honeybee. *Insect Mol. Biol.* **2006**, *15*, 615–636. [CrossRef] [PubMed]

35. Luo, W.C.; Zhang, Q. The effects of *Sophora alopecuroids* alkaloids on metabolic esterases of the diamondback moth. *Acta. Entomol. Sin.* **2003**, *46*, 122–125.

36. Habig, W.H. Assays for Differentiation of Glutathione S-Transferase. In *Method in Enzymology*; Willian, B.J., Ed.; Academic Press: New York, NY, USA, 1981; pp. 398–405.

37. Ding, S.Y.; Li, H.Y.; Li, X.F.; Zhang, Z.Y. Effects of two kinds of transgenic poplar on protective enzymes system in the midgut of larvae of American white moth. *J. For. Res. Jpn.* **2001**, *12*, 119–122.

38. Ali, S.; Huang, Z.; Ren, S.X. Media composition influences on growth, enzyme activity and virulence of the entomopathogen hyphomycete *Isaria fumosorosea*. *Entomol. Experiment. Appl.* **2009**, *131*, 30–38. [CrossRef]

39. Burhan, A.H.; Mohammad, R.A. Pathogenesis of *Paecilomyces lilacinus* against the immature stages of *Musca domestica* L. *J. Pharm. Sci. Res.* **2019**, *11*, 1595–1601.

40. Ali, S.; Huang, Z.; Ren, S.X. Production of cuticle degrading enzymes by *Isaria fumosorosea* and their evaluation as a biocontrol agent against diamondback moth. *J. Pest Sci.* **2010**, *83*, 361–370. [CrossRef]

41. Wu, J.; Li, J.; Zhang, C.; Yu, X.; Cuthbertson, A.G.S.; Ali, S. Biological impact and enzyme activities of *Spodoptera litura* (Lepidoptera: Noctuidae) in response to synergistic action of matrine and *Beauveria brongniartii*. *Front. Physiol.* **2020**, *11*, 584405. [CrossRef] [PubMed]

42. Bradford, M.M. A rapid and sensitive method for the quantitation of microgram quantities of protein utilizing the principle of protein dye binding. *Anal. Biochem.* **1976**, *72*, 248–254. [CrossRef]

43. Byrne, F.J.; Gorman, K.J.; Cahill, M.; Denholm, I.; Devonshire, A.L. The role of B-type esterases in conferring insecticide resistance in the tobacco whitefly, *Bemisia tabaci* (Genn). *Pest. Manag. Sci.* **2000**, *56*, 867–874.

44. Ellman, G.L.; Coutney, K.D.; Andre, V.; Featherstone, R.M. A new and rapid calorimetric determination of acetylcholinesterase activity. *Biochem. Pharmacol.* **1961**, *7*, 88–95. [CrossRef]

45. Beauchamp, C.; Fridovich, I. Superoxide dismutase: Improved assays and an assay applicable to acrylamide gels. *Anal. Biochem.* **1971**, *44*, 276–287. [CrossRef]

46. Beers, R.F.; Sizer, I.W. A spectrophotometric method for measuring the breakdown of hydrogen peroxide by catalase. *J. Biol. Chem.* **1952**, *195*, 133–140. [CrossRef]

47. Shannon, L.M.; Kay, E.; Lew, J.Y. Peroxidase isozymes from horseradish roots. I. Isolation and physical properties. *J. Biol. Chem.* **1966**, *241*, 2166–2172. [CrossRef]

48. SAS Institute. *SAS User's Guide*; Statistics SAS Institute: Cary, NC, USA, 2000.

Article

Application of an In Vitro Digestion Model for Wheat and Red Beetroot Bread to Assess the Bioaccessibility of Aflatoxin B$_1$, Ochratoxin A and Zearalenone and Betalains

Paula Llorens Castelló, Ana Juan-García *, Juan Carlos Moltó Cortés, Jordi Mañes Vinuesa and Cristina Juan García *

Laboratory of Food Chemistry and Toxicology, Faculty of Pharmacy, University of Valencia, 46100 Burjassot, Spain
* Correspondence: ana.juan@uv.es (A.J.-G.); cristina.juan@uv.es (C.J.G.)

Abstract: Nowadays, the bakery industry includes different bioactive ingredients to enrich the nutritional properties of its products, such as betalains from red beetroot (*Beta vulgaris*). However, cereal products are considered a major route of exposure to many mycotoxins, both individually and in combination, due to their daily consumption, if the cereals used contain these toxins. Only the fraction of the contaminant that is released from the food is bioaccessible and bioavailable to produce toxic effects. Foods with bioactive compounds vary widely in chemical structure and function, and some studies have demonstrated their protective effects against toxics. In this study the bioaccessibility and bioavailability of three legislated mycotoxins (AFB$_1$, OTA and ZEN), individual and combined, in two breads, one with wheat flour and the other with wheat flour enriched with 20% *Beta vulgaris*, were evaluated. Bioaccessibility of these three mycotoxins from wheat bread and red beet bread enriched individually at 100 ng/g was similar between the breads: 16% and 14% for AFB1, 16% and 17% for OTA and 26% and 22% for ZEN, respectively. Whereas, when mycotoxins were co-present these values varied with a decreasing tendency: 9% and 15% for AFB1, 13% and 9% for OTA, 4% and 25% for ZEN in wheat bread and in red beet bread, respectively. These values reveal that the presence of other components and the co-presence of mycotoxins can affect the final bioavailability; however, it is necessary to assess this process with in vivo studies to complete the studies.

Keywords: bioaccessibility; aflatoxin B1; ochratoxin A; zearalenone; bread; beetroot; LC-Q-TOF-MS

Key Contribution: Quantities of mycotoxins and betalains decrease all along the digested phases; If the mycotoxins are in a combined presence, the bioaccessibility decreases; ZEN combined with OTA and AFB1 exhibited the lowest bioaccessibility (4%); High bioaccessibility of OTA; AFB1 and ZEN in red beetroot breads was observed.

Citation: Llorens Castelló, P.; Juan-García, A.; Cortés, J.C.M.; Mañes Vinuesa, J.; Juan García, C. Application of an In Vitro Digestion Model for Wheat and Red Beetroot Bread to Assess the Bioaccessibility of Aflatoxin B$_1$, Ochratoxin A and Zearalenone and Betalains. *Toxins* **2022**, *14*, 540. https://doi.org/10.3390/toxins14080540

Received: 28 June 2022
Accepted: 1 August 2022
Published: 8 August 2022

Publisher's Note: MDPI stays neutral with regard to jurisdictional claims in published maps and institutional affiliations.

1. Introduction

Cereals and cereal-based products play a crucial role in the human diet and livestock feed, due to their valuable nutritional content, such as carbohydrates, proteins, fatty acids and vitamins [1]. However, these characteristics make cereals a good source of nutrients for the growth and appearance of fungi. In the case of mycotoxigenic fungi, it represents a toxicological risk that has to be controlled and, as far as possible, prevented. Mycotoxigenic fungi stand out for their ability to grow in a wide spectrum of climatic conditions. Fungal contamination of crops can be carried out at all stages of production; pre-harvest and harvest (in the field) or in the post-harvest stages (transport and storage). Good handling practices and a hazard analysis and critical control point (HACCP) system are essential in prevention of fungal contamination. However, in many cases, such systems are not applied and the presence of mycotoxins becomes possible. *Aspergillus* and *Fusarium* species colonize cereals and cereal products, and they are producers of the following three most toxicological mycotoxins: aflatoxins (*Aspergillus flavus* and *Aspergillus parasiticus*), ochratoxin-A (OTA)

(*Aspergillus ochraceus* and *Penicillum verucosum*) and zearalenone (ZEN) (*Fusarium graminearum, F. culmorum, F. equiseti,* and *F. verticillioides*). The intake of these mycotoxins could cause health effects, such as mutagenicity, dermato-toxicity, neurotoxicity, hepatotoxicity, teratogenicity, estrogenicity, carcinogenicity and immunosuppressive effects [2–5]. Furthermore, the IARC has classified aflatoxins (AFs) as agents of carcinogenicity. OTA is possibly carcinogenic to humans and ZEN is not classifiable as regards humans. Due to these effects in humans and animals, national and European legislative institutions have established maximum tolerable levels of AFB1 and OTA in bakery products to protect consumers from the health risks associated with their intake. It is established in EC Regulation No. 1881/2006 [6] that the highest legislative content for products derived from cereals, such as bread, is 2 µg/kg for AFB1, 3 µg/kg for OTA and 100 or 50 µg/kg for ZEN in cereal products or bread, respectively [6].

After ingestion of food contaminated by AFB1, OTA or ZEN, the quantities of mycotoxins that are absorbed gastro-intestinally and reach the circulation system determines their toxicological effect. In this sense, bioaccessibility is necessary to elucidate the risk assessment. Furthermore, previous studies have elucidated that the presence of other compounds, or the dietary source, influences the percentage of bioaccessibility. In wine, OTA presented 26% bioaccessibility [7]. Versantvoort et al., [8] evaluated the effect of a food-mix of peanut slurry, buckwheat and standard meal on the bioaccessibility of aflatoxin B1 (AFB1) and OTA, obtaining 84% for AFB1 and 86% for OTA in combined contamination, and 83% for AFB_1 and 100% for OTA in single contamination.

Red beetroot is rich in bioactive ingredients that ensure health-promoting effects and it is recognized as a functional food, due to its high nutritional value. Red beet is commonly consumed and used for manufacturing food coloring agents. Its consumption has witnessed a pronounced increase. In bakery, it is used to increase nutritional value of bakery products and also to obtain colorimetrically acceptable and tasty products. Furthermore, red beet is the precursor of betalains that have become popular because they have potent antioxidant, anticarcinogenic, hepatoprotective, antibacterial and anti-inflammatory activities, as well as intestinal and immune regulatory effects. In addition, they protect cells against peroxidation and DNA damage [9]. Recently, Penalva-Olcina et al., [10] observed the effective protection of beetroot extract in SH-SY5Y neuronal cells against Fumonisin B1, Ochratoxin A and their combination. Beetroot extract in bread or biscuits, as a product of daily consumption, can increase its intake and health effects.

Due to the health-related benefits, and as organoleptic ingredients, it would be a good protective strategy to attenuate the biological effects of AFB_1, OTA and ZEN in case they are ingested. However, to consider these effects it is first necessary to evaluate the bioaccessibility of both types of compounds. In this study, the bioaccessibility of three different legislated mycotoxins (AFs, OTA and ZEA), individually and combined, in two breads, one with wheat flour and the other with wheat flour enriched with *Beta vulgaris*, was evaluated. Furthermore, the bioaccessibility of some betalains, identified in enriched bread, all through the different digestive phases (salival, gastric and intestinal) were studied.

2. Results and Discussion

2.1. Fortification Levels of AFB_1, OTA and ZEN Analysis

European legislation has established maximum levels (MLs) for OTA, AFB_1 and ZEN in all products derived from unprocessed cereals, including processed cereal products and cereals intended for direct human consumption. Such levels are set at 3 and 2 µg/kg for OTA and AFB_1, respectively [6]. For unprocessed cereals the maximum level is set at 5 µg/kg for OTA. However, in processed cereal-based foods and baby foods for infants and young children the maximum is set at 0.5 and 0.1 µg/kg for OTA and AFB_1, respectively [6]. The highest values of ML for ZEN have been established as follows: 100 µg/kg for unprocessed cereals; 50 µg/kg in bread, including small bakery wares, pastries, biscuits, cereal snacks and breakfast cereals, excluding maize snacks and maize-based breakfast cereals; and 20 µg/kg for processed cereal-based foods (excluding processed maize-based foods) and baby foods for infants and young children. Despite these regulations and food controls, the mycotoxins have been detected in routine customs analysis and in products from various countries; in some cases, in bread from Guyana, Morocco, Portugal, Turkey and Malaysia, at levels 30 times above the maximum limits (MLs) [11–16]. Surprisingly, it is possible that many consumers have ingested these mycotoxins through one of the most consumed food products, bread.

In literature, different studies indicate that during the bakery process a reduction in the presence of mycotoxins by 7–85% and 6–40% for OTA and AFB_1, respectively, has been observed [17,18]. For this reason, this study was performed after the preparation of the bread and it was established that the studied levels of fortifications would be at two levels, 100 µg/kg and 10 µg/kg, before the digestion process, by adding 1 mL of the working solutions (100 µg/mL) in 10 g of wheat bread.

2.2. Bioaccessibility of OTA, AFB1 and ZEN in Studied Breads

Salival, gastric and intestinal phase digestions were evaluated. However, the bioaccessibility was calculated at the last phase, corresponding to the intestinal phase. These data were analyzed for both beet and wheat bread, comparing the differences in bioaccessibility between the levels of single mycotoxins and their combinations. The bioaccessibility value obtained was a percentage value, representing availability to be absorbed at the intestinal level, and it assumed the free fraction of the matrix. Therefore, the most relevant bioaccessibility results were those corresponding to the intestinal phase. Although, it should not be forgotten that a slight absorption of the components can occur throughout the gastrointestinal tract. For this reason, data from the other phases were collected and are discussed in this work.

Availability was calculated for all three phases, and it was calculated as the percentage of mycotoxins from the bread that was detected in extracts from each phase, using the formula: Availability = [(CDE)/(CB)] ∗ 100. (CDE: concentration in digested extract and CB: concentration in bread.

2.2.1. Availability of AFB1, OTA and ZEN in Salival, Gastric and Intestinal Phase in Both Breads

The percentage of ZEN progressively decreased during the digestion process at a fortification level of 100 ng/g in individual (51% to 22%) and combined (10% to 5%) presence in wheat (Figure 1(a1)). In red beet bread (Figure 1(b1)) a decrease in values in the intestinal phase was observed, but all values were above 23%. Regarding OTA and AFB_1, values presented variably along the digestion phases, reaching 15% in the intestinal phase in wheat bread for both mycotoxins (Figure 1(a1)), and 17% and 15% in red beet bread (Figure 1(b1)), respectively. OTA in the gastric phase presented higher percentages (20%, 42%, 31% and 25%) than in the saliva and intestinal phases, probably due to the fact that the bread was less digested, because in this phase amylase and pepsin activities are present. However, in the intestinal phase pancreatin and bile are active, so more components were loose in the bolus and OTA was degraded.

According to the results obtained, availability in each phase increased significantly, at the low level of fortification (10 ng/g), reaching the intestinal phase with values up to 65% for OTA and ZEA and, in the case of AFB_1, values of availability between 10–45% in wheat bread and slightly higher in beetroot bread with 25–45%. This fact might be due to adhesion mechanisms that favor its protection against the enzymes responsible for digestion. Contamination with these mycotoxins at low levels allows them to remain embedded in the food matrix, and, therefore, the percentage digested is lower, compared to those samples fortified at higher levels, where the mycotoxins are exposed to enzymatic action. Furthermore, the chemical structure is different, with AFB_1 presenting a higher amount of lactone and pentanone than OTA or ZEN. Based on the results obtained, ZEN is highly bioaccessible (100%) in wheat bread when fortified at 10 ng/g.

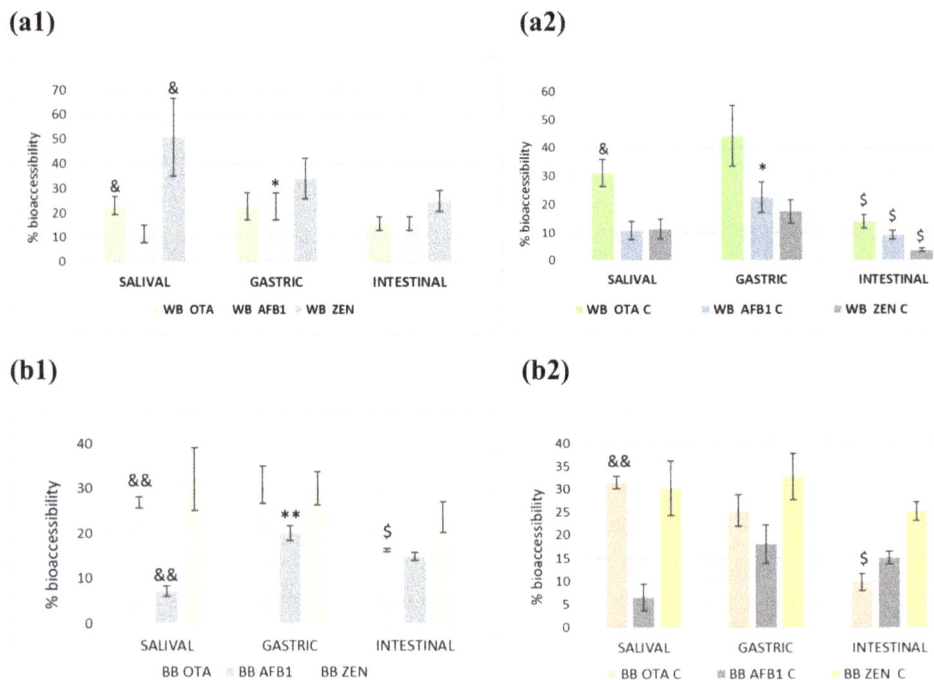

Figure 1. Percentage of digested mycotoxin (%) in three studied phases of: (**a**) wheat bread at 100 ng/g individual (**a1**) and combined fortification (**a2**); (**b**) beet bread at 100 ng/g individual (**b1**) and combined fortification (**b2**) (*: Salival-Gastric *** $p < 0.001$; ** $p < 0.01$; * $p < 0.05$; \$: Gastric-Intestinal \$\$\$ $p < 0.001$; \$\$ $p < 0.01$; \$ $p < 0.05$; and &: Salival-Intestinal &&& $p < 0.001$; && $p < 0.01$; & $p < 0.05$).

Comparing both breads it was observed that red beet bread presented high availability in the three phases in both fortifications (single Figure 1(b1) and combined Figure 1(b2)). However, AFB1 presented the lowest values (8–16% and 6–15%). No differences were observed between individual or combined fortifications.

The highest statistically significant differences for all three mycotoxins ($p < 0.05$) were observed in combined fortifications between gastric to intestinal phases in wheat flour bread (Figure 1(a2) and Figure 1(b2)).

2.2.2. Bioaccessibility of AFB1, OTA and ZEN in Both Breads

Bioaccessibility was calculated as the percentage of mycotoxins from the bread that were detected in intestinal phase extracts by the formula: Bioaccesibility = [(CDE)/(CB)] $*$ 100 in which CDE is the concentration in digested extract and CB is the concentration in bread.

The results were analyzed as individual and combined (AFB$_1$+OTA+ZEN) presence of mycotoxins in bread (Figure 2). The results showed that ZEN had high bioaccessibility in wheat flour bread as a single fortification (26%) and in red beetroot bread with both fortifications (22% for single and 26% combined with the other two mycotoxins). OTA presented values of 16% and 17% in wheat and beet bread, respectively, when fortified individually with 100 ng/g of OTA, and in combination AFB$_1$ + ZEN at 100 ng/g values of 14% in wheat bread and 10% in red beet bread.

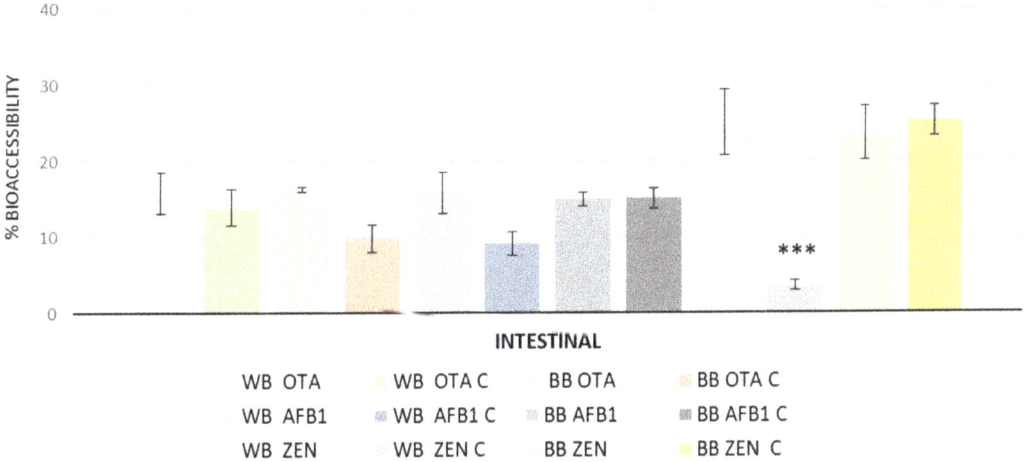

Figure 2. Individual (WB OTA, WB AFB$_1$, WB ZEN) and combined (WB OTAC, WB AFB$_1$C, WB ZENC) bioaccessibility results in wheat bread (WB) versus beet bread (BB) at 100 ng/g of mycotoxin (OTA, AFB$_1$, ZEN) (*** $p < 0.001$; ** $p < 0.01$; * $p < 0.05$).

Similar values were observed with AFB$_1$. However, in wheat flour bread combined with OTA+ZEN at 100 ng/g the bioaccessiblity was 9%, which was 8 units lower than wheat flour bread contaminated only with AFB1. The bioaccessibility of the three mycotoxins in bread were lower than those observed by other authors individually without meal. Sobral et al. [19] observed that AFB$_1$ and OTA were released along GI digestion, exhibiting a maximum cumulative isolated bioaccessibility of 51.6 and 72.4%, respectively, after intestinal digestion.

Regarding interactions, the release of AFB1 or OTA was significantly affected by the presence of the other mycotoxin.

2.2.3. Bioaccessibility of Betalains in Both Breads

Betanin is the only standard commercially available betalain and, therefore, it was the only compound of the betalain family of on which to perform further specific studies, such as qualitative and quantitative analyses. A tentative identification of betalains in dried beetroot powder was possible by LC-Q-TOF-MS which permitted the identification of seven betalains, two betacianins (betanin, betanidin) and four betaxanthins (Vulgaxanthin-III: Asparagine-betaxanthin; Vulgaxanthin-I: Glutamine-betaxanthin; Vulgaxanthin-II: Glutamic acid-betaxanthin; Tyrosine-betaxanthin). However, it was only possible to detect three betalains (betanin, betanidin and Vulgaxanthin-III) in the three phases of digestion and they were evaluated in digested red beetroot baked bread (Figure 3). In the literature, different studies indicate the total betalain content in red beet and derivative products [19,20]. Most recently, Sawicki et al., [21] studied thirteen varieties of red beets and reported betaxanthin levels ranging from 2.71 to 4.25 mg/g DW, whereas the content of betacyanins varied between 8.3–13.5 mg/g dry weight (DW).

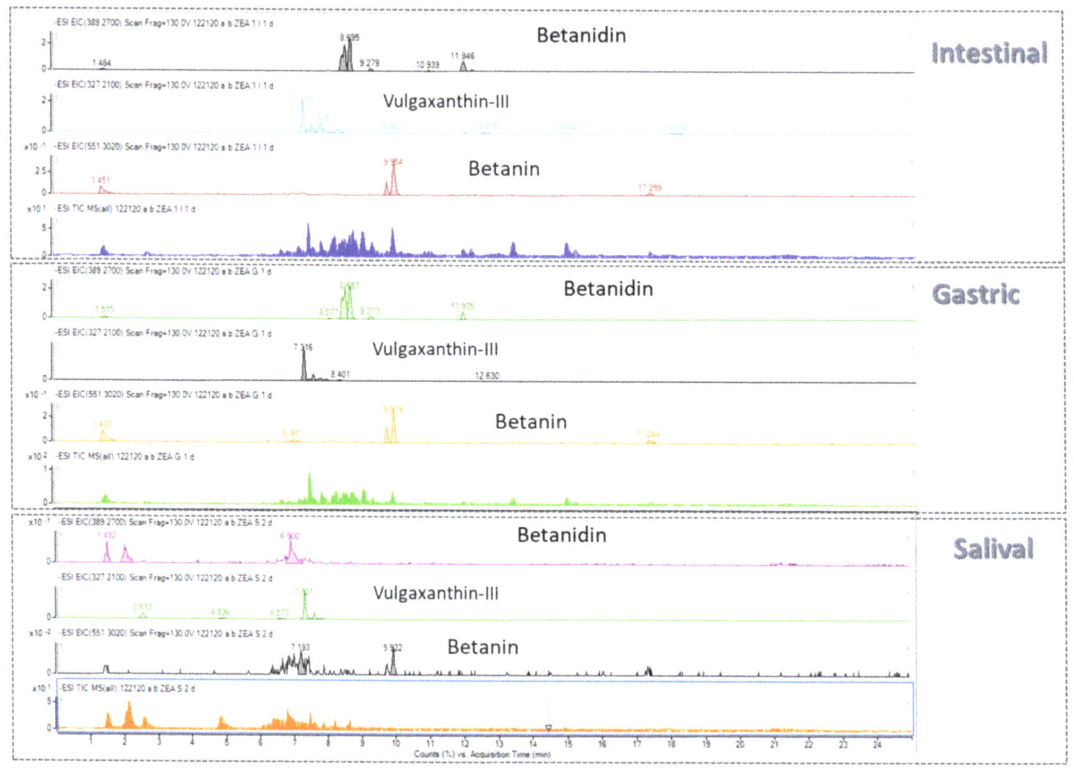

Figure 3. Total ion chromatograms (TIC) and extracted ion chromatograms (EIC) of betanidin, vulgaxanthin−III and betanin, from red beet bread digested phases (salival, gastric, intestinal).

Regarding availability, the three betalain compounds mentioned above presented high concentrations (80–90%) in the salival phase (Figure 4a), but during gastric and intestinal digestion the quantities decreased and the percentage in the intestinal phase in beet bread with a combination of AFB$_1$+OTA+ZEN (100 µg/kg) remained at 20%, 27% and 10% for betanin, betanidin and vulgaxanthin-III, respectively (Figure 4). Similar values were observed by Sawicki et al. (2020) for the gastric content after 2h of an intragastric administration of fermented red beetroot juice in rats for betanin and betanidin, obtaining values of 18% and 25%, respectively. These compounds in our study remained in intestinal bolus after digestion, so these amounts would be available to be absorbed in the intestinal tract. These results confirmed those shown by Sawicki et al. [22] in plasma and urine from healthy volunteers that consumed juice with a dose of 0.7 mg betalains/kg body weight, reaching the highest concentration in blood plasma (87.65 ± 15.71 nmol/L) and urine (1.14 ± 0.12 µmol) after the first and second weeks of juice intake. The study also indicated the presence of native betalains and their deglucosylated, decarboxylated, and dehydrogenated metabolites in human physiological fluids.

(a)

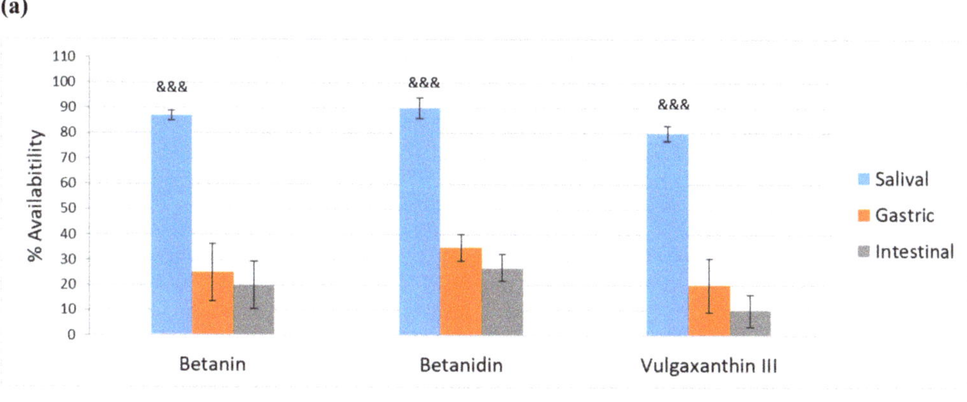

(b)

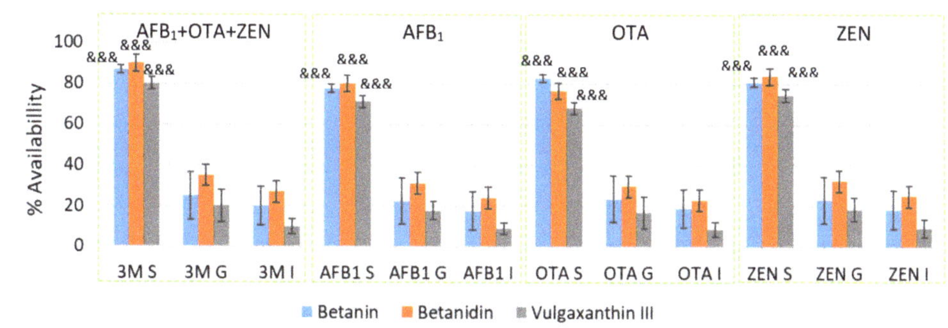

Figure 4. (**a**) Availability of three betalains (betanin, betanidin and vulgaxanthin-III) in beet-red bread. (**b**) Percentage (%) of three betalains (betanin, betanidin and vulgaxanthin-III) in three digested phases (S, G, I) of beet red bread contaminated with single (AFB$_1$ S, AFB$_1$ G, AFB$_1$ I, OTA S, OTA G, OTA I, ZEN S, ZEN G, ZEN I) and combined (3M S, 3M G, 3M I) mycotoxins at 100 µg/kg. (*: Salival-Gastric *** $p < 0.001$; ** $p < 0.01$; * $p < 0.05$; $: Gastric-Intestinal $$$ $p < 0.001$; $$ $p < 0.01$; $ $p < 0.05$; and &: Salival-Intestinal &&& $p < 0.001$; && $p < 0.01$; & $p < 0.05$).

3. Conclusions

According to the results reported, the quantities of the studied mycotoxins and betalains decreased all along the digested phases to the intestinal space. However, a high bioaccessibility in red beetroot breads, without significant differences in the values obtained, was observed. Whereas, if the mycotoxins were combined, the bioaccessibility decreased, being more significant for ZEN with values of 4%. Betalains presented similar bioaccessibility if the mycotoxins were in single or combined presence. It was observed that their availability decreased during the different phases of digestion and had bioaccessibility values of 20%, 27% and 10% for betanin, betanidin and vulgaxanthin-III, respectively.

These values revealed that the presence of other components, and mainly the copresence of mycotoxins, could affect the final bioavailability. However, it is necessary to assess this complete process with in vivo studies to fully understand the entire progression.

4. Materials and Methods

4.1. Chemicals, Reagents and Equipment

Standard ochratoxin A (OTA), aflatoxin B_1 (AFB$_1$) and zearalenone (ZEA) were purchased from Sigma-Aldrich (St. Louis, MO, USA). Individual stock solutions of mycotoxins were prepared in AcN at 500 μg/mL and formed solutions in MeOH at 100 μg/mL. The solutions were maintained at $-20\,^{\circ}$C in the dark. Betanin was purchased from Sigma-Aldrich and diluted to prepare solution in EtOH.

To prepare the gastrointestinal solutions, KCl 89.6 g/L, KSCN 20 g/L, NaH$_2$PO$_4$ 88.8 g/L, NaSO$_4$ 57 g/L, NaCl 175.3 g/L, NaHCO$_3$ 84.7 g/L, urea 20 g/L and Milli-Q water (Milli-Q water purification system (Millipore, Bedford, MA, USA)) were used. First, 145 mg α-amylase in 100 mL of Milli-Q water was prepared. A pepsin solution with 0.5 mg pepsin (1 g in 25 mL HCl (0.1 N) (Pepsin from porcine gastric mucosa, powder, $\geq$250 units/mg solid, P-7000, Sigma-Aldrich, St. Louis, MO, USA) was used. A solution of pancreatin: 1.10 mg (0,1 g Pancreatin (Pancreatin, from Porcine Pancreas, P1750, Sigma-Aldrich, St Louis, MO, USA) and 0.625 g bile salts (Bile extract porcine, B8631, Sigma-Aldrich, St. Louis, MO, USA) in 25 mL of NaHCO$_3$ (0.1 N) was prepared. The saliva solution was prepared mixing: 10 mL of KCl 89.6 g/L, 10 mL of KSCN 20 g/L, 10 mL of NaH$_2$PO$_4$ 88.8 g/L, 10 mL of NaSO4 57 g/L, 1.7 mL of NaCl 175.3 g/L, 20 mL of NaHCO$_3$ 84.7 g/L, 8 mL of urea (20g/L) and was completed to 500 mL with Milli-Q water.

4.2. Bread Preparation

Two different baked breads were prepared in the laboratory according to the following recipe: 300 g wheat flour, 175 mL water, 20 g fresh yeast (*Saccharomyces cerevisiae*), 10 g sugar and 5 g salt. All ingredients were mixed in a commercial bread maker. The bread was baked at 200 $^{\circ}$C for 40 min. Beetroot-enriched breads were prepared as detailed before regarding wheat flour bread but with slight modifications to obtain a bread with 20% beetroot. Beetroot dry powder was purchased from Agrosingularity S.L. (Murcia, Spain), made from dehydrated beetroot and blended.

Fortifications were made after the preparation of the bread, adding individual and mixed mycotoxins, before the digestion process, adding 1 mL of the working solutions (100 μg/mL) to 10 g of wheat bread.

4.3. In vitro Digestion Procedure

The procedure was the in vitro static model of Brodkorb et al. [23]. Three digestion phases were included: oral, gastric and duodenal. The assay was in triplicate for each bread, using an individually sterilized plastic bag (500 mL).

For the oral phase, 10 g of milled bread was combined for 5 min in a shaker with 6 mL of saliva solution (Section 2.1), 83 mL of milli-Q water, and 1 mL of -amylase solution. The Stomacher IUL Instrument homogenized the slurry for 30 s while simulating mastication (IUL S.A. Barcelona, Spain). An amount of 0.5 mL of pepsin solution for the stomach phase was then added to the bolus in an opaque Er-Lenmeyer flask. With HCl 6 N solution, the pH was raised to 2 to activate the stomach enzymes. The bolus was incubated for two hours at 37 $^{\circ}$C in a darkened room while being shaken at 100 rpm (orbital shaker, Infors AG CH-4103, Bottmingen, Switzerland). After that, 1.10 mL of the pancreatin/bile salts solution was added, and the pH was then raised to 6.5 for the duodenum phase using NaHCO3 1 N (intestinal). It was shaken for two hours at 37 $^{\circ}$C and 100 rpm. NaOH 0.5 N was used to bring the pH back to 7.2 after the incubation period.

After each stage of digestion, aliquots of each were obtained (5 or 10 mL), placed in an ice bath for 10 min to halt enzymatic activity, centrifuged for 5 min at 4500 rpm, 4 $^{\circ}$C, and then stored until the mycotoxins and duodenal bioaccessibility were determined. The remaining duodenal phase was centrifuged in a conical flask for 10 min at 4000 rpm, 4 $^{\circ}$C, and filtered afterwards. All collected supernatants were frozen at $-20\,^{\circ}$C. Additionally, enzyme inhibitors were utilized, It is strongly advised to use enough enzyme inhibitors to block the target digestion enzymes, as has already been demonstrated [24]. Trypsin and chymo-trypsin were both inhibited by a Bowman-Birk inhibitor (BBI; Sigma-Aldrich, cat. no. T9777). Amylase was inhibited by snap-freezing treatment and inactivation by extraction solvent and pepstatin A was employed to block pepsin (Sigma-Aldrich, cat. no. P5318). Pepstatin A was employed in a final concentration of 0.5–1.0 M in a total volume of 100 L of a BBI solution (0.05 g/L).

4.4. Sample Analysis

4.4.1. Extraction of AFB1, OTA and ZEN from Studied Bread

Spiked bread extraction was performed according to the validated method by Juan et al. (2016) [25] with slight modifications. First, homogenized and representative portions of 2 g were weighted into 50 mL polypropylene centrifuge tubes, spiked at the studied level of digestion. After that, 10 mL of acetonitrile/water (84:16, v/v) was added and shaken using a horizontal shaking device (IKA KS 260 basic) (250 shakes/min) for 1 h. Then, tubes were centrifuged for 5 min at 4500 rpm at 5 °C with an Eppendorf Centrifuge 5810R. The supernatant was filtered on Whatman filter paper No. 4 and 5 mL of the supernatant were evaporated to dryness at 38 °C under a gentle stream of nitrogen using a multi-sample TurboVap LV Evaporator (Zymark, Hoptkinton, MA, USA). The dried extract was prepared according to the analytical determination technique used.

4.4.2. Extraction of AFB$_1$, OTA and ZEN from the Simulated Physiological Fluids

At each stage of the digestion process, the mycotoxins were extracted into a simulated physiological fluid. Ethyl acetate was used to do a liquid-liquid extraction in accordance with El Jai et al. 's instructions with a few minor adjustments [26]. The amounts of 10 mL of intestinal fluid, 5 mL of saliva, and 5 mL of stomach juice were centrifuged, with the top layer being transferred to a falcon tube. Then, using an Eppendorf centrifuge 5810R, 5 mL of ethyl acetate was added twice, agitated, and centrifuged (Eppendorf, Hamburg, Germany). Using a multi-sample TurboVap LV Evaporator, the upper layer was deposited in 15 mL PTFE centrifuge tubes and evaporated to dryness at 35 °C with a mild stream of nitrogen (Zymark, Hoptkinton, MA, USA).

4.4.3. Extraction of Betalains

Betalains were analyzed in dried powdered beetroot, in prepared breads and in the simulated digestion solutions. An aliquot of dried powdered beetroot (0.1 g) or milled bread (5 g) were dissolved in 10 mL of 50% ethanol, agitated for 10 s and the homogenate centrifuged at 6000 rpm for 10 min. The simulated digestion solution (10 mL) was freeze-dried before adding the ethanol. After the centrifugation, the supernatant was collected and the same process was repeated twice to ensure maximum betalain extraction. The supernatant was evaporated to dryness using a multi-sample TurboVap LV Evaporator (Zymark, Hoptkinton, MA, USA) and redissolved with 1ml of a mixture of MeOH:H$_2$O (50:50, $v:v$), in order to be injected in LC-Q-TOF-MS.

4.4.4. Mycotoxin Analysis by LC–MS/MS

The dried residues were reconstituted to a final volume of 0.5 mL using methanol and water (70.30, v/v), and then filtered through a 13 mm/0.22 m nylon filter obtained from Analysis Vinicos S.L. (Tomelloso, Spain) before LC-MS/MS analysis.

A 3200 QTRAP®ABSCIEX, outfitted with a Turbo-VTM source (ESI) interface was attached to an LC-MS/MS system made up of an LC Agilent 1200 using a binary pump and an automated injector to carry out the analysis. At 25°C, the analytes were separated chromatographically using a reverse phase analytical column Gemini®NX-C18 (3 µm, 150 × 2 mm ID) and security guard cartridges Gemini-NX C18 (4 × 2 mm ID). Methanol was used as phase A of the mobile phase, along with 0.1 percent formic acid and 5 mM ammonium formate, and water was used as phase B (0.1 percent formic acid and 5 mM ammonium formate). The gradient used was as follows: equilibration for two minutes at 90 percent B, linear decrease to 20 percent of phase B in three minutes, maintenance of 20 percent of phase B for one minute, linear decrease from 20 to 10 percent of phase B in two minutes, maintenance of 10 percent of phase B for six minutes, linear decrease to 0 percent B in three minutes, maintenance of 100 percent A for one minute, and then linear increase from 0 to 50 percent B in three minutes, maintenance of initial conditions (90 percent B). In every phase, the flow rate was 0.25 mL/min. Runtime was 21 min in total. The injection had a 20 L volume.

The QTRAP system was employed as a triple quadrupole mass spectrometry detector (MS/MS) for mycotoxin analysis. The Turbo-VTM source was utilized in positive mode with the following source/gas parameter settings to investigate AFs, OTA and ZEN: 3.1 vacuum gauge (10×10^{-5} Torr), 20 curtain gas (CUR), 5500 ion spray voltage (IS), 450 °C source temperature, and 50 each of the ion source gases (GS1 and GS2). Table 1 displays the collision energy (CE), product ions (Q3), and precursor ions (Q1). For all analytes, the entrance potential (EP) was 10 V. Data collection and processing were carried out with the use of the Analyst®197 program, 1.5.2. The monitored fragments (retention duration, quantification ion, and confirmation ion), as well as the spectrometric parameters (declustering potential, collision energy, and cell exit potential), were carried out earlier by Juan et al. [27].

Table 1. Precursor ions (Q1), product ions (Q3), and collision energies (CE) used for the identification and quantification of the studied mycotoxins in bread samples.

Mycotoxin	Rt (min)	Quantitative Transition			Qualitative Transition		
		Q1 (m/z)	Q3 (m/z)	CE (eV)	Q1 (m/z)	Q3 (m/z)	CE (eV)
AFB$_1$	7.4	313	241	41	313	284	39
OTA	8.5	404	239	97	404	358	27
ZEN	8.5	319	301	10	319	282	10

4.4.5. Betalains Analysis by LC-Q-TOF-MS

Liquid chromatography with time-of-flight mass spectrometry (LC-Q-TOF-MS) analysis was carried out using an Agilent Technologies 1200 Infinity Series LC in conjunction with an Agilent Technologies 6540 UHD Accurate-Mass LC-Q-TOF-MS (Agilent Technologies, Santa Clara, CA, USA), equipped with an electrospray ionization ion source Agilent Technologies Dual Jet Stream (Dual AJS ESI). Chromatographic separation was performed with an Agilent InfinityLab Poroshell 120 EC-C18 (3×100 mm, 2.7 μm) column. The mobile phase consisted of 0.1% formic acid in milli-Q water (solvent A) and methanol (solvent B). The steps of the mobile phase gradient were applied as follows: 0-2 min, maintain 10% B; 2–5 min, get 70% B; 5–7 min, get 80% B; 7–8 min, get 90% B; kept 4 min at 90% B; 12–16 min, get 95% B; 16–18 min, get 50% B; 18–22 min, return to initial conditions 10% B; the total gradient run time was 25 min. The injection volume was 10 μL, obtaining the following chromatogram of the three studied betalains in red beetroot powder (Figure 5).

The N2 drying gas flow rate 12.0 L/min, the nebulizer pressure 45 psi, capillary voltage, 3500 V; frag-mentor voltage, 130 V; skimmer voltage 65 V and octopole RF peak, 750 V, and the gas drying temperature 370 °C gas drying temperature were the Q-TOF-MS conditions. Negative ions in the range of 100–1100 m/z for MS scans and 50–600 m/z for auto MS/MS scans were obtained using a dual AJS ESI interface in negative ionization mode at scan rates of 5 s/s for MS and 3 s/s for MS/MS, respectively. The collision energy levels used for MS/MS were 20 eV, 30 eV, and 40 eV in automated mode of acquisition. In order to enable internal mass correction, two reference masses at 121.0509 and 922.0098 m/z were used. Agilent MassHunter Workstation software B.08.00 was used for instrument control and data collecting. All the MS and MS/MS data of the validation standards were integrated by MassHunter Quantitative Analysis B.10.0 (Agilent Technologies).

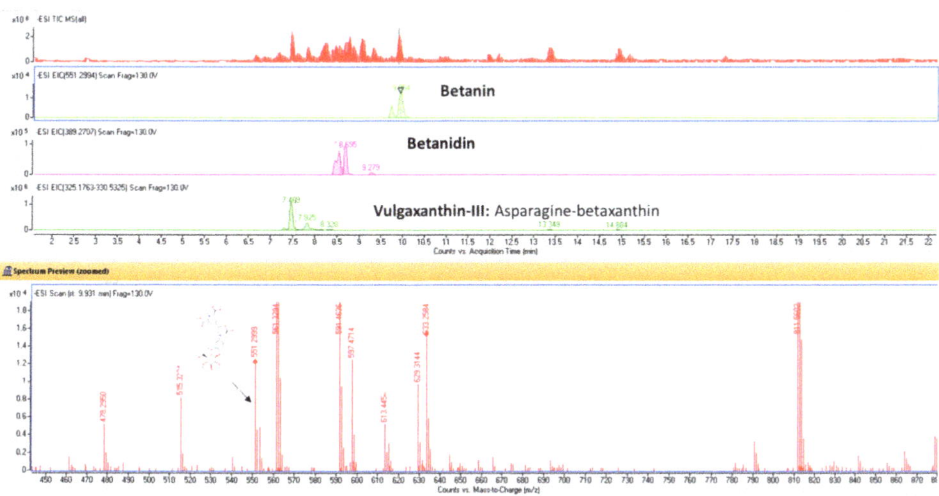

Figure 5. Total ion chromatograms (TIC) and extracted ion chromatograms (EIC) of betanin, betanidin and vulgaxanthin-III from red beetroot powder obtained with LC-Q-TOF-MS in negative electrospray ionization (ESI). Spectrum scan at elution time of betanin (9.931 min).

The main tools used for tentative identification of betalains were the interpretation of the observed MS/MS spectra (accurate mass, molecular formula and MS/MS fragmentation) in comparison with those found in the literature [28–30] and also by comparing their chromatographic and mass spectra characteristics with several online databases (Phenol-Explorer (Rothwell et al., 2012) [31], ChemSpider, MassBank, Spectral Database for Organic Compounds), and mass spectral data generated by authentic standards or related structural compounds (Table 2).

Table 2. Betalains tentatively identified by LC-Q-TOF-MS in red beet dried powder.

Betalains	Chemical Structure and Molecular Formula	RT (min)	Mass (Da)		ms/ms [a] [22]	Mass Error (ppm)
			Theorical	Observed		
Betanidin	$C_{18}H_{16}N_2O_8$	8.6	390.2552	390.2547	345	1.28
Betanin	$C_{24}H_{26}N_2O_{13}$	9.9	552.3083	552.3077	389	1.08
Vulgaxanthin-III	$C_{13}H_{15}N_3O_7$	7.3	328.2263	328.2258	277	1.52

[a] Identification with MS fragmentation of the standard (betanin) and database (betanidin and vulgaxanthin-III) [22].

4.5. Statistical Analysis

The statistical software program Microsoft 365 Excel 2015 was used to perform the statistical analysis of the data (correlation analysis, multiple linear regression analysis, and Student's t-test). Data from three separate experiments were expressed as (mean $\pm$ sd). The Student's t-test for paired samples was used to statistically analyze the findings. ANOVA and the Tukey HSD post hoc test for multiple comparisons were used to statistically assess differences from the control group; $p \leq 0.05$ was regarded as statistically significant.

Author Contributions: Conceptualization, C.J.G. and A.J.-G.; methodology, P.L.C. and C.J.G.; validation, P.L.C. and C.J.G.; formal analysis, P.L.C. and C.J.G.; data curation, P.L.C. and C.J.G.; writing—original draft preparation, P.L.C. and C.J.G.; writing—review and editing, C.J.G. and A.J.-G.; supervision, C.J.G., A.J.-G., P.L.C., J.M.V. and J.C.M.C. All authors have read and agreed to the published version of the manuscript.

Funding: This research received no external funding.

Institutional Review Board Statement: Not applicable.

Informed Consent Statement: Not applicable.

Data Availability Statement: Not applicable.

Acknowledgments: This project was financially supported by the Spanish Ministry of Science and Innovation PID2019-108070RB-I00ALI.

Conflicts of Interest: The authors declare no conflict of interest.

References

1. Castells, M.; Marin, S.; Sanchis, A.; Ramos, V.J. Distribution of fumonisins and aflatoxins in corn fractions during industrial cornflake processing. *J. Food Microbiol.* **2008**, *123*, 81–87. [CrossRef] [PubMed]
2. EFSA. *Effect on Dietary Exposure of an Increase in Total Aflatoxin Levels from 4 µg/kg to 10 µg/kg for Dried Figs*; EFSA Supporting Publications, EFSA: Parma, Italy, 2012; Volume 9, p. EN-311.
3. IARC. Some naturally occurring substances: Food items and constituents, heterocyclic aromatic amines and mycotoxins. In *IARC Monographs on the Evaluation of the Carcinogenic Risk of Chemicals to Humans*; World Health Organization: Geneva, Switzerland, 1993; Volume 56, p. 599.
4. Beltrán, Y.; Ibánez, M.; Sancho, J.V.; Hernández, F. Determination of mycotoxins in different food commodities by ultra-high-pressure liquid chromatography coupled to triple quadrupole mass spectrometry. *Rapid Commun. Mass Spectrom.* **2009**, *23*, 1801–1809. [CrossRef] [PubMed]
5. Copetti, M.; Iamanaka, T.; Pereira, J.L.; Lemes, D.; Nakano, F.; Taniwaki, M. Co-occurrence of ochratoxin-A and aflatoxins in chocolate marketed in Brazil. *Food Control* **2012**, *26*, 36–41. [CrossRef]
6. European Commission. Regulation (EC) No 1881/2006 of 19 December 2006 setting maximum levels for certain contaminants in foodstuffs. *Off. J. Eur. Union* **2006**, *L364*, 5.
7. González-Arias, C.A.; Piquer-García, I.; Marín, S.; Sanchís, V.; Ramos, A.J. Bioaccessibility of ochratoxin A from red wine in an in vitro dynamic gastrointestinal model. *World Mycotoxin J.* **2014**, *8*, 107–112. [CrossRef]
8. Versantvoort, C.H.M.; Van de Kamp, E.; Rompelberg, C.J.M. Development and Applicability of an in vitro Digestion Model in Assessing the Bioaccessibility of Contaminants from Food. RIVM Report 320102002/2004. 2004. Available online: http://www.rivm.nl/bibliotheek/rapporten/320102002.pdf (accessed on 27 July 2022).
9. Fu, Y.; Shi, J.; Xie, S.-Y.; Zhang, T.-Y.; Soladoye, O.P.; Aluko, R.E. Red Beetroot Betalains: Perspectives on Extraction, Processing, and Potential Health Benefits. *J. Agric. Food Chem.* **2020**, *68*, 11595–11611. [CrossRef]
10. Penalva-Olcina, R.; Juan, C.; Fernández-Franzón, M.; Juan-García, A. Effectiveness of beetroot extract in SH-SY5Y neuronal cell protection against Fumonisin B1, Ochratoxin A and its combination. *Food Chem. Tox.* **2022**, *165*, 113164. [CrossRef]
11. Morrison, D.M.; Ledoux, D.R.; Chester, L.F.B.; Samuels, C.A.N. Occurrence of aflatoxins in rice and in cassava (Manihot esculenta) products (meal, bread) produced in Guyana. *Mycotoxin Res.* **2019**, *35*, 75–81. [CrossRef]
12. Zinedine, A.; Mañes, J. Occurrence and legislation of mycotoxins in food and feed from Morocco. *Food Control* **2008**, *20*, 334–344. [CrossRef]
13. Paíga, P.; Morais, S.; Oliva-Teles, T.; Correia, M.; Delerue-Matos, C.; Duarte, S.C.; Pena, A.; Matos Lino, C. Extraction of ochratoxin A in bread samples by the QuEChERS methodology. *Food Chem.* **2012**, *135*, 2522–2528. [CrossRef]
14. Paíga, P.; Morais, S.; Oliva-Teles, T.; Correia, M.; Delerue-Matos, C.; Sousa, A.M.M.; Gonçalves, M.P.; Duarte, S.C.; Pena, A.; Matos Lino, C. Determination of Ochratoxin A in Bread: Evaluation of Microwave-Assisted Extraction Using an Orthogonal Composite Design Coupled with Response Surface Methodology. *Food Bioprocess Technol.* **2013**, *6*, 2466–2477. [CrossRef]

15. Kulahi, A.; Kabak, B. A preliminary assessment of dietary exposure of ochratoxin A in Central Anatolia Region, Turkey. *Mycotoxin Res.* **2020**, *36*, 327–337. [CrossRef] [PubMed]
16. Iqbal, S.Z.; Asi, M.R.; Jinap, S.; Rashid, U. Detection of aflatoxins and zearalenone contamination in wheat derived products. *Food Control* **2014**, *35*, 223–226. [CrossRef]
17. Vidal, A.; Morales, H.; Sanchis, V.; Ramos, A.J.; Marín, S. Stability of DON and OTA during the breadmaking process and determination of process and performance criteria. *Food Control* **2014**, *40*, 234–242. [CrossRef]
18. Keller Bol, E.; Araujo, L.; Fonseca Veras, F.; Welke, J.E. Estimated exposure to zearalenone, ochratoxin A and aflatoxin B1 through the consume of bakery products and pasta considering effects of food processing. *Food Chem. Toxicol.* **2016**, *89*, 85–91.
19. Slatnar, A.; Stampar, F.; Veberic, R.; Jakopic, J. HPLC-MSn Identification of Betalain Profileof Different Beetroot (Beta vulgaris L.ssp. vulgaris) Parts and Cultivars. *J. Food Sci.* **2015**, *80*, C1952–C1958. [CrossRef]
20. Sawicki, T.; Baczek, N.; Wiczkowski, W. Betalain profile, content and antioxidant capacity of red beetroot dependent on the genotype and root part. *J. Funct. Foods* **2016**, *27*, 249–261. [CrossRef]
21. Sobral, M.M.C.; Gonçalves, T.; Martins, Z.E.; Bäuerl, C.; Cortés-Macías, E.; Collado, M.C.; Ferreira, I.M. Mycotoxin Interactions along the Gastrointestinal Tract: In Vitro Semi-Dynamic Digestion and Static Colonic Fermentation of a Contaminated Meal. *Toxins* **2022**, *14*, 28. [CrossRef]
22. Sawicki, T.; Topolska, J.; Romaszko, E.; Wiczkowski, W. Profile and Content of Betalains in Plasma and Urine of Volunteers after Long-Term Exposure to Fermented Red Beet Juice. *J. Agric. Food Chem.* **2018**, *66*, 4155–4163. [CrossRef]
23. Brodkorb, A.; Egger, L.; Alminger, M.; Alvito, P.; Assunção, R.; Ballance, S.; Bohn, T.; Bourlieu-Lacanal, C.; Boutrou, R.; Carrière, F.; et al. INFOGEST static in vitro simulation of gastrointestinal food digestion. *Nat. Protoc.* **2019**, *14*, 991–1014. [CrossRef]
24. Llorens, P.; Pietrzak-Fie´cko, R.; Moltó, J.C.; Mañes, J.; Juan, C. Development of an Extraction Method of Aflatoxins and Ochratoxin A from Oral, Gastric and Intestinal Phases of Digested Bread by In Vitro Model. *Toxins* **2022**, *14*, 38. [CrossRef] [PubMed]
25. Juan, C.; Covarelli, L.; Beccari, G.; Colasante, V.; Mañes, J. Simultaneous analysis of twenty-six mycotoxins in durum wheat grain from Italy. *Food Control.* **2016**, *62*, 322–329. [CrossRef]
26. El Jai, A.; Juan, C.; Juan-García, A.; Mañes, J.; Zinedine, A. Multi-mycotoxin contamination of green tea infusion and dietary exposure assessment in Moroccan population. *Food Res. Int.* **2021**, *140*, 109958. [CrossRef] [PubMed]
27. Juan, C.; Oueslati, S.; Mañes, J.; Berrada, H. Multimycotoxin Determination in Tunisian Farm Animal Feed. *J. Food Sci.* **2019**, *84*, 3885–3893. [CrossRef]
28. Cai, Y.; Sun, M.; Corke, H. HPLC characterization of betalains from plants in the Amaranthaceae. *J. Chromatogr. Sci.* **2005**, *43*, 454–460. [CrossRef] [PubMed]
29. Li, H.; Deng, Z.; Liu, R.; Zhu, H.; Draves, J.; Marcone, M.; Tsao, R. Characterization of phenolics, betacyanins and antioxidant activities of the seed, leaf, sprout, flower and stalk extracts of three *Amaranthus* species. *J. Food Compos. Anal.* **2015**, *37*, 75–81. [CrossRef]
30. Deladino, L.; Alvarez, I.; De Ancos, B.; Sánchez-Moreno, C.; Molina-García, A.D.; Schneider Teixeira, A. Betalains and phenolic compounds of leaves and stems of Alternanthera brasiliana and Alternanthera tenella. *Food Res. Int.* **2017**, *97*, 240–249. [CrossRef]
31. Rothwell, J.A.; Urpi-Sarda, M.; Boto-Ordonez, M.; Knox, C.; Llorach, R.; Eisner, R.; Manach, C. Phenol-Explorer 2.0: A major update of the Phenol-Explorer database integrating data on polyphenol metabolism and pharmacokinetics in humans and experimental animals. *Database* **2012**, *2012*, bas031. [CrossRef]

 *toxins*

Article

Effects of Dietary Lanthanum Chloride on Growth Performance, Hematology and Serum Biochemistry of Juvenile *Clarias gariepinus* Catfish Fed Diets Amended with Mixtures of Aflatoxin B1 and Fumonisin B1

Bolade Thomas Adeyemo [1,*], Ndidi Gloria Enefe [2], Tanimomo Babatunde Kayode [1], Augustina Ezekwesili [1], Olatunde Hamza Olabode [3], Audu Zakariya [1], Gbenga Michael Oladele [4], Samson Eneojo Abalaka [5], Wesley Daniel Nafarnda [6] and Clement Barikuma Innocent Alawa [1]

[1] Department of Animal Health and Production, Faculty of Veterinary Medicine, University of Abuja, Gwagwalada 902001, Nigeria
[2] Department of Physiology and Biochemistry, Faculty of Veterinary Medicine, University of Abuja, Gwagwalada 902001, Nigeria
[3] Department of Veterinary Microbiology, Faculty of Veterinary Medicine, University of Abuja, Gwagwalada 902001, Nigeria
[4] Department of Veterinary Pharmacology, Faculty of Veterinary Medicine, University of Abuja, Gwagwalada 902001, Nigeria
[5] Department of Veterinary Pathology, Faculty of Veterinary Medicine, University of Abuja, Gwagwalada 902001, Nigeria
[6] Department of Veterinary Public Health and Preventive Medicine, Faculty of Veterinary Medicine, University of Abuja, Gwagwalada 902001, Nigeria
* Correspondence: bolade.adeyemo@uniabuja.edu.ng; Tel.: +234-8038049383

Citation: Adeyemo, B.T.; Enefe, N.G.; Kayode, T.B.; Ezekwesili, A.; Olabode, O.H.; Zakariya, A.; Oladele, G.M.; Abalaka, S.E.; Nafarnda, W.D.; Alawa, C.B.I. Effects of Dietary Lanthanum Chloride on Growth Performance, Hematology and Serum Biochemistry of Juvenile *Clarias gariepinus* Catfish Fed Diets Amended with Mixtures of Aflatoxin B1 and Fumonisin B1. *Toxins* **2022**, *14*, 553. https://doi.org/10.3390/toxins14080553

Received: 29 April 2022
Accepted: 28 June 2022
Published: 14 August 2022

Publisher's Note: MDPI stays neutral with regard to jurisdictional claims in published maps and institutional affiliations.

Abstract: This study aimed to determine the effects of dietary lanthanum chloride on the growth and health performance of juvenile *Clarias gariepinus* when fed diets experimentally contaminated with mixtures of aflatoxin B1 and fumonisin B1. A control diet, (mycotoxin free, diet A), mycotoxin contaminated (diet B), and two mycotoxin-contaminated diets amended with lanthanum chloride (200 mg/kg, diet C; and 400 mg/kg, diet D), were fed to 450 fish divided equally into five groups (each with three replicates) for 56 days. The fish were randomly sampled at the time points: day 7, 28 and day 56 for the zootechnical, hematological and serum biochemical evaluations. The fish fed the diets amended with lanthanum chloride exhibited significantly ($p < 0.05$) better performance indices compared with the fish fed only the mycotoxin-contaminated diet. Lanthanum chloride elicited significant ($p < 0.05$) increases in erythrocytes and leucocytes count and significant ($p < 0.05$) reduction in serum transaminase, alkaline phosphatase, lactate dehydrogenase activities, urea and uric acid concentrations in the fish fed the diets contaminated with mixtures of aflatoxin B1 and fumonisin B1. The study indicates that juvenile *Clarias gariepinus* may be beneficially cultured with mycotoxin-contaminated grains amended with 200 to 400 mg/kg lanthanum chloride.

Keywords: aflatoxin B1; fumonisin B1; lanthanum chloride; *Clarias gariepinus*; hematology; serum biochemistry

Key Contribution: The dietary inclusion of lanthanum chloride at 200–400 mg/kg diet improves the production performance of juvenile *Clarias gariepinus* catfish when fed diets contaminated with mixtures of aflatoxin B1 and fumonisin B1. Dietary lanthanide chloride at 200–400 mg/kg in the diet ameliorates the hematological and serum biochemical derangements produced by dietary exposures to mixtures of aflatoxin B1 and fumonisin B1.

1. Introduction

Aquaculture production represents a significant source of animal protein to millions of people worldwide. According to the Food and Agriculture Organization of the United Nations [1], the total world aquaculture production has risen to 51.4 million tons by volume and \$60.0 billion by value. The increase in the production of cultured fish has also led to a significant decrease in the landings of several capture fisheries, a direct consequence of the use of fish as the source of animal proteins in aquafeeds [2–4]. To mitigate this negative trend, plant-based proteins are increasingly being used as sustainable alternatives to fish-meal-based proteins in aquafeeds [5].

In Africa, the quality of the plant products used in fish feed formulations is said to be a limiting factor in the progressive increase in aquaculture productivity; in addition, these feed ingredients are, very often, ideal substrates for the growth of fungi, such as *Aspergillus* spp. and *Fusarium* spp. [6,7]. These fungi, under favorable conditions (that is, those generally present in the tropical regions of the world), may result in the synthesis of mycotoxins, such as aflatoxin B1 and fumonisin B1 [4].

The aflatoxins are a group of mycotoxins produced by the blue-green molds, *Aspergillus flavus* and *Aspergillus parasiticus* [8,9]. These molds are common contaminants in the feed ingredients of agricultural origin (such as cotton seed, ground nut, maize, wheat, soya bean and the respective by-products from the agricultural processing of these commodities). The aflatoxins have also been reported in fish meal. Four major aflatoxins (AFB_1, AFB_2, AFG_1 and AFG_2) have been reported to be direct contaminants of feed ingredients and formulated agricultural and aquacultural feeds [10,11]. Of the aflatoxins, AFB_1 is reported to be the most prevalent, most potent and the most carcinogenic [12–14], and has been classified as a group 1 carcinogen by the International Agency for Research on Cancer [15].

Aflatoxicosis, a disease state caused by the effects of aflatoxins, has been noted to be common in aquaculture [16]. According to [17], *Oncorhynchus mykiss* fed cotton seed meals contaminated with aflatoxins developed liver tumors and exhibited a mortality pattern of up to 85%. Other information available on the effects of AFB_1 in the cultivable species of fin and non-fin fishes, i.e., *Oncorhynchus mykiss* [18]; *Ictalurus punctatus* [19,20]; *Oreochromis niloticus* [21]; *Labeo rohita* [22] and *Penaeus monodons* [23], suggests that fish exhibit a wide plasticity in the susceptibility to AFB_1, and that cold water species are more sensitive when compared to warm water fishes [18,20]. These species-specific differences in sensitivity to AFB_1 have been attributed to the differences in the metabolism of aflatoxin B1 in the liver and the affinity of AFB_1-derived metabolites to hepatic macromolecules [14,24].

Fusarium verticillioides, the mold primarily associated with the production of the fumonisins, is prevalent in hot–humid regions of the world, and its occurrence has been related to the presence of invading insects [24,25]. Though maize is most frequently contaminated by fumonisins, these mycotoxins have been found at high concentrations in wheat, asparagus, tea and cowpea [26]. According to [27], it is very difficult to obtain uncontaminated maize, even if the contamination level is not significant. In most investigations, FB_1 is the most prevalent toxin, with a co-occurrence of FB_2 and FB_3 [25]. Several countries in Africa, North and South America, Asia and Europe have reported FB_1 in cereals at levels from 0.02 to 25.9 ng/kg, and FB_2 at levels from 0.05 to 11.3 ng/kg [26,28]. FB_1 is stable in acetonitrile-water (1:1), at food-processing temperatures and light, but unstable in methanol [26].

Feed additives are used world-wide for many different reasons. Some help to cover the need for essential nutrients, and others help to increase animal performance, feed intake and thereby optimize feed utilization. The rare earth elements (REE), composed of about 15 elements with atomic numbers ranging from 57 (lanthanum) to 71 (lutetium), are a promising set of feed additives in animal production [29,30]. The REE are reported as being used as performance enhancers in animal production, without affecting the quality of the final produce [31,32].

The growth-promoting effects of REE are reported to be based on the type and concentration of the REE applied [33,34]. It is reported that the application of REE additives, with concentrations ranging from 100 to 200 mg/kg, in the diets of 40–50 days old piglets, significantly improved the daily body weight gain [35]. It is reported that diets supplemented with mixtures of lanthanum chloride at 100, 200 and 300 mg/kg per diet improved the activities of proteinase, lipase and amylase in the liver and pancreas of the adults and fry of carp (*Cyprinus carpio*) [36]. Furthermore, dietary lanthanum at 75 mg/kg has been reported to result in a 2–5% increase in body weight and a 7% increase in the feed conversion ratio of piglets [31].

Clarias gariepinus, also called the African sharp tooth fish, is widely farmed in the West African sub-region; this fish is cultured based on the aquafeeds produced from maize and soybean cake [37,38], resulting in the risk of inadvertent dietary AFB_1 and/or FB_1 exposures. In a previous study, we observed a marked decrease in weight gain following dietary exposures to mixtures of AFB_1 and FB_1, and reported the tolerable limits for dietary exposures to mixtures of AFB_1 and FB_1 in juvenile *Clarias gariepinus* to be 17.6 µg AFB_1/kg and 24.5 mg FB_1/kg [39].

The addition of rare earth metals to animal diets as growth promoters is considered to be a promising alternative to the use of antibiotics and other chemicals [29,40,41]. There are reports on the potential for the use of lanthanum chloride (LC), either as immunostimulants and/or growth-promoters in agriculture, as well as in aquaculture [31,42]. The present study was set up to determine the effects of lanthanum chloride on the growth performance, hematology and serum chemistry of the juvenile *Clarias gariepinus*, when fed diets contaminated with mixtures of aflatoxin B1 and fumonisin B1.

2. Results

The nutrient composition, proximate composition and the mycotoxin analysis of the experimental diets are shown in Table 1. It shows that the concentrations of AFB_1 and FB_1 in the produced diets are significantly higher than the concentrations of the purified mycotoxins added to the experimental diets at the time of the diet formulation and feed production. Hence, the concentrations of the mycotoxins in the diets would, hereafter, be appropriately quantified as follows: Diet A, the control diet, 2.0 µg AFB_1; 3.0 mg FB_1/kg diet; Diets B, C and D, 19.7 µg AFB_1; 28.5 mg FB_1/kg diet. Table 1 also shows that there were no variations ($p > 0.05$) in the percentage of crude protein, metabolizable energy, digestible energy and the total lipids contents of the formulated feeds.

Table 1. Ingredients and Proximate Composition of Diets Amended with Lanthanum chloride, Bentonite clay and Mixtures of Aflatoxin B_1 and Fumonisin B_1.

Parameter	Diet A	Diet B	Diet C	Diet D
[Y] Fish meal (%)	19.0	19.0	19.0	19.0
Soybean cake (%)	38.00	38.00	38.00	38.00
Maize (%)	32.22	32.24	32.23	32.23
Palm oil (%)	1.00	1.00	1.00	1.00
Fish oil (%)	6.00	6.00	6.00	6.00
Vit/min premix (%)	0.50	0.50	0.50	0.50
Bone meal (%)	1.00	1.00	1.00	1.00
NaCl (%)	0.23	0.23	0.23	0.23
LC (mg/kg)	0.0	0.0	200.00	400.00
AFB_1 (µg/kg): FB_1 (mg/kg)	2.0; 3.0	19.7; 28.5	19.7; 28.5	19.7; 28.5
[v] Starch binder (%)	2.00	2.00	2.00	2.00
Crude protein (%)	40.04	40.00	40.02	40.01
Gross energy (kj/g)	20.00	20.01	19.97	20.02
Digestible energy (kj/g)	12.00	12.04	12.01	12.07
Total lipids (%)	9.0	9.0	9.0	9.0
Moisture (%)	2.27	2.33	2.33	2.33
Ash (%)	9.65	9.70	9.71	9.65

[Y] Fish meal of 72% crude protein; [v] Cassava starch binder; LC = Lanthanum chloride; BC = Bentonite clay. Diet A (control 2.0 µg AFB_1/kg + 3.0 mg FB_1/kg); Diet B (19.7 µg AFB_1/kg + 28.5 mg FB_1/kg); Diet C (lanthanum chloride 200.0 mg/kg and 19.7 µg AFB_1/kg + 28.5 mg FB_1/kg); Diet D (lanthanum chloride 400.0 mg/kg and (19.7 µg AFB_1/kg + 28.5 mg FB_1/kg).

2.1. Effect on Growth Performance

Table 2 shows the results obtained for the zootechnical assessments of the fish fed the various experimental diets. The one way analysis of variance (ANOVA) shows that there were no significant variations ($p > 0.05$) in the body weights of the fish in the treatment groups compared with the body weight of the fish in the control group at the start of the feeding study. However, there were significant variations ($p < 0.05$) in the final weight of the fish in the control group when compared with the final weight of the fish in the treatment groups. The fish fed the control diet (2.0 µg AFB_1; 3.0 mg FB_1/kg diet) had the highest final body weight, while the fish fed diet B (19.7 µg AFB_1; 28.5 mg FB_1/kg diet) had the lowest weight gain. The Tukey's post hoc assessment revealed the weight gained by the fish fed diet D (lanthanum chloride 400.0 mg/kg and 19.7 µg AFB_1/kg + 28.5 mg FB_1/kg) was significantly ($p < 0.05$) higher than the weight gained by the fish fed diet C (lanthanum chloride 200.0 mg/kg and 19.7 µg AFB_1/kg + 28.5 mg FB_1/kg).

Table 2. Growth response of juvenile *Clarias gariepinus* catfish fed diets amended with lanthanum chloride and mixtures of aflatoxin B1 and fumonisin B1 for 56 days.

	Diet A (Control)	*Diet B*	*Diet C*	*Diet D*
Mean initial body weight (g)	85.04 ± 1.17 [a]	85.00 ± 0.21 [a]	85.02 ± 1.21 [a]	85.09 ± 1.01 [a]
Mean final body weight (g)	168.0 ± 4.99 [a]	116.2 ± 2.07 [b]	131.4 ± 6.35 [c]	138.5 ± 3.19 [d]
Mean weight gained (g)	81.00 ± 1.11 [a]	31.5 ± 1.12 [b]	47.90 ± 1.09 [c]	51.2 ± 1.25 [d]
Feed fed (g)	197.38 ± 1.42 [a]	102.94 ± 1.11 [b]	94.91 ± 1.13 [c]	105.04 ± 1.08 [d]
FCR	2.437 ± 0.11 [a]	3.268 ± 0.05 [b]	1.981 ± 0.07 [c]	2.052 ± 0.07 [c]
FCE	41.04 ± 2.97 [a]	30.60 ± 1.66 [b]	50.47 ± 1.14 [c]	49.21 ± 1.12 [d]
SGR (%/day)	1.19 ± 0.03 [a]	0.54 ± 0.01 [b]	0.77 ± 0.01 [cd]	0.83 ± 0.66 [cd]
Survival (%)	100.00	100.00	100.00	100.00

Diet A (control 2.0 µg AFB_1/kg + 3.0 mg FB_1/kg); Diet B (19.7 µg AFB_1/kg + 28.5 mg FB_1/kg); Diet C (lanthanum chloride 200.0 mg/kg and 19.7 µg AFB_1/kg + 28.5 mg FB_1/kg); Diet D (lanthanum chloride 400.0 mg/kg and 19.7 µg AFB_1/kg + 28.5 mg FB_1/kg). Rows with different superscripts are significantly different ($p < 0.05$).

The quantity of feed consumed by the experimental fish are shown in Table 2. The one way analysis of variance (ANOVA) shows that the quantity of feed consumed by the fish fed diet A (the control diet) was significantly ($p < 0.05$) more than the quantities of feed consumed by the fish fed the treatment diets (diets B, C and D). The Tukey's post hoc analysis reveals that the quantities of feed consumed by the fish in the various treatment groups varied significantly ($p < 0.05$) from one another, and that the fish fed diet C (lanthanum chloride 200.0 mg/kg and 19.7 µg AFB_1 + 28.5 mg FB_1/kg), consumed the lowest quantity of fed.

In addition, the results obtained for the feed conversion ratio (FCR) are shown in Table 2. The one way analysis of variance (ANOVA) shows that the FCR of the fish fed the control diet (2.0 µg AFB_1; 3.0 mg FB_1/kg diet) differed significantly ($p < 0.05$) compared with the FCR of the fish fed the treatment diets (diets B, C and D). The fish fed diet B (19.7 µg AFD_1 + 28.5 mg FB_1/kg) had the highest FCR (3.268) while the fish fed diet C had the lowest FCR (1.981). The variations in the FCR of the fish fed diets C (lanthanum chloride 200.0 mg/kg and 19.7 µg AFB_1 + 28.5 mg FB_1/kg) and diet D (lanthanum chloride 400.0 mg/kg and 19.7 µg AFB_1 + 28.5 mg FB_1/kg) were also not significant ($p > 0.05$).

The results obtained for the feed conversion efficiency of the fish fed the experimental diets are shown in Table 2. The one way analysis of variance (ANOVA) shows the feed conversion efficiency (FCE) of fish fed the control diet differed significantly ($p < 0.05$) compared with the FCE of fish fed the treatment diets (diets B, C and D). Furthermore, Tukey's post hoc evaluation shows there were no significant variations ($p > 0.05$) in the FCE of the fish fed diet C compared with the FCE of the fish fed diet D. The fish fed the diets containing lanthanum chloride had the highest FCE (50.47), while the fish fed diet B had the lowest FCE (30.60).

The data for the specific growth rate (SGR) shows that the fish fed diet A had the highest SGR (1.195) and the fish fed diet B had the lowest SGR (0.535). The one way analysis of variance (ANOVA) of this data shows that the fish fed diet A (the control diet) had significantly ($p < 0.05$) higher SGR values compared with the fish fed the treatment diets (diets B, C and D). Tukey's post hoc evaluation reveals there were no significant variation in the SGR of fish fed diet C and diet D.

2.2. Effects on Hematology

At 7 days post commencement of the trials (Table 3), the one way analysis of variance (ANOVA) showed that there were significant ($p < 0.05$) decreases in the erythrocytes, leucocytes and the hematocrit values in the fish fed the treatment diets (diets B, C and D) compared with the fish fed the control diet (diet A). Table 3 further shows that the fish fed diet B had the lowest erythrocytes count (1.43×10^9 cells/mm^3), leucocytes count (1.95×10^6 cells/mm^3) and hematocrit values (16.24%), while the fish fed the control diet had the highest values for these parameters; there were no clear patterns in these parameters in the fish fed the lanthanum chloride.

Table 3. Hematological profile of juvenile *Clarias gariepinus* catfish fed diets amended with lanthanum chloride and mixtures of aflatoxin B1 and fumonisin B1 for 7 days.

Parameter	Diet A (Control)	Diet B	Diet C	Diet D
Erythrocytes (10^9 cells/mm^3)	2.21 ± 0.01 [a]	1.43 ± 0.04 [b]	1.68 ± 0.01 [c]	1.60 ± 0.02 [d]
Leucocytes (10^6 cells/mm^3)	2.30 ± 01.01 [a]	1.95 ± 0.01 [b]	2.04 ± 0.03 [c]	2.10 ± 0.01 [d]
Hematocrit (%)	29.85 ± 0.12 [a]	16.42 ± 0.36 [b]	22.76 ± 1.88 [c]	25.19 ± 1.17 [d]
Hemoglobin concentration (g/dl)	9.00 ± 0.59 [a]	7.79 ± 0.64 [b]	8.39 ± 0.51 [c]	8.50 ± 0.39 [a]

Diet A (control 2.0 µg AFB$_1$/kg + 3.0 mg FB$_1$/kg); Diet B (19.7 µg AFB1/kg + 28.5 mg FB1/kg); Diet C (lanthanum chloride 200.0 mg/kg and 19.7 µg AFB$_1$/kg + 28.5 mg FB$_1$/kg); Diet D (lanthanum chloride 400.0 mg/kg and 19.7 µg AFB1/kg + 28.5 mg FB$_1$/kg). Rows with different superscripts are significantly different ($p < 0.05$).

In this same period, the one way analysis of variance (ANOVA) also shows that there were significant ($p < 0.05$) variations in the hemoglobin concentration in the fish fed the treatment diets (diets B, C and D) compared with the hemoglobin concentrations of the fish fed the control diet. The Tukey post hoc analysis shows there were no significant ($p > 0.05$) variations in the hemoglobin concentrations of the fish fed diet A (the control diet) and the fish fed diet D.

There was a significant ($p < 0.05$) reduction in the erythrocytes count, the leucocytes count, and the hematocrit volume of the fish fed the treatment diets (diets B, C and D) compared with the corresponding values in fish fed the control diet (diet A) on day 28 (Table 4) and day 56 (Table 5) of the trial. The Tukey's post assessments reveal there were no significant ($p > 0.05$) variations in the values obtained for fish fed diet C and diet D on day 28 and day 56.

Table 4. Hematological profile of juvenile *Clarias gariepinus* catfish fed diets amended with lanthanum chloride and mixtures of Aflatoxin B1 and fumonisin B1 for 28 days.

Parameter	Diet A(Control)	Diet B	Diet C	Diet D
Erythrocytes (10^9 cells/mm^3)	2.18 ± 0.07 [a]	1.21 ± 0.06 [b]	1.72 ± 0.01 [c]	1.69 ± 0.01 [c]
Leucocytes (10^6 cells/mm^3)	2.26 ± 0.01 [a]	1.14 ± 0.01 [b]	2.11 ± 0.02 [c]	2.13 ± 0.01 [d]
Hematocrit (%)	30.17 ± 1.39 [a]	18.31 ± 4.12 [b]	25.49 ± 1.09 [c]	24.83 ± 1.00 [d]
Hemoglobin concentration (g/dl)	9.97 ± 0.59 [a]	6.31 ± 1.47 [b]	7.83 ± 0.98 [cd]	8.00 ± 0.81 [cd]

Diet A (control 2.0 µg AFB$_1$/kg + 3.0 mg FB$_1$/kg); Diet B (19.7 µg AFB1/kg + 28.5 mg FB1/kg); Diet C (lanthanum chloride 200.0 mg/kg and 19.7 µg AFB$_1$/kg + 28.5 mg FB$_1$/kg); Diet D (lanthanum chloride 400.0 mg/kg and 19.7 µg AFB1/kg + 28.5 mg FB$_1$/kg). Rows with different superscripts are significantly different ($p < 0.05$).

Table 5. Hematological profile of juvenile *Clarias gariepinus* catfish fed diets amended with lanthanum chloride and mixtures of Aflatoxin B1 and fumonisin B1 for 56 days.

Parameter	Diet A (Control)	Diet B	Diet C	Diet D
Erythrocytes (10^9 cells/mm^3)	2.44 ± 0.01 [a]	1.25 ± 0.06 [b]	2.06 ± 0.01 [c]	1.99 ± 0.01 [d]
Leucocytes (10^6 cells/mm^3)	2.26 ± 0.01 [a]	1.09 ± 0.01 [b]	2.03 ± 0.02 [c]	2.01 ± 0.01 [d]
Hematocrit (%)	30.00 ± 2.03 [a]	16.18 ± 6.04 [b]	25.61 ± 2.39 [c]	25.70 ± 4.05 [c]
Hemoglobin concentration (g/dl)	10.05 ± 0.16 [a]	6.72 ± 1.39 [b]	8.53 ± 0.74 [c]	8.37 ± 0.55 [c]

Diet A (control 2.0 µg AFB$_1$/kg + 3.0 mg FB$_1$/kg); Diet B (19.7 µg AFB$_1$/kg + 28.5 mg FB$_1$/kg); Diet C (lanthanum chloride 200.0 mg/kg and 19.7 µg AFB$_1$/kg + 28.5 mg FB$_1$/kg); Diet D (lanthanum chloride 400.0 mg/kg and 19.7 µg AFB$_1$/kg + 28.5 mg FB$_1$/kg). Rows with different superscripts are significantly different ($p < 0.05$).

2.3. Effects on the Erythrocytic Indices

Figure 1 shows the mean corpuscular volume (mcv) of the fish fed the control diet and the contaminated diets amended with lanthanum chloride for 56 days. At day 7 of the feeding study, the mcv of the fish fed the diet contaminated with the mixtures of the AFB$_1$ and FB$_1$ (diet B) was significantly ($p < 0.05$) reduced compared with the mcv of the fish fed the control diet. The Tukey's post hoc analysis reveals that, at this same interval, the mcv of the fish fed the mixtures of AFB$_1$- and FB$_1$-contaminated diets, amended with 400 mg/kg lanthanum chloride, increased significantly ($p < 0.05$) compared with the mcv of the fish fed the control diet (Figure 1). At day 28, the mcv values of the fish fed the treatment diets were significantly higher than those of the fish fed the control diet. The post hoc evaluations showed that there were no significant ($p > 0.05$) variations in the mcv of the fish fed the mycotoxin-contaminated diets (diet B) compared with the mcv of the fish fed the contaminated diet amended with 200 mg/kg lanthanum chloride (diet C). The post hoc assessment further shows that the mcv values obtained for the fish fed diet C were not significantly different ($p > 0.05$) compared with the mcv values of the fish fed diet D. At day 56 of the feeding trial, the one way analysis of variance (ANOVA) showed that the mean mcv of the fish fed the treatment diets were significantly higher ($p < 0.05$) than the mean mcv values of the fish fed the control diet. The post hoc assessment shows that there were no significant variations in the mcv values obtained for the fish fed the treatment diets ($p > 0.05$).

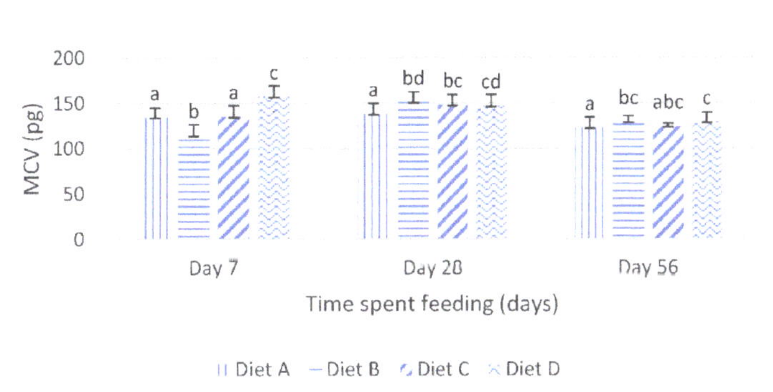

Figure 1. Mean corpuscular volume juvenile *Clarias gariepinus* fed diets amended with lanthanum chloride and mixtures of aflatoxin B1 and fumonisin B1 for 56 days. Bars with different superscripts are significantly different ($p < 0.05$) at specified periods of feeding.

The results obtained for the mean corpuscular hemoglobin (mch) of the fish fed the experimental diets are shown in Figure 2. They show that, at day 7 of the feeding study, there were significant ($p < 0.05$) increases in the mch values of the fish fed the treatment diets compared to the mch values of the fish fed the control diet. Figure 2 also shows the significant ($p < 0.05$) reduction in the mch values in the fish fed the contaminated diets, amended with lanthanum chloride, at day 28 and day 56 of the feeding trial. The post hoc assessment shows that there were no significant ($p > 0.05$) variations in the mch values of the fish fed the contaminated diets, amended with 200 mg/kg lanthanum chloride and 400 mg/kg lanthanum chloride, and the control diet at these sampling intervals.

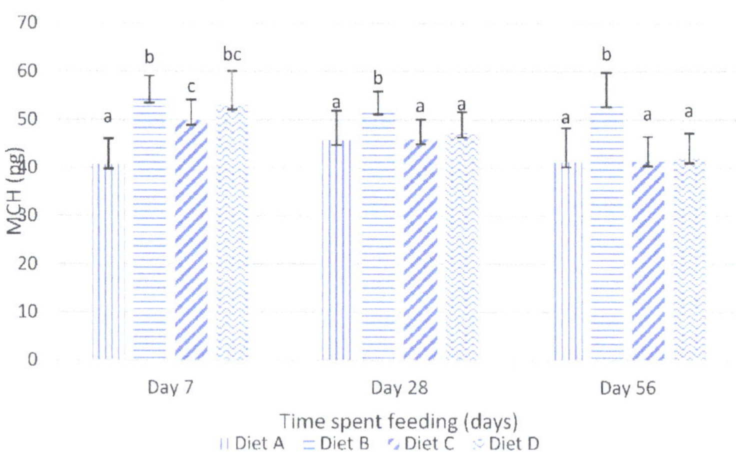

Figure 2. Mean corpuscular hemoglobin of juvenile *Clarias gariepinus* fed diets amended with lanthanum chloride and mixtures of aflatoxin B1 and fumonisin B1 for 56 days. Bars with different superscripts are significantly different ($p < 0.05$) at specified periods of feeding.

The mean corpuscular hemoglobin concentration (mchc) of the fish fed the experimental diets are shown in Figure 3. It shows that, at day 7 of the feeding, the mchc values of the fish fed the diets contaminated with mixtures of AFB_1 and FB_1 were significantly ($p < 0.05$) increased compared with the mchc values of the fish fed the contaminated diets amended with lanthanum chloride. There were no significant variations in the mchc values of the fish fed the contaminated diet B and the fish fed the contaminated diet C, amended with 200 mg/kg lanthanum chloride, at days 28 and day 56; furthermore, the mchc values of the fish fed the contaminated diet D, amended with 400 mg/kg lanthanum chloride, was not significantly ($p > 0.05$) different from the mchc values of the fish fed the control diet, at days 28 and 56 of the study (Figure 3).

The results obtained for the serum biochemistry evaluations are shown in Tables 6 and 7. At 7 days post commencement of the trial, there were no significant variations ($p > 0.05$) in the serum total protein concentrations of the fish fed diet A (the control diet) compared with those of the fish fed the treatment diets (diets B, C and D). The serum albumin concentration of the fish fed diet D (the contaminated diet amended with 400 mg/kg lanthanum chloride) was significantly ($p < 0.05$) higher compared with those of the fish fed diet A (the control diet). The Tukey's post hoc assessment shows the variations in the serum albumin concentrations of fish fed diet C and diet D were not significant ($p > 0.05$). Table 6 also shows there were significant increases ($p < 0.05$) in the serum transaminase activities of fish fed the treatment diets compared with those of fish fed the control diets. The Tukey's post hoc assessment shows that there were no significant variations ($p > 0.05$) in the alanine amino transferase and aspartate amino transferase values of the fish fed 200 mg/kg and 400 mg/kg lanthanum chloride.

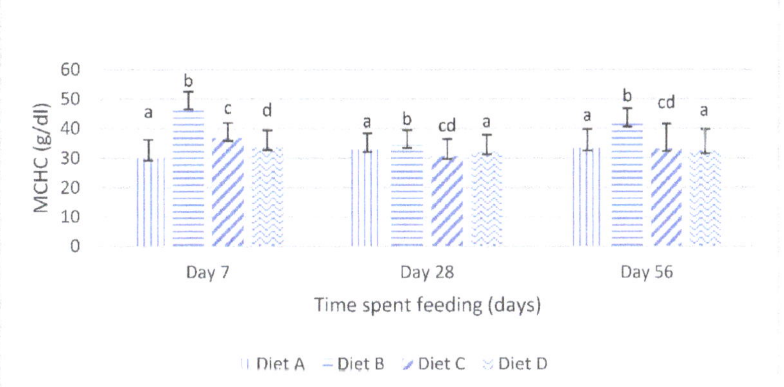

Figure 3. Mean corpuscular hemoglobin concentration of juvenile *Clarias gariepinus* fed diets amended with lanthanum chloride and mixtures of aflatoxin B1 and fumonisin B1 for 56 days. Bars with different superscripts are significantly different ($p < 0.05$) at specified periods of feeding.

Table 6. Serum biochemical profile of juvenile *Clarias gariepinus* catfish fed diets amended with lanthanum chloride and mixtures of aflatoxin B1 and fumonisin B1 for 7 days.

Parameter	Diet A (Control)	Diet B	Diet C	Diet D
Total Protein (g/dL)	6.01 ± 1.08 [a]	6.00 ± 1.27 [a]	5.94 ± 1.12 [a]	6.01 ± 1.13 [a]
Albumin (g/dL)	3.81 ± 0.23 [a]	3.93 ± 0.17 [a]	4.10 ± 0.11 [ab]	4.17 ± 0.88 [b]
Globulin (g/dL)	2.20 ± 0.01 [a]	2.07 ± 0.03 [b]	1.84 ± 0.04 [cd]	1.84 ± 0.02 [cd]
ALT (u/L)	12.34 ± 10.18 [a]	76.08 ± 11.29 [bc]	81.00 ± 26.08 [cd]	78.11 ± 13.21 [bd]
AST (u/L)	79.18 ± 5.39 [a]	144.03 ± 12.01 [b]	84.27 ± 4.19 [c]	80.59 ± 6.13 [ac]
ALP (u/L)	154.22 ± 7.07 [a]	248.76 ± 9.26 [b]	175.08 ± 6.15 [cd]	172.31 ± 8.33 [cd]
Creatinine (μmol/L)	2.19 ± 0.16 [a]	8.00 ± 0.6 [b]	5.61 ± 0.13 [cd]	6.02 ± 0.67 [cd]
Urea (mg/dL)	18.95 ± 3.11 [a]	33.17 ± 1.94 [b]	23.09 ± 2.13 [c]	27.12 ± 3.15 [d]
Uric Acid (μmol/L)	2.00 ± 0.08 [a]	6.03 ± 1.33 [b]	3.04 ± 0.99 [cd]	3.11 ± 1.04 [cd]
LDH (u/L)	607.19 ± 10.11 [a]	854.47 ± 21.16 [b]	677.16 ± 29.17 [c]	644.82 ± 11.1 [d]

Diet A (control 2.0 μg AFB_1/kg + 3.0 mg FB_1/kg); Diet B (19.7 μg AFB_1/kg + 28.5 mg FB_1/kg); Diet C (lanthanum chloride 200.0 mg/kg and 19.7 μg AFB_1/kg + 28.5 mg FB_1/kg); Diet D (lanthanum chloride 400.0 mg/kg and 19.7 μg AFB_1/kg + 28.5 mg FB_1/kg). Rows with different superscripts are significantly different ($p < 0.05$).

Table 7. Serum biochemical profile of juvenile *Clarias gariepinus* catfish fed diets amended with lanthanum chloride and mixtures of aflatoxin B1 and fumonisin B1 for 56 days.

Parameter	Diet A (Control)	Diet B	Diet C	Diet D
Total Protein (g/dL)	5.63 ± 0.55 [a]	6.08 ± 0.23 [abcd]	5.99 ± 0.28 [bcd]	5.85 ± 0.34 [abc]
Albumin (g/dL)	3.76 ± 0.03 [a]	4.06 ± 0.12 [be]	3.93 ± 0.06 [c]	3.82 ± 0.02 [d]
Globulin (g/dL)	1.87 ± 0.01 [a]	2.02 ± 0.03 [be]	2.06 ± 0.01 [c]	2.03 ± 0.02 [d]
ALT (u/L)	13.61 ± 0.31 [a]	54.93 ± 2.17 [b]	25.00 ± 1.10 [c]	27.09 ± 1.15 [d]
AST (u/L)	83.24 ± 1.88 [a]	159.07 ± 9.01 [b]	92.8 ± 2.19 [cd]	94.01 ± 3.24 [cd]
ALP (u/L)	162.01 ± 10.02 [a]	243.91 ± 6.24 [b]	183.2 ± 7.1 [cd]	185.3 ± 6.5 [cd]
Creatinine (μmol/L)	2.24 ± 0.19 [a]	6.39 ± 0.01 [b]	3.25 ± 0.17 [cd]	3.15 ± 0.43 [cd]
Urea (mg/dL)	20.38 ± 2.17 [a]	31.68 ± 1.81 [b]	26.39 ± 3.14 [c]	24.28 ± 1.19 [d]
Uric Acid (μmol/L)	1.83 ± 0.11 [a]	5.07 ± 0.19 [b]	2.91 ± 0.01 [c]	3.06 ± 0.13 [d]
LDH (u/L)	593.82 ± 12.19 [a]	965.5 ± 34.28 [b]	617.8 ± 19.0 [ce]	637.9 ± 18.1 [d]

Diet A (control 2.0 μg AFB_1/kg + 3.0 mg FB_1/kg); Diet B (19.7 μg AFB_1/kg + 28.5 mg FB_1/kg); Diet C (lanthanum chloride 200.0 mg/kg and 19.7 μg AFB_1/kg + 28.5 mg FB_1/kg); Diet D (lanthanum chloride 400.0 mg/kg and 19.7 μg AFB_1/kg + 28.5 mg FB_1/kg). Rows with different superscripts are significantly different ($p < 0.05$).

At this same time interval, the serum creatinine, urea and uric acid concentrations of the fish fed the experimental diets (diets B, C and D) were significantly ($p < 0.05$) higher compared with those of the fish fed the control diet; Tukey's post hoc evaluations at this interval also show the serum concentrations of the creatine, urea and uric acid obtained for the fish fed the diets amended with 200 mg/kg (diet C) and 400 mg/kg lanthanum chloride (diet D) were significantly ($p < 0.05$) lower than those obtained for the fish fed only the contaminated diet (diet B); furthermore, the differences obtained in these values for the fish fed 200 mg/kg lanthanum chloride and 400 mg/kg lanthanum chloride were not significant ($p > 0.05$).

The results obtained for the serum biochemical evaluations at day 56 of the feeding trial are shown in Table 7. It shows that the serum total protein concentrations of the fish fed the treatment diets were significantly ($p < 0.05$) higher compared with those of the fish fed the control diet. The Tukey's post hoc assessment reveals that there were no significant ($p > 0.05$) variations in the serum total protein concentration of the fish fed diet C and diet D. Table 7 also shows that the serum albumin concentration of the fish fed the control diet differed significantly ($p < 0.05$) compared with those of the fish fed diets B, C and D.

The serum globulin concentrations of the fish fed diets B, C and D were significantly ($p < 0.05$) higher compared with those of the fish fed diet A (the control diet) at 56 days post commencement of the trial (Table 7). The Tukey's post hoc evaluation reveals the difference in the serum globulin concentrations of the fish fed diet B and diet D, at 56 days of the feeding trial, were not significant ($p > 0.05$).

The serum transaminases (alanine aminotransferase and aspartate aminotransferase activities), and the alkaline phosphatase activities increased significantly ($p < 0.05$) in the fish fed diets B, C and D compared with their corresponding values in the fish fed the control diet (Table 7). The Tukey's post hoc evaluation shows the serum alanine aminotransferase (ALT) activities of the fish fed diet B, and diet D differed significantly ($p < 0.05$) at 56 days of feeding, with the fish fed diet B having the lowest ALT activity (25.00 u/L) and the fish fed diet D having the highest ALT activity (33.51 u/L).

The serum concentrations of creatinine, urea and uric acid and the serum activity of lactate dehydrogenase (LDH) increased significantly ($p < 0.05$) in the fish fed the treatment diets (diets B, C and D) compared to their corresponding values in the fish fed the control diet (diet A) at 7 days of feeding (Table 6) and at 56 days of feeding (Table 7).

3. Discussion

The water quality parameters of the culture tanks in the present study were determined to be within the range recommended for the culture of clariid catfishes [39]; thus, may not have contributed to the pathophysiological observations recorded in this study. The proximate and mycotoxin ($AFB_1 + FB_1$) content of the feed was not altered by the addition of the lanthanum chloride; furthermore, the AFB_1 and FB_1 content of the final feed was higher than the respective concentrations of the purified mycotoxins introduced into the diets at formulation. This is an expected result, as it has been previously noted that agricultural products are often contaminated with various mycotoxins, and that these mycotoxins occur at varying concentrations; hence, the difference in the AFB_1 and FB_1 contents of the produced diets reflects the concentrations of these mycotoxins in the agricultural materials used in the production of the feed [9,43].

The fish fed diet A (the control diet) consumed the most quantity of feed, while the fish fed diet B (19.7 µg AFB_1 + 28.5 mg FB_1/kg) consumed the lowest quantities of fed. This is an expected result, as mycotoxins, especially aflatoxins, are reported to cause a reduction in feeding, or an outright feed refusal, with a consequent decrease in the performance in the animals [41,44]. This agrees with the findings of [32], who observed that dietary rare earth elements improve the body weight gain and feed conversion ratio, without increasing feed intake. The results of the present study also show that the fish fed diets containing 400 mg/kg lanthanum chloride consumed more feed compared with the fish fed 200 mg/kg lanthanum chloride; thus, indicating that the feed consumption increased with the dietary concentration of lanthanum chloride.

The fish fed the diets containing AFB_1 and FB_1 exhibited the lowest weight gain. This is similar to the findings of our earlier study [39], where poor growth performance was recorded in the juvenile *Clarias gariepinus* catfish when fed diets amended with doses of mixtures AFB_1 and FB_1. The aflatoxins are reported to cause gastrointestinal dysfunctions marked by significant changes in the gut morphology, reduced digestive ability and a disruption of the digestive enzymes and intestinal innate immunity [23]. The fumonisins are reported to negatively influence growth performance by their abilities to interfere with cellular growth and cell–cell interactions [20,45]. The poor growth recorded for fish when fed diets containing mixtures of AFB_1 and FB_1 may be as a consequence of the combined activities of the two mycotoxins.

The fish fed diet D (400 mg/kg lanthanum chloride) exhibited a superior weight gain compared with the fish fed diet C (200 mg/kg lanthanum chloride), suggesting that the weight gain in the juvenile *Clarias gariepinus*, fed diets contaminated with mixtures of aflatoxin B1 and fumonisin B1, was influenced by the concentration of the dietary inclusion of lanthanum chloride. The exact mechanisms of the growth promotion by lanthanum chloride are yet to be described. It is, however, reported that lanthanum chloride may promote weight gain in animals by improving the utilization of dietary nutrients, such as total energy, crude protein and fat [46]. It is also reported that dietary lanthanum chloride increases the secretion of gastric juices in the exposed animals [47]; thus, the increased weight gain observed in the present study may be a function of the increased activities of gastric enzymes in the exposed fish [47] and, since the fish fed the diets containing lanthanum chloride at 400 mg/kg diet consumed more feed compared with those fed the diets containing lanthanum chloride at 200 mg/kg diet, it is therefore reasonable for them to gain better weight.

The fish fed diet B (19.7 µg AFB_1 + 28.5 mg FB_1/kg diet) exhibited the highest (3.268), and the lowest feed conversion ratio (30.60 ± 1.60) compared with the fish fed the other diets. This is a result of the deleterious effects of the mixed mycotoxins in the diets [13,40,48,49]. The feed conversion ratio and the feed conversion efficiency were significantly improved by the addition of the lanthanum chloride into the diets. The results of the present study further show that fish fed diet C (lanthanum chloride 200 mg/kg diet), exhibited the lowest (1.981 ± 0.07) feed conversion ratio. Hence, the lanthanum chloride at 200 mg/kg inclusion rates produced the best nutrient utilization in the juvenile *Clarias gariepinus* fed diets contaminated with mixtures of aflatoxin B1 and fumonisin B1.

The evaluations of the hematological parameters of the fish are required in the physiological assessment of the effects of exposure to sub-chronic concentrations of contaminants [50] and/or the physiological response to the dietary intake of essential nutrients [51]. This is because the determination of the erythrocytes count, the hematocrit values, and the hemoglobin concentration in the fish aids in the assessment and prognostication of anemias [52].

The results of the present study show significant decreases in the erythrocytes counts, the hematocrit values and the hemoglobin concentrations of the fish fed the diets contaminated with mixtures of AFB_1 and FB_1 compared with the corresponding values for the fish fed the control diets at days 7, 28 and 56 of the trial. This is similar to the findings of [8], who reported a disruption in the protein digestion and absorption in Nile tilapia following dietary exposures to AFB_1.

There were no significant variations in the erythrocytes counts, the hematocrit values and the hemoglobin concentrations of the fish fed the low (200 mg/kg) or high (400 mg/kg) concentrations of lanthanum chloride; however, the fish fed the mycotoxin-contaminated diets amended with lanthanum chloride exhibited significantly higher erythrocytes counts, hematocrit values and hemoglobin concentrations compared with their corresponding values in the fish fed diets contaminated with the mixtures of AFB_1 + FB_1 only. This indicated that dietary lanthanum chloride may ameliorate the depression of erythropoiesis induced by dietary exposures to mixtures of AFB_1 and FB_1. This finding may be a result of the increases in feed consumption and improved utilization of dietary nutrients [53,54], or due to the anti-oxidative effects of lanthanum chloride, wherein lanthanum chloride is able to protect the oxidation of dietary fatty acids, such as omega-3 fatty acids, thereby making it more available and/or enhancing their absorption [55].

There were significant and sustained leukocytopenia in the fish fed the diets contaminated with AFB_1 and FB_1 throughout the duration of the study. This agrees with the reports [21,45,48,49], where it was reported that the dietary mycotoxins elicit a suppression of the immune response of exposed animals. Furthermore, the fish fed the diets contaminated with AFB_1 and FB_1 and containing lanthanum chloride exhibited significantly higher leucocytes counts compared with the fish fed diets contaminated AFB_1 and FB_1 alone, indicating lanthanum chloride may have some ameliorative effect on the leucocytes counts of juvenile *Clarias gariepinus* fed diets contaminated with mixtures of AFB_1 and FB_1. The mechanism for these immunoprotective effects may not be unconnected with the antioxidant activities of lanthanum chloride [53,55] and/or the increased nutrient absorption and utilization effects of lanthanum chloride [56,57].

Although the results of the present study show that the dietary lanthanum chloride elicited significant changes in the erythrocytic indices (the mean corpuscular volume, the mean corpuscular hemoglobin values and the mean corpuscular hemoglobin concentration) of juvenile *Clarias gariepinus* fed with the contaminated diets, the observed changes were well within the scope of the hematological reference intervals for juvenile *Clarias gariepinus* [39]. It is probable that these changes may have been more pronounced if the duration of the study had extended beyond the 56 days duration, as it is generally reported that the effects of dietary exposures to mycotoxins are dependent on the duration of the exposure and the concentration of the mycotoxins [5,13,20].

The serum total proteins, consisting of the albumin and globulin concentrations, provide critical information reflecting the functional statuses of various organs and/or systems; since they are involved in the specific immune responses of the fish and participate in the maintenance of the acid-base balance [58,59], the serum proteins are also involved in the protection of the cellular integrity of cells, such as the erythrocytes, hepatocytes and the nephrocytes [50]. The serum total proteins also provide an easy and readily available source of energy in emergencies, such as that obtained in situations of feed deprivation [60–62].

The serum total proteins increase in cases of generalized chronic inflammation and in inflammatory disorders affecting the liver and the kidneys [60]. The present study was marked by hyperproteinemia (observed 56 days post dietary exposure) in the fish fed the diets contaminated with mixtures of AFB_1 and FB_1; this is an expected result as both of the mycotoxins have been reported to elicit hepatic and nephrotic syndromes in exposed fish [45,63]. There were no significant differences in the serum total protein of the fish fed the diets amended with lanthanum chloride or bentonite clay. This may be the consequence of ingested free mycotoxins, especially of the fumonisins [64].

The serum albumins are produced in the liver. Therefore, the synthetic capacity of the liver (which is an estimate of the protein losing nephropathy) may be estimated by the determination of the serum albumin concentration [65]. According to [66], malnutrition, increased protein catabolism, enteropathy and/or chronic nephropathy are marked by a reduced serum albumin concentration (termed hypoalbuminemia). The fish fed the diets contaminated with the mixtures of AFB_1 and FB_1 exhibited significantly elevated serum albumin concentrations. These may be a result of the combined effects of AFB_1 and FB_1 [63]. As observed for the serum total proteins, the inclusion of the lanthanum chloride in the diets elicited a marginal but significant reduction in the serum albumin concentration compared with those of the fish fed the diets contaminated with only the mixtures of AFB_1 and FB_1. This may indicate that lanthanum chloride may have some hepatoprotective and nephroprotective properties in juvenile *Clarias gariepinus* fed diets contaminated with mixtures of AFB_1 and FB_1.

The hepatic enzymes (transaminase and alkaline phosphatase) are liberated into the serum in situations of hepatocellular or cholestatic liver injuries [67]. In hepatopathies, such as those seen in hepatocellular degenerations, aspartate aminotransferase (AST) and alanine aminotransferase (ALT) are liberated into the serum, while alkaline phosphatase

(ALP) is liberated into the serum in hepatic cholestasis [66]. The results of the present study shows significant elevations in the serum activities of these enzymes in the fish fed the diets containing the mixtures of AFB_1 and FB_1. The highest values for the AST, ALT and ALP were obtained in the fish fed diet B (contaminated with AFB_1 and FB_1). The AST, ALT and the ALP activities were significantly lower in the fish fed the diets containing lanthanum chloride and bentonite clay, compared with the corresponding values in the fish fed AFB_1- and FB_1-contaminated diets only. The fumonisins are reported to cause cellular rupture and necrosis by the inhibition of mitochondrial respiration and the complete deregulation of calcium homeostasis [68]; meanwhile, by its ability to preferentially bind to calcium, the lanthanum chloride may prevent tissue necrosis via this mechanism and, hence, reduce the elaboration of these enzymes into the serum, indicating some erythrocyte, hepatic and kidney protective effects of the dietary lanthanum chloride [24,69].

The serum creatinine concentration is increased significantly in the skeletal muscle necrosis and/or atrophy, as well as in chronic nephropathies [58,70]. In the present study, there were significant elevations of the serum creatinine concentration in the fish fed the diets containing the mixtures of AFB_1 and FB_1. The highest values for the serum creatinine concentration were obtained in the fish fed diet B (contaminated with AFB_1 and FB_1), indicating significant skeletal muscle necrosis and/or atrophy, as well as probable chronic nephropathy [66]. The serum creatinine concentrations of the fish fed AFB_1 and FB_1-contaminated diets, containing lanthanum chloride or bentonite clay, were significantly lower compared with those of the fish only fed the AFB_1- and FB_1-contaminated diets. This may be indicative of some liver and kidney protective effects of the lanthanum chloride in the *Clarias gariepinus* fed the diets contaminated with the mixtures of AFB_1 and FB_1 [24,71].

The serum urea nitrogen and the uric acid concentrations are critical analytes required in the assessments of the functional status of the kidney [67]. It is reported that decreases in the blood urea nitrogen concentrations are usually observed in hepatic insufficiencies and in cases of malnutrition, while an increased blood urea nitrogen concentration is commonly reported in renal disease, shock and in cardiac insufficiencies [71,72]. The results obtained from the present study show the serum urea and uric acid concentrations of the fish fed the diets contaminated with the mixtures of AFB_1 and FB_1 were significantly higher than those of the fish fed the control diet, indicating a significant impact on the kidneys [8,41,69]. The fish fed the diets contaminated with the mixtures of AFB_1 and FB_1 containing lanthanum chloride exhibited significantly lowered serum urea and uric acid concentrations compared to the fish fed the diets contaminated with only mixtures of AFB_1 and FB_1, indicating that the dietary lanthanum chloride in juvenile *Clarias gariepinus* may have some ameliorating effects on the AFB_1- and FB_1-induced kidney toxicities [70,73].

Lactate dehydrogenase (LDH) is an enzyme found in several tissues/organs (such as the muscles, liver, heart, kidneys and the blood vessels. It catalyzes the reversible transformation of pyruvate into lactate [74]. The increased serum activity of LDH is indicative of degenerative changes in any of the aforementioned tissues/organs [66,67]. The results of our study show that the serum LDH activity of the fish fed the diets contaminated with only mixtures of AFB_1 and FB_1 were significantly higher compared with those of the fish fed the control diet, indicating a significant impact on the kidneys and/or the other aforementioned organs [67,70]. The fish fed the diets contaminated with the mixtures of AFB_1 and FB_1 containing lanthanum chloride exhibited significantly lowered serum urea and uric acid concentrations compared to the fish fed the diets contaminated with only mixtures of AFB_1 and FB_1, indicating that the dietary lanthanum chloride in juvenile *Clarias gariepinus* may have some ameliorating effects on the toxicities of the mixtures of AFB_1 and FB_1 [40,75].

4. Conclusions

Under the present culture conditions, the dietary lanthanum chloride at 200 or 400 mg/kg feed could promote the growth performance, nutrient utilization and ameliorate the hematological and serum biochemical derangements produced by the dietary exposures to the mixtures of AFB_1 and FB_1 in the juvenile *Clarias gariepinus* catfish. A further study is needed to determine and confirm the exact dietary concentration of lanthanum chloride needed for the optimization of the growth and health performance of this fish species, under the challenge of inadvertent dietary exposures to mixtures of aflatoxin B1 and fumonisin B1.

5. Materials and Methods

5.1. Experimental Fish and Experimental Design

Three hundred and sixty (360) juvenile *C. gariepinus* catfish were acquired from a commercial catfish hatchery. The fish were allowed to acclimatize to laboratory conditions for 21 days before the commencement of the experiment.

Fifteen (12) 1000 L capacity tanks retrofitted with water inflow and outflow devices were divided into five groups (each consisting of a triplicate set of tanks, with each tank containing 30 juvenile *C. gariepinus catfish*), as described in [75].

The experiment adopted a complete randomized design with a triplicate of each treatment. Four groups, consisting of one control and three treatment groups, were used for this study. Group A (control group, were fed the control diet, i.e., no lanthanum chloride, no mycotoxin); Group B (treatment 1, were fed the basal diet amended with mixtures of AFB_1 and FB_1 at an inclusion rate of 19.7 µg AFB_1/kg diet and 28.5 mg FB_1/kg diet); Group C (treatment 2, were fed the basal diet amended with lanthanum chloride 200.0 mg/kg diet and mixtures of AFB_1 and FB_1 at an inclusion rate of 19.7 µg AFB_1/kg diet and 28.5 mg FB_1/kg diet); Group D (treatment 3, were fed the basal diet amended with lanthanum chloride 400.0 mg/kg diet and mixtures of AFB_1 and FB_1 at an inclusion rate of 19.7 µg AFB_1/kg diet and 28.5 mg FB_1/kg diet). The groups of fish were fed any of the diets A, B, C or D in triplicates for 56 days.

5.2. Experimental Feeds

The experimental feeds were produced at the University of Abuja feed mill, following the procedures reported by [39], with slight adjustments. The basal diet was formulated using the following ingredients (fish meal 19%, soybean cake 37%, maize 32.25%, palm oil 1.0%, fish oil 6.0%, starch binder 2.0%, vitamin/mineral premix 0.5%, bone meal 1.0%, salt 0.25%), to meet the nutritional requirements of the juvenile clariid fish [37].

The mixtures of the AFB_1 and FB_1 diets were produced by adding 1 mg crystalline AFB_1 (Sigma Chemicals, St Louis, MO, USA) to 1 mL chloroform (to produce 1 mg: 1000 µL aliquot of AFB_1). The quantity of the solution required to produce the chosen concentration of the AFB_1 in the mixed mycotoxin diets was then pipetted using an automated adjustable pipette into 100 mL volumetric flasks. This volume was then made up to the 100 mL mark with methanol. The FB_1 contents of the respective diets were added by careful measurements of the desired quantities of FB_1 from a pre-produced aliquot of 1 g FB_1 dissolved in 1000 µL of acetonitrile-water (v/v) (resulting in 1 µL:1 mg solution of fumonisin B1).

The ingredients for the basal diets were weighed, completely mixed, and added to the liquid mixture of AFB_1 and FB_1 (i.e., 17.0 µg AFB_1; 23.0 FB_1/kg diet), 200.0 mg lanthanum chloride, 400.0 mg lanthanum chloride or 400 mg/kg. Diet A (control 0.0 µg AFB_1/kg + 0.0 mg FB_1/kg diet); diet B (17.0 µg AFB_1; 23.0 FB_1/kg diet); diet C (lanthanum chloride 200.0 mg/kg and 17.0 µg AFB_1; 23.0 FB_1/kg diet) and diet D (lanthanum chloride 400.0 mg/kg and 17.0 µg AFB_1; 23.0 FB_1/kg diet).

The mixtures were subsequently blended and placed in a hot air oven for the methanol to evaporate. These were then pelletized with an extruder pelletizer, after the addition of weighted portions of the starch binders. The mycotoxins content of the compounded diets were then analyzed using the multi-mycotoxin LC-MS/MS method [76,77] and, thereafter, individually packed in airtight polyethylene bags and stored in a deep freezer (at 2–4 °C) until use. According to the results obtained following the mycotoxin assessments of the produced diets, the AFB_1 and FB_1 concentrations of the produced experimental diets were consequently adjusted, as shown in Table 1.

At time points day 7, 28 and day 56, five (5) fish were randomly selected (by the aid of a handheld sampling net) from each tank for hematological determinations. The sampled fish were weighed, length measured and then bled via caudal veni-puncture, using a 23 G needle fitted on a 5 mL syringe pre-rinsed with ethylene diamine tetra acetic (EDTA) solution (for hematological assessments), and without EDTA (for serum biochemical determinations).

5.3. Hematological Analysis

The hematological parameters, such as the hemoglobin concentration (Hb), total erythrocytes count (RBC) and total leucocytes count (WBC), packed cell volume (PCV), mean corpuscular hemoglobin (MCH), mean corpuscular hemoglobin concentration (MCHC) and mean corpuscular volume (MCV), were analyzed.

The Hb concentration was determined by the cyanmethemoglobin method [50]. A Neubauer hemocytometer was used for the determination of the total red and white blood corpuscle count, by using Natt and Herricks dilution fluid. The cell count was estimated as per [78], using the following formulae:

$$\text{RBC count} = \text{number of cells counted} \times \text{dilution factor} \times \text{depth of chamber}/\text{area counted}$$

where the dilution factor is one in 200; the depth is 1/10 mm; and the area counted is $80/400 = 1/5$ sq.

$$\text{WBC count} = \text{number of cells counted} \times \text{blood dilution} \times \text{chamber depth}/\text{number of chambers counted:}$$

$$\text{WBC}/\text{mm}^3 = \text{total white blood cells} \times 200 \times 10/4$$

The hematocrit (PCV) was determined by the following methods of [50] briefly, where micro-hematocrit capillaries were centrifuged at 12,000 g for 5 min in a micro-hematocrit centrifuge (Ocean Med, London, UK). The erythrocytic indices were determined mathematically, following the methods of [58].

5.4. Serum Biochemical Parameters

The serum was obtained from the blood (collected without anticoagulant), and kept in the narrow-bored glass test tubes after centrifugation at $5000 \times g$ for 5 min in a laboratory centrifuge (Uniscope model SM 112, Surgifield Medicals, Devon, UK. After centrifugation, the liquid fraction (the serum) was eluted from the top of the tube, using a clean and sterile 1 mL syringe. The collected serum was then transferred into a fresh Eppendorf tube and stored at -20 °C until used for biochemical analyses.

From the serum, the following biochemical parameters were determined: total protein and albumin (in g dL^{-1}) using the methods of [17]; alanine amino transaminase and aspartate aminotransferase (UL^{-1}), alkaline phosphatase (UL^{-1}), lactase dehydrogenase (UL^{-1}), creatinine kinase (μmol/L) following the methods of [67]. The serum urea (mg dL^{-1}) and uric acid concentrations (μmol/L)) were determined, using the methods of [79]. The serum biochemistry evaluations were carried out, using a laboratory spectrophotometer (SpectrumLab 750, Huddinge, Sweden) and Dialab serum diagnostic kits (Dialab, Neudorf, Austria). The globulin values were calculated mathematically by subtracting the albumin values from the values obtained for serum total protein. All of the serum biochemical parameters were obtained by following strictly the guidelines of the manufacturers of the diagnostic kits.

5.5. Zootechnical Evaluations

The fish in each group were fed the respective feed at 5% of the biomass (the feed was divided into two portions, given at 0800 h and at 1800 h) daily. All of the unconsumed feed was siphoned, dried and weighed in order to determine the exact amount of feed consumed per group. At the set time points, the fish were individually weighed (to the nearest 0.00, using a sensitive laboratory scale) and their length measured (to the nearest

0.00 cm using a ruler). At the close of the feeding trial, the following growth indices were determined mathematically according to [80]:

(a)   Growth rate (g/d) = $W_f - W_i / T_f - T_i$ Where W_f = final weight W_i = initial weight $T_f - T_i$ = Time (in days) spent feeding between W_f and W_i;

(b)   Specific growth rate (%/day) = $100 \times [(\ln W_f - \ln W_i)/(T_f - T_i)]$;

(c)   Feed conversion ratio (FCR) = weight of feed consumed (g)/$W_f - W_i$ (W_f = final weight; W_i = initial weight);

(d)   Feed conversion efficiency (%) = [$W_f - W_i$ weight of feed consumed (g)] $\times$ 100

5.6. Statistical Evaluation

Data were expressed as the mean $\pm$ Standard Deviation of the mean of each group (n = 30) and analyzed by a one way analysis of variance (ANOVA). The variant means were then separated by the Duncan's multiple range post hoc test using the Statistical Package of Social Sciences (SPSS) for Windows version 20.0 (IBM Corporation, Armonk, USA). Differences in the means were considered significant at $p < 0.05$.

Author Contributions: Conceptualization, B.T.A. and N.G.E.; methodology, B.T.A., N.G.E.; A.Z. and S.E.A.; validation, N.G.E.; G.M.O. and S.E.A.; formal analysis, T.B.K., A.E. and C.B.I.A.; investigation, A.E.; resources, T.B.K.; data curation, O.H.O. and G.M.O.; writing—original draft preparation, B.T.A. and N.G.E.; writing—review and editing, W.D.N., B.T.A. and O.H.O.; visualization, A.Z.; supervision, W.D.N. and C.B.I.A.; funding acquisition, B.T.A. All authors have read and agreed to the published version of the manuscript.

Funding: This research was funded by The University of Abuja Institutional Based Research Grant, grant number TETFUND/DESS/UNI/ABUJA/RP/VOL. 2 and the APC was funded by TETFUND/DESS/UNI/ABUJA/RP/VOL. 2.

Institutional Review Board Statement: Not applicable.

Informed Consent Statement: Not applicable.

Data Availability Statement: The data presented in this study are available on request from the corresponding author.

Acknowledgments: This work was conducted at the Veterinary Teaching and Research Farm, Faculty of Veterinary Medicine, University of Abuja. We therefore use this opportunity to appreciate the supporting staff of the farm for their efforts in the daily feeding of the experimental fish, and the Dean Faculty of Veterinary Medicine (Sani Saka), for granting permission to the facility. We also want to thank Alani and Lilian Uchenna for proofreading the manuscript.

Conflicts of Interest: The authors declare no conflict of interest.

References

1.   Food and Agricultural Organization (FAO) of the United Nations. The state of the World Fisheries and Aquaculture. In *Contributing to Food Security and Nutrition for All*; FAO: Rome, Italy, 2016; 200p.

2.   Olsen, R.L.; Hasan, M.R. A limited supply of fishmeal: Impact on future increases in global aquaculture. *Trends Food Sci. Technol.* **2012**, *27*, 120–128. [CrossRef]

3.   Encarnacao, P. Recent updates on the effects of mycotoxins in aquafeeds. *Int. Aquafeed* **2011**, *14*, 10–13.

4.   Santos, G.A.; Rodrigues, I.; Naechrer, K.; Encarnacao, P. Mycotoxins in Aquaculture: Occurrence in Feed Components and Impact on Animal Performance. *Aquac. Eur.* **2010**, *35*, 6–10.

5.   Spring, P.; Fegan, D.F. Mycotoxin—A rising threat to aquaculture? In *Nutritional Biotechnology in the Feed and Food Industries, Proceeding of the Alltech's 21st Annual Symposium, Lexington, KY, USA, 22–25 May 2005*; Lyons, T.P., Jacque, K.A., Eds.; Alltech UK: Stamford, UK, 2005; pp. 323–332.

6.   Rodrigues, I.; Naehrer, K. A three-year survey on the worldwide occurrence of mycotoxins in feedstuffs and feed. *Toxins* **2012**, *4*, 663–675. [CrossRef]

7.   Bankole, S.A.; Mabekoje, O.O. Occurrence of aflatoxins and fumonisins in preharvest maize from south-western Nigeria. *Food Addit. Contam.* **2004**, *21*, 251–255. [CrossRef]

8.   Abdelhamid, F.M.; Elshopakey, G.E.; Aziza, A.E. Ameliorative effects of dietary *Chlorella vulgaris* and β-glucan against diazinon-induced toxicity in Nile tilapia (*Oreochromis niloticus*). *Fish Shellfish. Immunol.* **2020**, *96*, 213–222. [CrossRef]

9.   Oyedele, O.A.; Ezekiel, C.N.; Sulyok, M.; Adetunji, M.C.; Warth, B.; Atanda, O.O.; Krska, R. Mycotoxin risk assessment for consumers of groundnut in domestic markets in Nigeria. *Int. J. Food Microbiol.* **2017**, *251*, 24–32. [CrossRef]

10.   Food and Drug Administration (FDA); US Food and Drug Administration; Centre for Food Safety and Applied Nutrition; Centre for Veterinary Medicine. Guidance for Industry 2001. Fumonisin levels in Human Foods and Animal Feeds, November 9. 2001. Available online: https://www.fda.gov/regulatory-information/search-fda-guidance-documents/guidance-industry-fumonisin-levels-human-foods-and-animal-feeds (accessed on 13 January 2019).

11.   Food and Agriculture Organization of the United Nations (FAO). *Worldwide Regulation of Mycotoxins in Food and Feed in 2004*; FAO Food and Nutrition paper No 81, FAO: Rome, Italy, 2004; pp. 1728–3264.

12.   Puschner, B. Mycotoxins. *Vet. Clin. North Am. Small Anim. Pract.* **2002**, *32*, 409–419. [CrossRef]

13.   Sepahdar, A.; Ebrahimzadeh, H.A.; Sharifpour, I.; Khorsaivi, A.; Motallebi, A.A.; Mohseni, M.; Kakoolaki, S.; Pourali, H.R.; Hallajian, A. Effects of Different Dietary Levels of AFB$_1$ on Survival Rate and Growth Factors of Beluga (*Huso huso*). *Iran. J. Fish. Sci.* **2009**, *9*, 141–150.

14.   Dey, D.K.; Chang, S.N.; Kang, S.C. The inflammation response, and risk associated with Aflatoxin B1 contamination was minimized by insect peptide CopA3 treatment and act towards the beneficial health outcomes. *Environ. Pollut.* **2020**, *268*, 115713. [CrossRef]

15.   International Agency for Research on Cancer (IARC). *Chemical Agents and Related Occupations: A Review of Human Carcinogens*, International Agency for Research on Cancer (IARC): Lyon, France, 2012.

16.   Spring, P. Mycotoxins—A rising threat to aquaculture. *Feedmix* **2005**, *13*, 1–5.

17.   Royes, J.B.; Yanong, R.P.E. Molds in Fish and Aflatoxicosis. 2002. Available online: http://edis.ifas.ufe.edu/fa095 (accessed on 22 March 2022).

18.   Han, X.Y.; Huang, Q.C.; Li, W.F.; Xu, Z.R. Changes in growth performance digestive enzyme activities and nutrient digestibility of Cherry valley ducks in response to aflatoxin B1levels. *Livest. Sci.* **2008**, *119*, 216–220. [CrossRef]

19.   Jantrarotai, W.; Lovell, R.T. Subchronic toxicity of dietary aflatoxin B1 to channel catfish. *J. Aquat. Anim. Health* **1990**, *2*, 248–254. [CrossRef]

20.   Manning, B.B.; Li, M.H.; Robinson, E.A. Aflatoxins from moldy corn cause no reduction in channel catfish (*Ictalurus puntatus*) performance. *J. World Aquac.* **2005**, *23*, 167–179.

21.   Tuan, N.A.; Manning, B.B.; Lovell, R.T.; Rottinghaus, G.E. Responses of Nile Tilapia (*Oreochromis niloticus*) fed diets containing different concentrations of moniliformin of fumonisin B1. *Aquaculture* **2003**, *217*, 515–528. [CrossRef]

22.   Sahoo, P.K.; Mukherjee, S.C. Immunosuppressive effects of afflation B1 in Indian major carp (*Labeo rohita*). *Comp. Immunol. Microb. Infec. Dis.* **2001**, *24*, 143–149. [CrossRef]

23.   Boonyaratpalin, M.; Supamattaya, K.; Verakunipiraya, V.; Supraset, D. Effects of aflatoxin B1 on growth performance blood component, immune function and histopathological changes in black tiger shrimp (*Peanus monodons*. Fabricius). *Aquacuture Res.* **2001**, *32*, 388–398. [CrossRef]

24.   Mughal, M.J.; Peng, X.; Kamboh, A.A.; Zhou, Y.; Fang, J. Aflatoxin B1 induced systemic toxicity in poultry and rescue effects of selenium and zinc. *Biol. Trace Elem. Res.* **2017**, *178*, 292–300. [CrossRef]

25.   Zychowski, K.E.; Hoffmann, A.R.; Ly, H.J.; Pohlenz, C.; Buentello, A.; Romoser, A.; Gatlin, D.M.; Phillips, T.D. The effect of aflatoxin-B1 on red drum (*Sciaenops ocellatus*) and assessment of dietary supplementation of NovaSil for the prevention of aflatoxicosis. *Toxins* **2013**, *5*, 1555–1573. [CrossRef]

26.   Maja, S. Fumonsins and their effect on animal with a brief review. *Vet. Achiv* **2001**, *71*, 299–320.

27.   Walter, F.O.; Maracas, W.F. O Discovery and occurrence of the fumonsins historical perspectives. *Environ. Health* **2001**, *109*, 239–243.

28.   Voss, K.A.; Smith, G.W.; Hashcek, W.M. Fumonisins: Toxicokinetics, mechanism of action and toxicity. *Anim. Feed. Sci. Technol.* **2007**, *137*, 299–325. [CrossRef]

29.   Slamova, R.; Trckova, M.; Vondruskova, H.; Zraly, Z.; Pavlik, I. Applied Clay Science Clay minerals in animal nutrition. *Appl. Clay Sci.* **2011**, *51*, 395–398. [CrossRef]

30.   Stanely, V.G.; Gray, C.; Daley, M.; Krueger, W.F.; Sefton, A.E. An Alternative to Antibiotic-Based Drugs in Feed for Enhancing Performance of Broilers Grown on *Eimeria* spp.-Infected Litter. *Poult. Sci.* **2001**, *83*, 39–44. [CrossRef]

31.   Casewell, M.; Friis, C.; Marco, E.; McMullin, P.; Phillips, I. The European ban on growth-promoting antibiotics and emerging consequences for human and animal health. *J. Antimicrob. Chemother.* **2003**, *52*, 159–161. [CrossRef]

32.   Rambeck, W.A.; He, M.L.; Wehr, U. Influence of the alternative growth promoter "rare earth elements" on meat quality in pigs. In Proceedings of the British Society of Animal Science Pig and Poultry Meat Quality: Genetic and Non-Genetic Factors, Krakow, Poland, 14–15 October 2004.

33.   He, M.L.; Wehr, U.; Rambeck, W.A. Effect of low doses of dietary rare earth elements on growth performance of broilers. *J. Anim. Physiol. An. N.* **2009**, *94*, 86–92. [CrossRef]

34.   Prause, B.; Gebert, S.; Wenk, C.; Rambeck, W.A.; Wanner, M. Impact of rare earth elements on metabolism and energy-balance of piglets. In Proceedings of the 9th Congress of the European Society of Veterinary and Comparative Nutrition, Grugliasco, Italy, 22–24 September 2005.

35.   He, M.L.; Ranz, D.; Rambeck, W.A. Study on performance enhancing effect of rare earth elements in growing and fattening pigs. *J. Anim. Physiol. Anim. Nutr.* **2001**, *85*, 263–270. [CrossRef]

36. Fan, C.; Zhang, F.; Wu, Y.; Zhang, G.; Ren, J. A study of feeding organic REE additives on growing pork pigs. *QingHai J. Husb. Vet.* **1997**, *27*, 23–24.
37. Shaw, H.; Jiang, Z.; Zong, Z.; Liu, H.; Liu, W. Effect of dietary Rare Earth Elements on some enzyme activities in liver, pancreas and blood of carp. *Chin. Feed.* **1999**, *1*, 21–22.
38. Ayinla, O.A. *Analysis of Feeds and Fertilizers for Sustainable Development in Nigeria*; Technical Paper No. 83; Nigeria Institute of Oceanography and Marine Research Lagos Nigeria: Lagos, Nigeria, 2007; 13p.
39. Tiamiyu, L.O.; Solomon, G.S.; Oketa, E.J. Effects of different boiling periods of soybean (*Glycine max* (L) Merill) on growth, performance of tilapia (*Oreochromis niloticus*) fingerlings. *J. Aquat. Sci.* **2006**, *21*, 15–18.
40. Adeyemo, B.T.; Tiamiyu, L.O.; Ayuba, V.O.; Musa, S.; Odo, J. Effects of dietary mixed aflatoxin B1 and fumonisin B1 on growth performance and haematology of juvenile *Clarias gariepinus* catfish. *Aquaculture* **2018**, *491*, 190–196. [CrossRef]
41. Ayyat, M.S.; Ayyat, A.M.N.; Al-Sagheer, A.A.; El-Hais, A.E.M. Effect of some safe feed additives on growth performance, blood biochemistry, and bioaccumulation of aflatoxin residues of Nile tilapia fed aflatoxin-B1 contaminated diet. *Aquaculture* **2018**, *495*, 27–34. [CrossRef]
42. Hussain, D.; Mateen, A.; Gatlin, D.M. Alleviation of aflatoxin B1 (AFB$_1$) toxicity by calcium bentonite clay: Effects on growth performance, condition indices and bioaccumulation of AFB1 residues in Nile tilapia (*Oreochromis niloticus*). *Aquaculture* **2017**, *475*, 8–15. [CrossRef]
43. Tang, Y.; Zhang, B.; Wang, L.; Li, J. Effect of dietary rare earth elements-amino acid compounds on growth performance of carp and rainbow trout. *Chin. J. Fish.* **1997**, *10*, 88–90.
44. Rofiat, A.S.; Fanelli, F.; Atanda, O.; Sulyok, M.; Cozzi, G.; Bavaro, S.; Krska, R.; Logrieco, A.F.; Ezekiel, C.N. Fungal and bacterial metabolites associated with natural contamination of locally processed rice (*Oryza sativa* L.) in Nigeria. *Food Addit. Contam. Part A* **2015**, *32*, 950–959. [CrossRef]
45. Rushing, B.R.; Selim, M.I. Aflatoxin B1: A review on metabolism, toxicity, occurrence in food, occupational exposure, and detoxification methods. *Food Chem. Toxicol.* **2019**, *124*, 81–100. [CrossRef]
46. Yildirim, M.; Manning, B.B.; Lovell, R.T.; Grizzler, J.M. Toxicity of Moniliformin and Fumonisin B1 Fed Singly and in Combination in Diets for Young Catfish *Ictalurus punctatus*. *J. World Aquac. Soc.* **2000**, *31*, 599–608. [CrossRef]
47. Xie, J.; Wang, Z. The effect of organic rare-earth compounds on production performance of chicken. In Proceedings of the 2nd International Symposium on Trace Elements and Food Chain, Wuhan, China, 12–15 November 1998; p. 74.
48. Xu, X.; Xia, H.; Rui, G.; Hu, C.; Yuan, F. Effect of lanthanum on secretion of gastric acid in stomach of isolated mice. *J. Rare Earths* **2004**, *22*, 427–464.
49. Gonçalves, R.A.; Cam, T.D.; Tri, N.N.; Santos, G.A.; Encarnação, P.; Hung, L.T. Aflatoxin B1 (AFB1) reduces growth performance, physiological response, and disease resistance in Tra catfish (*Pangasius hypophthalmus*). *Aquac. Int.* **2018**, *26*, 921–936. [CrossRef]
50. Anater, A.; Mayes, L.; Meca, G.; Ferrer, E.; Luciano, F.B.; Pimpão, C.T.; Font, G. Mycotoxins and their consequences in aquaculture: A review. *Aquaculture* **2016**, *451*, 1–10. [CrossRef]
51. Campbell, T.W. Haematology of fish. In *Veterinary Haematology and Clinical Chemistry*; Thrall, M.A., Weiser, G., Allison, R., Eds.; John Wiley and Sons: Ames, IA, USA, 2012; pp. 293–312.
52. Fazio, F. Fish haematology analysis as an important tool of aquaculture: A review. *Aquaculture* **2019**, *500*, 237–242. [CrossRef]
53. Witeska, M. Erythrocytes in teleost fishes: A review. *Zool. Ecol.* **2013**, *23*, 275–281. [CrossRef]
54. Zeng, Z.; Jiang, W.; Wu, P.; Liu, Y.; Zeng, Y.; Jiang, J.; Kuang, S.; Tang, L.; Zhou, X.; Feng, L. Dietary aflatoxin B1 decreases growth performance and damages the structural integrity of immune organs in juvenile grass carp (*Ctenopharyngodon idella*). *Aquaculture* **2019**, *500*, 1–17. [CrossRef]
55. Ou, X.; Guo, Z.; Wang, J. The effects of rare earth element additive in feed on piglets. *Livest. Poult. Ind.* **2000**, *4*, 21–22.
56. Harper, A.F.; Estienne, M.J.; Meldrum, J.B.; Harrell, R.J.; Diaz, D.E. Assessment of a hydrated sodium calcium aluminosilicate agent and antioxidant blend for mitigation of aflatoxin- induced physiological alterations in pigs. *J. Swine Health Prod.* **2010**, *18*, 282–289.
57. Wang, Y.; Guo, F.; Yuan, K.; Hu, Y. Effect of lanthanum chloride on apoptosis of thymocytes in endotoxemia mice. *Acta Acad. Med. Jiangxi* **2003**, *43*, 31.
58. Wang, H.; Sun, H.; Chen, Y.; Wang, X. The bioaccumulation of rare earth elements in the internal organs of fish and their effect on the activities of enzymes in liver. *China Environ. Sci.* **1991**, *19*, 141–144.
59. Hrubec, T.C.; Cardinale, J.L.; Smith, S.A. Haematology and Plasma Chemistry Reference Intervals for Cultured Tilapia (Oreochromis Hybrid). *Vet. Clin. Pathol.* **2000**, *29*, 7–12. [CrossRef]
60. Lusková, V. Annual cycles and normal values of hematological parameters in fishes. *Acta Sci. Nat. Brno* **1997**, *31*, 1–70.
61. Mantantzis, K.; Schlaghecken, F.; Sünram-Lea, S.I.; Maylor, E.A. Sugar rush or sugar crash? A meta-analysis of carbohydrate effects on mood. *Neurosci. Biobehav. Rev.* **2019**, *101*, 45–67. [CrossRef] [PubMed]
62. Hemre, G.I.; Mommsen, T.P.; Krogdahl, A. Carbohydrates in fish nutrition: Effects on growth, glucose metabolism and hepatic enzymes. *Aquac. Nutr.* **2002**, *8*, 175–194. [CrossRef]
63. Metcalfe, N.B.; Monaghan, P. Compensation for a bad start: Grow now, pay later? *Trends Ecol. Evol.* **2001**, *16*, 254–260. [CrossRef]
64. Santacroce, M.P.; Conversano, M.; Casalino, E.; Lai, O.; Zizzadoro, C.; Centoducati, G.; Crescenzo, G. Aflatoxins in aquatic species: Metabolism, toxicity and perspectives. *Rev. Fish Biol. Fish.* **2008**, *18*, 99–130. [CrossRef]

65. Carraro, A.; De Giacomo, A.; Giannossi, M.L.; Medici, L.; Muscarella, M.; Palazzo, L.; Quaranta, V.; Summa, V. Clay minerals as adsorbents of aflatoxin M1 from contaminated milk and effects on milk quality. *Appl. Clay Sci.* **2014**, *88*, 92–99. [CrossRef]
66. McCue, M.D. Starvation physiology: Reviewing the different strategies animals use to survive a common challenge. *Comp. Biochem. Physiol.* **2010**, *156*, 1–18. [CrossRef]
67. Wagner, T.; Congleton, J.L. Blood chemistry correlates of nutritional condition, tissue damage, and stress in migrating juvenile chinook salmon (*Oncorhynchus tshawytscha*). *Can. J. Fish. Aquat. Sci.* **2004**, *61*, 1066–1074. [CrossRef]
68. Lemarie, P.; Drai, P.; Mathieu, A.; Lemaire, S.; Carriere, S.; Giudicelli, S.; Lafaurie, M. Changes with different diets in plasma enzymes (GOT, GPT, LDH, ALP) and plasma lipids (cholesterol, triglycerides) of seabass (*Dicentrarchus labrax*). *Aquaculture* **1991**, *93*, 63–75. [CrossRef]
69. Domijan, A.M.; Abramov, A.Y. Fumonisin B1 Inhibits Mitochondrial Respiration and Deregulates Calcium Homeostasis—Implication to Mechanism of Cell Toxicity. *Int. J. Biochem. Cell Biol.* **2011**, *43*, 897–904. [CrossRef]
70. Indresh, H.; Devegowda, G.; Ruban, S.; Shivakumar, M. Effects of high-grade bentonite on performance, organ weights and serum biochemistry during aflatoxicosis in broilers. *Vet. World* **2013**, *6*, 313. [CrossRef]
71. Bucci, T.Y.; Howard, P.C.; Tollerson, W.H.; LaBorde, J.B.; Hansen, D.K. Renal effects of fumonisin mycotoxins in animals. *Toxicol. Pathol.* **1998**, *26*, 160 161. [CrossRef]
72. Waisbren, S.E.; Gropman, A.L. Improving long-term outcomes in urea cycle disorders. *J. Inherit. Metab. Dis.* **2016**, *39*, 573. [CrossRef]
73. Bezerra, R.F.; Soares, M.C.F.; Santos, A.J.G.; Maciel-Carvalho, E.V.; Coelho, L.C.B.B. Secondary indicators of seasonal stress in the Amazonian pirarucu fish (*Arapaima gigas*). In *Advances in Environmental Research*; Daniels, J.A., Ed.; Nova Science Publishers: New York, NY, USA, 2013; pp. 233–244.
74. Naiel, M.A.E.; Ismael, N.E.M.; Shehata, S.A. Ameliorative effect of diets supplemented with rosemary (*Rosmarinus officinalis*) on aflatoxin B1 toxicity in terms of the performance, liver histopathology, immunity and antioxidant activity of Nile Tilapia (*Oreochromis niloticus*). *Aquaculture* **2019**, *511*, 734–764. [CrossRef]
75. Knowles, S.; Hrubec, T.C.; Smith, S.A.; Bakal, R.S. Haematology and plasma chemistry reference intervals for cultured short Nose Sturgeon (Acipenser brevirostrum). *Vet. Clin. Pathol.* **2006**, *35*, 434–440. [CrossRef]
76. Adeyemo, B.T.; Oloyede, L.T.; Ogeh, A.V.; Orkuma, C.J. Growth performance and serum lipids profile of *Clarias gariepinus* catfish following experimental dietary exposure to fumonisin B1. *Open J. Vet. Med.* **2016**, *6*, 127–138. [CrossRef]
77. Monbaliu, S.; Van Poucke, C.; Detaverneir, C.; Dumoulin, F.; Van de Velde, M.; Schoeters, E.; Van Dyek, S.; Averkieva, O.; Van Peteghem, C.; De Saeger, S. Occurrence of Mycotoxins in Feed as Analyzed by a Multi-Mycotoxin LC-MS/MS Method. *J. Agric. Food Chem.* **2010**, *58*, 66–71. [CrossRef]
78. Campbell, T.W. Haematology of fish. In *Veterinary Haematology and Clinical Chemistry*; Troy, D.B., Ed.; Lippincott. Williams and Wilkins: Baltimore, MD, USA, 2004; pp. 277–289.
79. Myburgh, J.G.; Botha, C.J.; Booyse, D.G.; Reyers, F. Provisional clinical chemistry parameters in the African Sharptooth catfish (*Clarias gariepinus*). *J. S. Afr. Vet. Assoc.* **2008**, *79*, 156–160. [CrossRef]
80. Arnason, T.; Björnsson, B.; Steinarsson, A.; Oddgeirsson, M. Effects of Temperature and Body Weight on Growth Rate and Feed Conversion Ratio in Turbot (*Scohthalmus maximus*). *Aquaculture* **2009**, *295*, 218–225. [CrossRef]

Article

Spaceflight Changes the Production and Bioactivity of Secondary Metabolites in *Beauveria bassiana*

Youdan Zhang [1,2], Xiaochen Zhang [1,2], Jieming Zhang [1,2], Shaukat Ali [1,2,*] and Jianhui Wu [1,2,*]

1   Key Laboratory of Bio-Pesticide Innovation and Application, Guangzhou 510642, China
2   Engineering Research Center of Biological Control, Ministry of Education and Guangdong Province, South China Agricultural University, Guangzhou 510642, China
*   Correspondence: aliscau@scau.edu.cn (S.A.); jhw@scau.edu.cn (J.W.)

Abstract: Studies on microorganism response spaceflight date back to 1960. However, nothing conclusive is known concerning the effects of spaceflight on virulence and environmental tolerance of entomopathogenic fungi; thus, this area of research remains open to further exploration. In this study, the entomopathogenic fungus *Beauveria bassiana* (strain SB010) was exposed to spaceflight (ChangZheng 5 space shuttle during 5 May 2020 to 8 May 2020) as a part of the Key Research and Development Program of Guangdong Province, China, in collaboration with the China Space Program. The study revealed significant differences between the secondary metabolite profiles of the wild isolate (SB010) and the spaceflight-exposed isolate (BHT021, BH030, BHT098) of *B. bassiana*. Some of the secondary metabolites/toxins, including enniatin A2, brevianamide F, macrosporin, aphidicolin, and diacetoxyscirpenol, were only produced by the spaceflight-exposed isolate (BHT021, BHT030). The study revealed increased insecticidal activities for of crude protein extracts of *B. bassiana* spaceflight mutants (BHT021 and BH030, respectively) against *Megalurothrips usitatus* 5 days post application when compared crude protein extracts of the wild isolate (SB010). The data obtained support the idea of using space mutation as a tool for development/screening of fungal strains producing higher quantities of secondary metabolites, ultimately leading to increased toxicity/virulence against the target insect host.

Keywords: space exposure; entomopathogenic fungi; pathogenicity; secondary metabolites

Key Contribution: The study revealed significant differences between the secondary metabolite profiles of the wild isolate (SB010), and the spaceflight-exposed isolate (BHT021, BH030, BHT098) of *B. bassiana*. The study revealed lower median lethal concentrations (LC_{50}) of crude protein extracts of *B. bassiana* spaceflight mutants against *M. usitatus* 5 days post application when compared LC_{50} of crude protein extracts of the wild isolate (SB010).

Citation: Zhang, Y.; Zhang, X.; Zhang, J.; Ali, S.; Wu, J. Spaceflight Changes the Production and Bioactivity of Secondary Metabolites in *Beauveria bassiana*. *Toxins* **2022**, *14*, 555. https://doi.org/10.3390/toxins14080555

Received: 27 July 2022
Accepted: 13 August 2022
Published: 15 August 2022

Publisher's Note: MDPI stays neutral with regard to jurisdictional claims in published maps and institutional affiliations.

1. Introduction

An increase in space exploration has resulted in an increase in studies on understanding the changes in physiology of living organisms under spaceflight conditions [1]. Spaceflight conditions include long-term exposure to microgravity, radiation, and isolation [2]. The higher costs, limited number of launches, and complexity of experimental design under space habitats are the main difficulties in performing studies under true spaceflight conditions, and consequently, many studies have also been performed under ground-based simulated microgravity [3]. Different investigations have explored the effects of space conditions on plants, animals, as well as microorganisms. However, microorganisms can be considered strong candidates to study the responses to variations in environmental conditions because of their rapid life cycle, easy handling, and stability [4,5]. The explorations regarding influences of spaceflight on growth and metabolism of microorganisms have few significant implications. Firstly, microorganisms can impact (positively or negatively) human, animal and plant life [6]. Secondly, they are known for producing secondary metabolites, which are used as medicine, biopesticides, plant growth-promoting agents, etc. [7,8]. Several studies have explained the relationship between changed environmental conditions during spaceflight and variations in morphological as well as physiological characteristics of microorganisms, including germination, virulence, host pathogen relationship, secondary metabolite production, and gene expression [2,5,9].

Metabolism is the sum of all reactions in a living cell required for maintenance, development, and division. Microbial metabolism is comprised of primary metabolites (the intracellular molecules that enable growth and proliferation) and secondary metabolites (predominantly extracellular molecules that facilitate a microbe's interaction and adaptation with its environment) [10,11]. Secondary metabolites produced by microorganisms are predominantly low-molecular-weight extracellular compounds [12]. They are usually produced in late-exponential and stationary phases and are not directly associated with growth, development, or division of microorganisms [10]. These specialized products are most notable for their use in healthcare settings as antimicrobial, antiparasitic, and pest-control agents [13–17]. Many of the intermediates in primary metabolism are precursors of secondary metabolites, and cells have evolved complex molecular switches linking primary and secondary metabolic pathways. These include high expression of the secondary metabolism genes at specific times in the cell cycle and controlling the flow of primary metabolites (carbon and nitrogen) through different pathways by feedback regulation [12,13,18]. Thus far, studies on secondary metabolism have focused on only a few microorganisms (mainly *Streptomycetes*, *Escherichia coli*, and *Bacillus*) and are mostly limited to one or a few metabolites per study. These studies have suggested altered secondary metabolite production levels, but the specific responses have been unique to each species. Furthermore, previous studies have been either limited to already-known metabolites or have focused on microorganisms that are already well-known metabolite producers. However, space vehicles hold diverse species whose behaviors are unstudied and could have responses under microgravity beyond prediction. Additionally, understanding microbes at a global metabolomics level could provide more comprehensive knowledge about the overall responses exhibited under microgravity.

Filamentous fungi are a major group of microorganisms critical to the production of different commercial enzymes, biopesticides, and organic compounds [19,20]. Several species of filamentous fungi belonging to the phylum Ascomycota (Subkingdom Dikarya) are known for their pathogenicity against insects [21]. Fungi belonging to genus *Beauveria* (Hypocreales: Cordycipitaceae) are one of the most common insect pathogenic fungal species. *Beauveria* species are known to cause widespread epizootics of insect populations because of their saprophytic behavior [22]. Studies regarding responses of microorganisms to spaceflight date back to 1960, but the responses to microgravity and its analogs have been investigated in a few microorganisms (bacteria as well as fungi), showing plausible but conflicting results for cellular growth rates and secondary metabolism under spaceflight and simulated microgravity experiments [23–26]. To date, limited concrete studies on the influences/utilization of spaceflight on production of secondary metabolites by entomopathogenic fungi make this area of research open for detailed investigations.

In this study, the entomopathogenic fungus *Beauveria bassiana* (SB010) was sent to the Taingong space station for exposure to spaceflight conditions on the ChangZheng 5 space shuttle during 5 May 2020 to 8 May 2020 as a part of the Key Research and Development Program of Guangdong Province, China, in collaboration with the China Space Program. The aims of this study were to (i) extract and characterize the changes in mycelial protein extract/secondary metabolites profiles of different *B. bassiana* spaceflight mutants and (ii) undertake toxicity assays of mycelial extract/secondary metabolites *B. bassiana* spaceflight mutants against *Megalurothrips usitatus* (Thysanoptera: Thripidae).

2. Results

Secondary metabolite production was quantified from the ethyl acetate extracts of wild isolate (SB010) and spaceflight mutants (BHT021, BHT030, BHT098) of *B. bassiana*. The total protein concentration of ethyl acetate extract differed significantly among all four isolates ($F_{3,8}$ = 41.32, $p < 0.001$). The highest protein contents were obtained from the spaceflight mutant BHT021, with a mean value of 1.87 ± 0.035 mg/mL. For the wild isolate (SB010), the concentration of protein observed was 1.26 ± 0.03 mg/mL. The lowest protein content was produced by the space mutant BHT098, with a mean value of 1.05 ± 0.05 mg/mL (Figure 1).

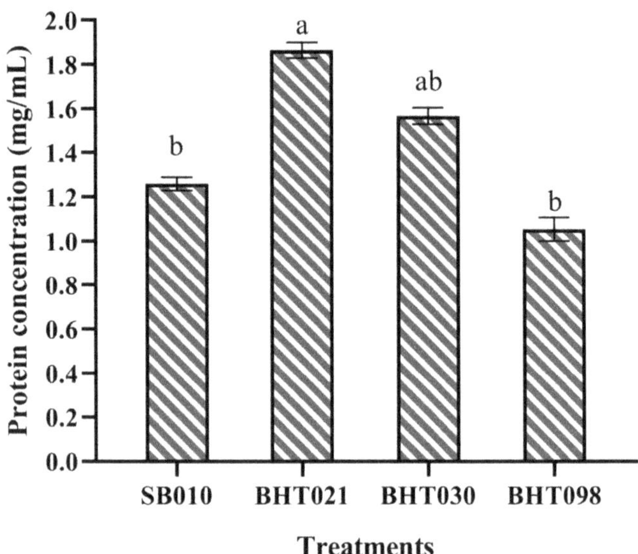

Figure 1. Total protein content of the crude protein extracts of the wild isolate (SB010) and spaceflight mutants (SCPH0535) of *Beauveria bassiana*. Error bars indicate standard error of means. Bars having different letters are significantly different from each other (Tukey's test at 5% level of significance).

As determined by LC-MS analysis, several differences were observed between the metabolite profiles of the four isolates (Figure 2; Supplementary file S1). A section of peaks between retention time of 8–9 min and 14.5–17 min was evident in the LC-MS profile of the spaceflight-exposed mutants (BHT021, BHT030, and BHT098) when compared with the wild isolate (SB010). In addition, a broader pattern of peaks between retention time of 10–11 min was observed for the spaceflight mutant BHT021 when compared with the wild isolate (SB010). Analysis of fragmentation pattern of peaks from the LC-MS profiles of the four isolates revealed the production of 43, 79, 44, and 47 secondary metabolites known for insecticidal activity by the SB010, BHT021, BHT030, and BHT098, respectively (Supplementary file S1). Some of the secondary metabolites/toxins, including enniatin A2, brevianamide F, macrosporin, aphidicolin, and diacetoxyscirpenol, were only produced by the spaceflight mutants BHT021 and HT030 (Supplementary file S1).

The FTIR analysis of ethyl acetate extracts of *B. bassiana* wild isolate (SB010) and spaceflight mutants (BHT021, BHT030, and BHT098) showed marked variations for functional groups of secondary metabolites (Figure 3). The FTIR profiles of the wild isolate (SB010) showed a broad peak at 3386.79 cm^{-1} (due to overlap of O-H or N-H stretching) along with sharp peaks at 2924.24 cm^{-1} (aliphatics (C-H), methylene functional group), 1647.51 cm^{-1} (carbonyl stretch amide linkage), and 641.42 cm^{-1} (aromatics).

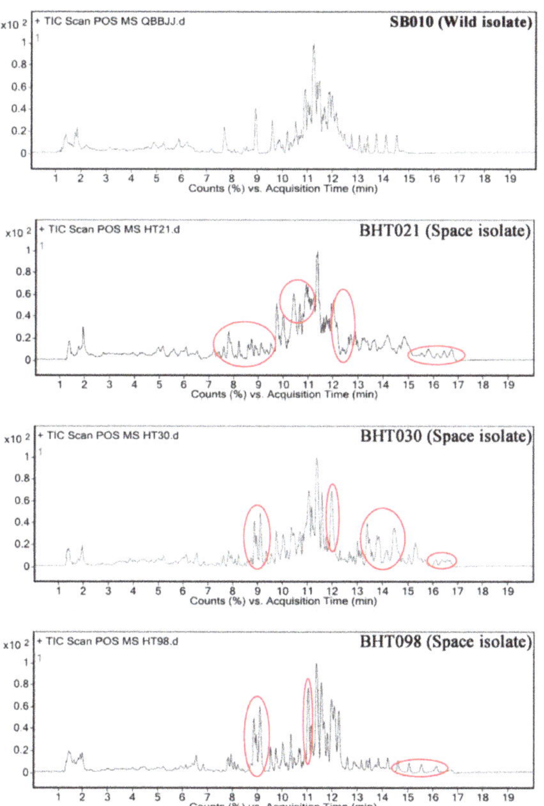

Figure 2. LC-MS analysis of the crude protein extracts from the wild isolate and spaceflight mutants of *Beauveria bassiana*. The indicated peaks (circle) show the variation in secondary metabolite profile of spaceflight mutants from the wild isolates.

In the case of *B. bassiana* spaceflight mutant (BHT021), the O-H or N-H stretching was observed at 3308.91 cm^{-1}. The methylene (C-H) peak was observed at 2955.88 cm^{-1}, whereas amide linkage peaks were observed at 1649.24 and 1623.96 cm^{-1}, respectively. Furthermore, four sharp peaks of ester (C-O) functional group were observed at 1385.16, 1329.16, 1298.90, and 1241.14 cm^{-1}, respectively. Five strong peaks representing the presence of aromatics were observed between 766.32–559.74 cm^{-1}.

The FTIR profiles of *B. bassiana* spaceflight mutant (SB030) showed a broad peak at 3311.40 cm^{-1} (due to overlap of O-H or N-H stretching) along with peaks at 2924.24 cm^{-1} (aliphatics (C-H), methylene functional group), 1648.98 cm^{-1} (carbonyl stretch amide linkage). Furthermore, four sharp peaks of ester (C-O) functional group were observed at 1384.34, 1330.16, 1306.46 and 1241.58 cm^{-1}, respectively. Five strong peaks representing the presence of aromatics were observed between 766.66–559.74 cm^{-1} (Figure 3).

In the case of *B. bassiana* spaceflight mutant (BHT098), the O-H or N-H stretching was observed at 34.44.52 cm^{-1}. The methylene (C-H) peak was observed at 2924.49 cm^{-1}, whereas amide linkage peaks were observed at 1647.75 cm^{-1}, respectively. Furthermore, a single sharp peak of ester (C-O) functional group was observed at 138,400 cm^{-1}. Two strong peaks representing the presence of aromatics were observed at 637.34 and 562.01 cm^{-1} (Figure 3).

The results showed the presence of additional ester (C-O) and aromatic functional groups in ethyl acetate extracts of the *B. bassiana* spaceflight mutants when compared to the FTIR profile of *B. bassiana* wild isolate (SB010).

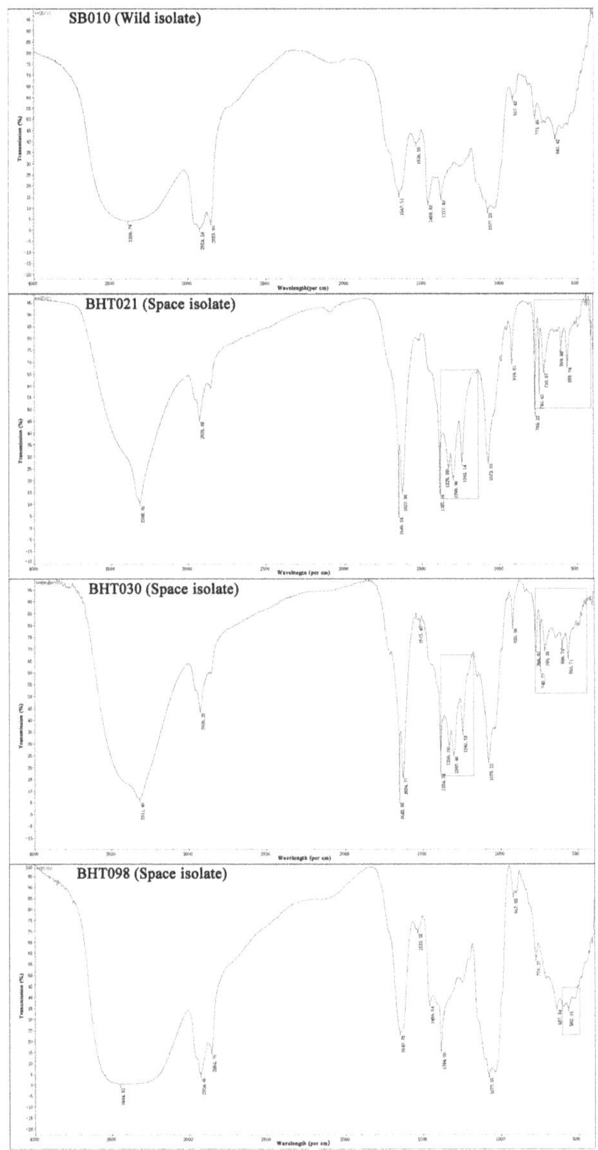

Figure 3. Fourier transformation infrared (FTIR) spectroscopy of the crude protein extracts of wild isolate (SB010) and spaceflight mutants (BHT021, BHT030, and BHT098) of *Beauveria bassiana*. The indicated peaks (circle) show the variation in secondary metabolite profile of spaceflight mutants from the wild isolates.

2.1. Nuclear Magnetic Resonance (NMR)

The characterization of protein extracts of *B. bassiana* wild isolate SB010 through NMR analysis revealed bands of high resonance in allylic, pyrrolidine ring and NeCH (2.0–2.6 ppm and 2.7–3.5 ppm), and ether linkage (3.5–4.3 ppm) regions. The NMR spectrograms of metabolic compounds produced by *B. bassiana* space mutants (BHT021, BHT030, and BHT098) showed high-resonance bands in aliphatic (0.9–1.4 ppm), allylic, pyrrolidine ring (2.0–2.6 ppm), NeCH (2.1–3.5 ppm), and ether linkage (3.5–4.3 ppm) regions (Figure 4).

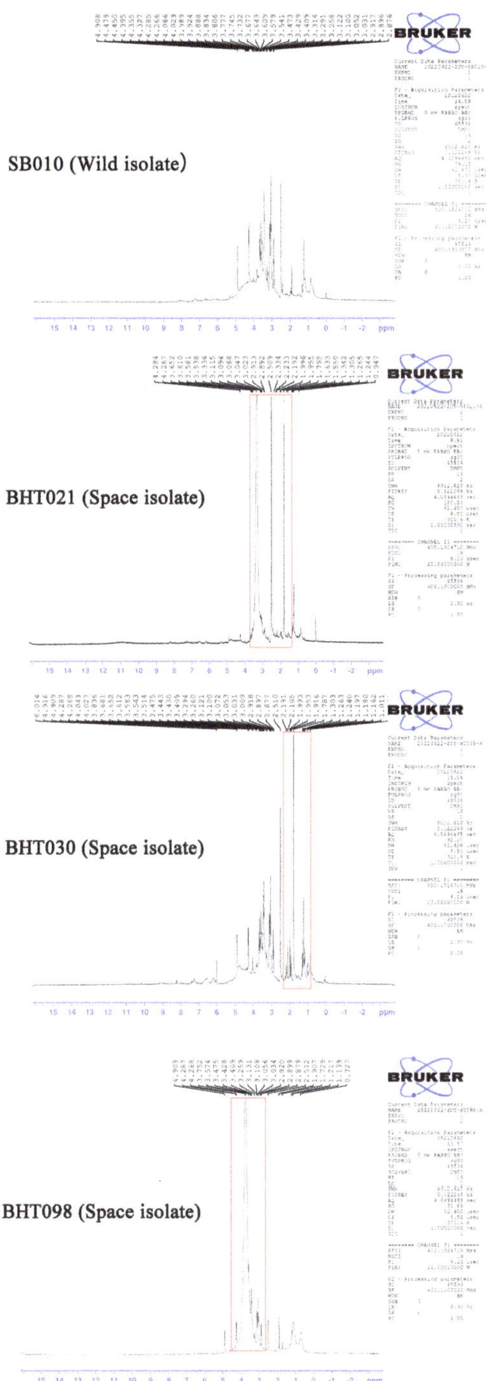

Figure 4. Nuclear magnetic resonance (NMR) resonance analysis of the crude protein extracts of wild isolate (SB010) and spaceflight mutants (BHT021, BHT030, and BHT098) of *Beauveria bassiana*. The indicated peaks (circle) show the variation in secondary metabolite profile of spaceflight mutants from the wild isolates.

2.2. Toxicity of Metabolites against M. usitatus

The insecticidal activities of crude protein extracts from *B. bassiana* wild isolate (SB010) and space mutants (BHT021, BHT030, and BHT098) applied at different concentrations against *M. usitatus* adults differed significantly among different isolates and their concentrations ($F_{15,48}$ = 23.81; p = 0.0034). The different concentrations of the crude protein extracts of *B. bassiana* spaceflight mutants (BHT021 and BHT030) had higher efficacy against *M. usitatus* adults 5 days post application compared to the wild isolate (SB010) (Table 1). The observed mortality (%) of *M. usitatus* adults following treatment with different concentration of the crude protein extracts of spaceflight mutants (BHT098) were lower than the mortality values observed for the crude protein extracts of wild isolate (SB010).

Table 1. Dose–mortality responses of *M. usitatus* adults to crude protein extracts of wild isolate (SB010) and spaceflight mutants (BHT021, BHT030, and BHT098) of *Beauveria bassiana*.

Treatments	Concentration (μg/mL)	Mortality (%)
	ddH$_2$O	5.00 j
	100	88.33 ab
SB010	75	80.00 c
	50	73.33 d
	25	65.00 ef
	12.5	50.00 h
	ddH$_2$O	5.00 j
	100	100.00 a
BHT021	75	91.67 ab
	50	83.33 bc
	25	70.00 e
	12.5	63.33 f
	ddH$_2$O	5.00
	100	95.00 a
BHT030	75	83.33
	50	75.00 d
	25	63.33 f
	12.5	56.67 g
	ddH$_2$O	5.00 j
	100	81.67 c
BHT098	75	73.33 d
	50	61.67 f
	25	55.00 g
	12.5	46.67 i

The difference between the means (±SE) followed by various letters is significant (Tukey's p < 0.05).

2.3. Transmission Electron Microscopic Examination of M. usitatus Midgut following Treatment with Mycelial Extracts from B. bassina Wild Isolate and Space Mutants

The ultrastructural changes of the fat body and somatic cells of the treated group were observed by transmission electron microscopy after feeding the toxin for 72 h compared with the control. The microvilli of the midgut cells in the control were abundant and neatly arranged with a fenestrated morphology, and after 72 h of toxin feeding, the microvilli of the midgut of adult thrips were swollen, shortened and thinned, vacuolated, and partially dislodged (Figure 5). Compared with the control, the microvilli in the lumen of the midgut cells of *M. usitatus* treated with mycelial extracts of space mutants (BHT012 and BHT030) were sparse, disorganized, and vacuolated, while some of them were even completely lost. The nuclei of the blank group were flat, but after treatment with mycelial extracts of space mutants (BHT021 and BHT030), the fat body cells of *M. usitatus* adults were severely damaged, and their nuclei appeared expanded, folded, and detached, and the lipid droplet membrane structure became transparent and vacuolated (Figure 5).

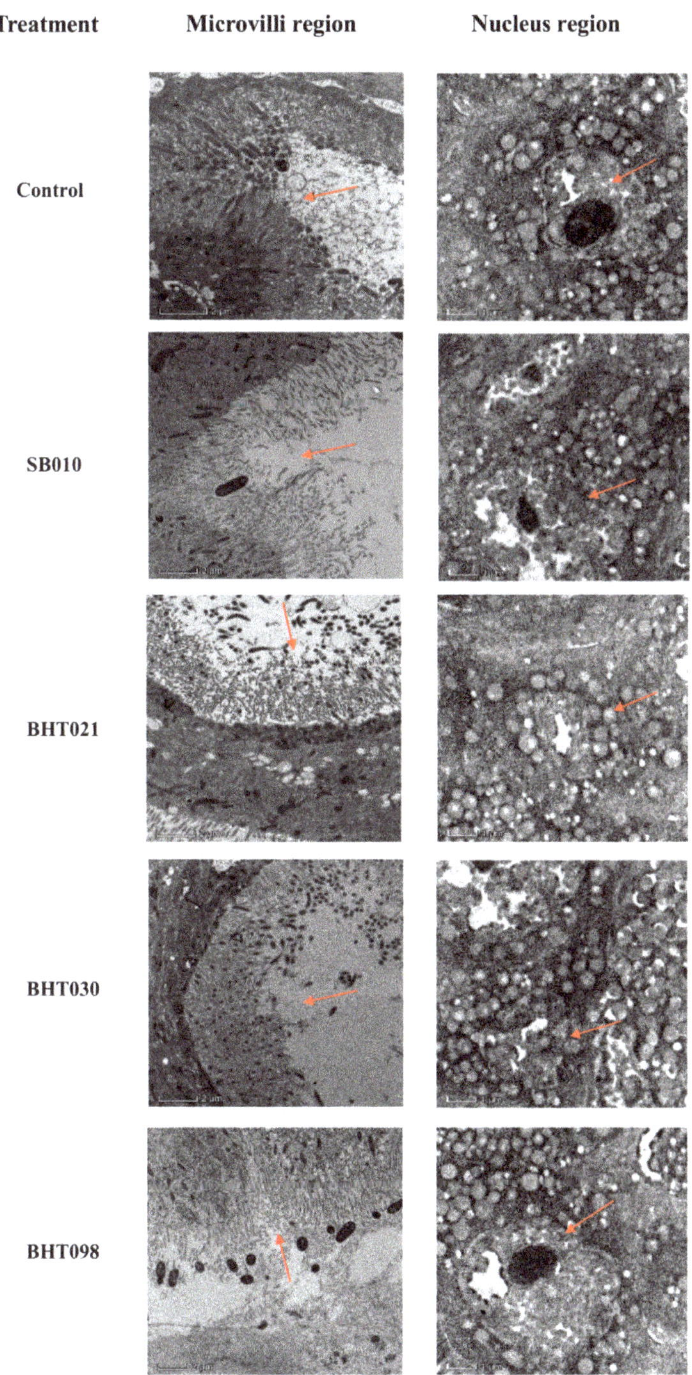

Figure 5. Transmission electron microscopic examination of *M. usitatus* following treatment with mycelial extracts from *Beauveria bassiana* wild isolate and space mutants. Arrow marks points with variations in *M. usitatus* midgut anatomy following different treatment.

3. Discussion

Studies investigating the influences of spaceflight on secondary metabolite production by microorganisms would be interesting, as the biosynthesis and concentration of microbial secondary metabolites are sensitive to extracellular environmental cues (nutrients availability, temperature, and osmotic stress) [27,28]. The results showed significant changes in total yield and secondary metabolite profiles of *B. bassiana* spaceflight mutants and the wild isolate. These findings are consistent with Lam et al. [29], who observed increased production of actinomycin D by *Streptomyces plicatus* WC56452 after spaceflight on the U.S. Space Shuttle mission STS-80. Similarly, Luo et al. [30] observed 13–18% higher production of nikkomycins by *Streptomyces ansochromogenus* during 15 days of spaceflight. In a different study, Nikkomycins-producing strains of *Streptomyces ansochromogenus* were investigated to understand the biological response to production onboard a satellite for 15 days. The production of Nikkomycins in nearly all strains was reported to be increased by 13–18%, with increases specifically in Nikkomycin X and Z [2,30]. Similarly, a study on *Cupriavidus metallidurans* under simulated microgravity showed increased production of the polyester polymer poly-b-hydroxybutyrate after 24 h but not after 48 h compared with ground controls [7]. All these results suggest that effects of microgravity on secondary metabolism may be specific depending on the strain, growth condition, pathway utilized, or time course analyzed.

Based on already-reported data and the results of this study as well as complexity of regulatory mechanisms of secondary metabolites production, the observed changes in secondary metabolite profiles of *B. bassiana* in response to spaceflight exposure can be strain, growth media, and secondary metabolism production pathway, depicting an inconsistent trend. These changes are induced by the indirect physical influences of spaceflight (such as variations in fluid dynamics or the extracellular transport of metabolites) [2]. Furthermore, different extracellular environmental cues can induce or enhance secondary metabolisms in different microorganisms, followed by the transfer of extracellular stresses to the downstream responsive genes by a cascade of complex signal transduction steps [29]. Therefore, although all of the *B. bassiana* space flight mutants went through same sort of extracellular stress, the amount and number of secondary metabolites produced by each mutant were different. In short, the changes in secondary metabolites production in *B. bassiana* following spaceflight exposure were induced by the variations in microenvironments around the fungal cells [2]. However, space vehicles hold diverse species whose behaviors are unstudied and could have responses under microgravity beyond prediction. Additionally, understanding microbes at a global metabolomics level could provide more comprehensive knowledge about the overall responses exhibited under microgravity since entomopathogenic fungi offer a wealth of potential for the discovery of new and important microbial products for pest control. Broadening the horizon of fungal species and understanding the altered levels under microgravity could offer unique advantages for the biopesticides industry. There remains much to discover about the nature of diverse secondary metabolisms in such stressful environments of spaceflight. The changes of environmental factors, such as temperature, oxygen availability, and diffusion limitations, under microgravity can provide a condition that can be harnessed in the best way possible to be used for engineered microorganisms to generate useful metabolites. Therefore, understanding the specific cause-and-effect mechanisms of fungal responses to microgravity at the molecular level could provide ground-breaking discoveries for space applications and biopesticides development. However, in order to highlight the microbial responses to microgravity for long periods of time, different technologies (such as low-shear-modeled microgravity (LSMMG)) have been developed to mimic real microgravity [31], and these technologies can be used more often (to mimic microgravity and ground level) to induce higher production of secondary metabolites by insect pathogenic fungi.

The examination of the biological activities of fungal crude protein extracts from *B. bassiana* wild isolate SB010 and space mutants (BHT021, BHT030, and BHT098), applied at different concentrations against *M. usitatus* adults, showed significantly different rates of insect mortality. The crude protein extracts of *B. bassiana* space mutants BHT021 and BHT030 (used at 100 µg/mL) induced higher mortality of *M. usitatus* adults when compared with the mortality caused by *B. bassiana* wild isolate SB010. On the other hand, *M. usitatus* mortality observed in response to treatment with crude protein extracts of *B. bassiana* space mutant BHT098 was lower than mortality caused by *B. bassiana* wild isolate SB010. Our results are similar to the already-reported previous studies on changes to microbial pathogens (*Bacillus subtilis; Escherichia coli; Pseudomonas aeruginosa; Staphylococcus aureus*) induced by spaceflight [29]. The transmission electron microscopy of the midgut of fourth instar *M. usitatus* adults showed nuclei enlargement, folding of nuclear membrane, and detachment of microvilli from midgut cells. The secretion of secondary metabolites is important for pathogenic processes, and variations in secondary metabolite profiles of spaceflight mutants compared to the wild isolate may have led to variations in biological activities of crude protein extracts against *M. usitatus*. These data further explain that increased secondary metabolism can increase the virulence of microorganisms [2].

4. Conclusions

The results obtained from our studies provide substantial evidence that spaceflight exposure can alter the secondary metabolites production and biological activities of *B. bassiana*, which can serve as a baseline information for the studies on the effects of microgravity on insect pathogenic fungi. However, the specific reasons or mechanisms regulating the above-mentioned changes are unclear. Therefore, further studies on corresponding gene expression/regulation and characterization of micro-environments around the fungal cells should be emphasized in near future for identification as well as application of secondary metabolites produced by *B. bassiana*.

5. Materials and Methods

5.1. Insect Cultures

M. usitatus adults were reared on cowpea pods following the method of Espinosa et al. [32] and Du et al. [33] for multiple generations in an artificial climate chamber (Model PYX-400Q-A, Shaoguan City Keli Instrument Co., Ltd., Ningbo, China). Freshly emerged *M. usitatus* adult females were used in subsequent studies.

5.2. Fungal Preparations

Beauveria bassiana isolate SB010 (deposited at Guangdong Microbiological Research Centre repository under accession number GDMCC NO. 60359) was used for the study. The fugal isolate was cultured on potato dextrose agar (PDA) plates for 15 d followed by preparation of a basal conidial concentration (1×10^8 conidia/mL) using the method of Ali et al. [34].

5.3. Exposure of Beauveria bassiana to Spaceflight Conditions

One milliliter of *B. bassiana* conidial suspension (1×10^6 conidia/mL) grown on PDA broth was individually loaded into polypropylene (PE) centrifuge tubes, which were then sealed with parafilm M (Bemis, Neenah, WI, USA). Four PE centrifuge tubes were exposed to simulated microgravity in a 3D rotating experimental device (temperature: 20 °C; speed: 9 rpm/min) for 72 h at Shenzhou Space Biology Science and Technology Corporation, Ltd., Beijing, China.

Polypropylene centrifuge tubes (4 tubes) having *B. bassiana* conidial suspension were exposed to spaceflight conditions by the following method. The PE tube samples were pooled and placed into experimental boxes. The experimental boxes were placed in the ChangZheng 5 space shuttle. The samples were then flown to space within the shuttle during 5 May 2020 to 8 May 2020. The samples stayed in higher earth orbit (altitude 3000–8000 km) for 67 h and faced the Van Allen radiation belt (high-energy particle radiation belt) several times during the spaceflight. The sample box was retrieved from the returning space capsule and opened after 10 days, with the PE tubes being stored at $-20\,^{\circ}\text{C}$ until further use.

Aliquots (200 µL) of *B. bassiana* conidial suspension exposed to spaceflight conditions were cultured on PDA plates (15 plates each) following the method of Zhao et al. [35]. The fastest-growing colonies from spaceflight-exposed *B. bassiana* conidial suspensions were selected and named as BHT021, BHT030, and BHT098.The fugal isolate was cultured on PDA plates for 15 d followed by preparation of a basal conidial concentration $(1 \times 10^8 \text{ conidia/mL})$ using the method of Ali et al. [34].

5.4. Production and Characterization of Crude Protein Extracts of Wild Isolate (SB010) and Spaceflight Mutants (BHT021, BHT030, and BHT098) of Beauveria bassiana

Sterilized growth medium, 100 mL, (containing/L: glucose 30 g, yeast extract 3 g, KH_2PO 0.39 g, $Na_2HPO_4 \cdot 12H_2O$ 1.42 g, NH_4NO_3 0.70 g, and KCl 1.00 g), added to Erlenmeyer flasks (250 mL), was inoculated with 5 mL of conidial suspension $(1 \times 10^8 \text{ conidia/mL})$ of wild isolate (SB010) and spaceflight mutants (BHT021, BHT030, and BHT098) of *B. bassiana* followed by incubation at 150 rpm and $27\,^{\circ}\text{C}$ for 5 days. After 5 days of growth, cultures were centrifuged in an Eppendorf 5804R centrifuge (Eppendorf, Framingham, MA, USA) at 10,000 rpm, $4\,^{\circ}\text{C}$ for 10 min, and the resultant supernatant was extracted with ethyl acetate (1: 1 v/v ratio) following Wu et al. [36]. Three individual samples were run for each treatment as biological replicate.

The total protein concentration of the extracts was quantified by Bradfords' method using bovine albumin serum as standard [37].

The liquid chromatography–mass spectrophotometry (LC-MS) analysis of obtained protein extracts was carried out by the method of Wu et al. [36] using LC Agilent 1200 LC-MS/MS system. The detailed description of LC-MS protocol can be seen in the supplementary materials (Supplementary file 2).

Fourier-transformed infrared spectroscopy analysis was performed by using MIR8035 FTIR spectrometer (Thermo Fisher Scientific, Germany). All measurements were made at a resolution of $4\,\text{cm}^{-1}$ over a frequency range of 400 to 4000 cm^{-1}. The liquid sample was loaded directly, and the spectra were recorded at room temperature.

Nuclear magnetic resonance (NMR) was performed using a Bruker advance III-HD 600 NMR spectrometer (Bruker, Karlsruhe, Germany) by following the method of Wu et al. [36].

5.5. Toxicity of Crude Protein Extracts of Wild Isolate (SB010, and Spaceflight Mutants (BHT021, BHT030, and BHT098) of Beauveria bassiana against Megalurothrips usitatus

The concentration mortality response of crude protein extracts of wild isolate (SB010) and spaceflight mutants (BHT021, BHT030, and BHT098) of *B. bassiana* against *M. usitatus* adult females was studied by following the method Du et al. [26]. Briefly, centrifuge tubes (9 mL) and bean pods (1 cm) were individually immersed in different crude protein extracts of different concentrations (100, 75, 50, 25, and 12.5 µg/mL) followed by drying under sterile conditions. Centrifuge tubes and bean pods immersed in ddH$_2$O with 0.05% Tween-80 served as control. Adult females of *M. usitatus* (100 individuals) were inoculated to treated bean pods with camel hair brush, and bean pods were placed in a treated centrifuge tube. Each centrifuge tube was sealed with a cotton plug to prevent the thrips from escaping and was placed at $26 \pm 1\,^{\circ}\text{C}$, $70 \pm 5\%$ R.H., and 16:8 L:D photoperiod. The insects were observed on a daily basis for 5 days to record mortality data as outlined by Du et al. [33]. The treatments were repeated three times with fresh batches of insects.

Changes in the appearance of the infected *M. usitatus* midgut were directly monitored at 5 days post treatment under a JEM1011 transmission electron microscope (Nikon Co. Ltd., Tokyo, Japan) following Du et al. [26]. The treated *M. usitatus* adults were sampled at 5 d post treatment and were dissected under stereo microscope (Stemi 508, Carl-ZEISS, Jena, Germany) to obtain midgut samples. The samples were fixed overnight in 2.5% glutaraldehyde + 2% paraformaldehyde solution at $4\,^{\circ}\text{C}$ followed by rinsing with PBS buffer (0.1 M). The samples were then stained overnight with 1% uranyl acetate at $4\,^{\circ}\text{C}$ followed by dehydration with gradient concentrations of ethanol. The dehydrated tissues were embedded in silica gel blocks, and sections were cut using automatic microwave tissue processing instrument (EM AMW, Leica Microsystems, Wetzlar, Germany) and cryo-ultramicrotome (EM UC7/FC7, Leica Microsystems, Wetzlar, Germany).

5.6. Data Analysis

Data regarding total protein concentration were subjected to ANOVA-1. Means were compared by Tukey's HSD test ($p < 0.05$). Mortality (%) data were arcsine transformed before further analysis. The mortality (%) data were subjected to ANOVA-2, and significance between means was also tested by Tukey's HSD test ($p < 0.05$). SAS 9.2 was used for all statistical analysis [38].

Supplementary Materials: The following supporting information can be downloaded at: https://www.mdpi.com/article/10.3390/toxins14080555/s1, Supplementary file S1: Detailed LCMS analysis of the crude protein extracts from wild isolate and spaceflight mutants of *Beauveria bassiana*. Supplementary file S2: Supplementary methods.

Author Contributions: Conceptualization, J.W. and S.A.; methodology, J.W. and S.A.; software, X.Z.; validation, Y.Z. and J.Z.; formal analysis, Y.Z.; investigation, Y.Z. and X.Z.; resources, J.W.; data curation, Y.Z. and J.Z.; writing—original draft preparation, Y.Z. and S.A.; writing—review and editing, S.A.; supervision, J.W.; project administration, J.W. and S.A.; funding acquisition, J.W. and S.A. All authors have read and agreed to the published version of the manuscript.

Funding: This research was funded by grants from The Key Area Research and Development Program of Guangdong Province (2018B020205003) and Science and technology program of Guangdong Province (2021A1515011058).

Institutional Review Board Statement: Not applicable.

Informed Consent Statement: Not applicable.

Data Availability Statement: The raw data supporting the conclusion will be made available by the corresponding author on request.

Acknowledgments: The authors want to thank the handling editor and anonymous reviewers for their constructive comments and suggestions.

Conflicts of Interest: The authors declare no conflict of interest.

References

1. Gilbert, R.; Torres, M.; Clemens, R.; Hateley, S.; Hosmani, R.; Wade, W.; Bhattacharaya, S. Spaceflight and simulated microgravity conditions increase virulence of *Serratia marcescens* in the *Drosophila melanogaster* infection model. *Npj Microgravity* **2020**, *6*, 4. [CrossRef] [PubMed]
2. Huang, B.; Li, D.G.; Huang, Y.; Liu, C.T. Effects of spaceflight and simulated microgravity on microbial growth and secondary metabolism. *Milit. Med. Res.* **2018**, *5*, 18. [CrossRef] [PubMed]
3. Barrila, J.; Radtke, L.A.; Crabbe, A.; Saker, S.F.; Herbst-Kralovetz, M.M.; Ott, C.M.; Nickerson, C.A. Organotypic 3D cell culture models: Using the rotating wall vessel to study host–pathogen interactions. *Nat. Rev. Microbiol.* **2010**, *8*, 791–801. [CrossRef] [PubMed]
4. Taylor, P.W. Impact of space flight on bacterial virulence and antibiotic susceptibility. *Infect. Drug Resist.* **2015**, *8*, 249–262. [CrossRef]
5. Senatore, G.; Mastroleo, F.; Leys, N.; Mauriello, G. Effect of microgravity and space radiation on microbes. *Future Micobiol.* **2018**, *13*, 831–847. [CrossRef]
6. Horneck, G.; Kluas, D.M.; Mancinelli, R. Space microbiology. *Microbiol. Mol. Biol. Rev.* **2010**, *74*, 121–156. [CrossRef]
7. Benoit, M.R.; Li, W.; Stodieck, L.S.; Lam, K.S.; Winther, C.L.; Roane, T.M.; Klaus, D.M. Microbial antibiotic production aboard the international Space Station. *Appl. Microbiol. Biotechnol.* **2006**, *70*, 403–411. [CrossRef]
8. Rosenzweig, J.A.; Abogunde, O.; Thomas, K.; Lawal, A.; Nguyen, Y.U.; Sodipe, A.; Jejelowo, A. Spaceflight and modelled microgravity effects on microbial growth and virulence. *Appl. Microbiol. Biotechnol.* **2010**, *85*, 885–891. [CrossRef]
9. Abshire, C.F.; Prasai, K.; Soto, I.; Shi, R.; Concha, M.; Baddoo, M.; Flemington, E.K.; Ennis, D.G.; Scott, R.S.; Harrison, L. Exposure of *Mycobacterium marinum* to low-shear modelled microgravity: Effect on growth, the transcriptome and survival under stress. *Npj Microgravity* **2016**, *2*, 16038. [CrossRef]
10. Seyedsayamdost, M.R. Toward a global picture of bacterial secondary metabolism. *J. Ind. Microbiol. Biotechnol.* **2019**, *46*, 301–311. [CrossRef]

11. Horak, I.; Engelbrecht, G.; van Rensburg, P.J.J.; Claassens, S. Microbial metabolomics: Essential definitions and the importance of cultivation conditions for utilizing *Bacillus* Species as bionematicides. *J. Appl. Microbiol.* **2019**, *127*, 326–343. [CrossRef] [PubMed]
12. Sharma, G.; Curtis, P.D. The impacts of microgravity on bacterial metabolism. *Life* **2022**, *12*, 774. [CrossRef] [PubMed]
13. Sanchez, S.; Demain, A.L. Metabolic regulation and overproduction of primary metabolites. *Microb. Biotechnol.* **2008**, *1*, 283–319. [CrossRef] [PubMed]
14. Milojevic, T.; Weckwerth, W. Molecular mechanisms of microbial survivability in outer space: A systems biology approach. *Front. Microbiol.* **2020**, *11*, 923. [CrossRef]
15. Singh, R.; Kumar, M.; Mittal, A.; Kumar, P. Microbial metabolites in nutrition, healthcare and agriculture. *3 Biotech.* **2017**, *7*, 15. [CrossRef]
16. Craney, A.; Ozimok, C.; Pimentel-Elardo, S.M.; Capretta, A.; Nodwell, J.R. Chemical perturbation of secondary metabolism demonstrates important links to primary metabolism. *Chem. Biol.* **2012**, *19*, 1020–1027. [CrossRef] [PubMed]
17. Li, T.; Chang, D.; Xu, H.; Chen, J.; Su, L.; Guo, Y.; Chen, Z.; Wang, Y.; Wang, L.; Wang, J. Impact of a short-term exposure to spaceflight on the phenotype, genome, transcriptome and proteome of *Escherichia coli*. *Int. J. Astrobiol.* **2015**, *14*, 435–444. [CrossRef]
18. Komatsu, M.; Uchiyama, T.; Omura, S.; Cane, D.E.; Ikeda, H. Genome-minimized *Streptomyces* host for the heterologous expression of secondary metabolism. *Proc. Natl. Acad. Sci. USA* **2010**, *107*, 2646–2651. [CrossRef]
19. Su, X.Y.; Schmitz, G.; Zhang, M.L.; Mackie, R.I.; Cann, I.K.O. Heterologous gene expression in filamentous fungi. *Adv. Appl. Microbiol.* **2012**, *81*, 1–61.
20. Wang, Q.; Zhong, C.; Xiao, H. Genetic engineering of filamentous fungi for efficient protein expression and secretion. *Front. Bioeng. Biotechnol.* **2020**, *8*, 293. [CrossRef]
21. Baron, N.C.; Rigobelo, E.C.; Zied, D.C. Filamentous fungi in biological control: Current status and future perspectives. *Chil. J. Agri. Res.* **2019**, *79*, 307–315. [CrossRef]
22. Wu, J.H.; Yu, X.; Wang, X.S.; Tang, L.D.; Ali, S. Matrine enhances the pathogenicity of *Beauveria brongniartii* against *Spodoptera litura* (Lepidoptera: Noctuidae). *Front. Microbiol.* **2019**, *10*, 1812. [CrossRef] [PubMed]
23. Leys, N.M.E.J.; Hendrickx, L.; De Boever, P.; Baatout, S.; Mergeay, M. Space flight effects on bacterial physiology. *J. Biol. Regul. Homeost. Agents* **2004**, *18*, 193–199. [PubMed]
24. Nickerson, C.A.; Ott, C.M.; Wilson, J.W.; Ramamurthy, R.; Pierson, D.L. Microbial responses to microgravity and other low-shear environments. *Microbiol. Mol. Biol. Rev.* **2004**, *68*, 345–361. [CrossRef]
25. Demain, A.L.; Fang, A. Secondary metabolism in simulated microgravity. *Chem. Rec.* **2001**, *1*, 333–346. [CrossRef]
26. Benoit, M.R.; Klaus, D.M. Microgravity, bacteria, and the influence of motility. *Adv. Space Res.* **2007**, *39*, 1225–1232. [CrossRef]
27. Bibb, M.J. Regulation of secondary metabolism in *Streptomycetes*. *Curr. Opin. Microbiol.* **2005**, *8*, 208–215. [CrossRef]
28. Viollier, P.H.; Kelemen, G.H.; Dale, G.E.; Nguyen, K.T.; Buttner, M.J.; Thompson, C.J. Specialized osmotic stress response systems involve multiple SigB-like sigma factors in *Streptomyces coelicolor*. *Mol. Microbiol.* **2003**, *47*, 699–714. [CrossRef]
29. Lam, K.; Gustavson, D.R.; Pirnik, D.L.; Pack, E.; Bulanhagui, C.; Mamber, S.W.; Forenza, S.; Stodieck, L.S.; Klaus, D.M. The effect of space flight on the production of actinomycin D by *Streptomyces plicatus*. *J. Ind. Microbiol. Biotechnol.* **2002**, *29*, 299–302. [CrossRef]
30. Luo, A.; Gao, C.; Song, Y.; Tan, H.; Liu, Z. Biological responses of a *Streptomyces* strain producing- nikkomycin to space flight. *Space Med. Med. Eng.* **1998**, *11*, 411–414.
31. Herranz, R.; Anken, R.; Boonstra, J.; Braun, M.; Christianen, P.C.M.; de Geest, M.; Hauslage, J.; Hilbig, R.; Hill, R.J.A.; Lebert, M.F.; et al. Ground-based facilities for simulation of microgravity: Organism-specific recommendations for their use, and recommended terminology. *Astrobiology* **2013**, *13*, 1–17. [CrossRef] [PubMed]
32. Espinosa, P.J.; Contreras, J.; Quinto, V.; Grávalos, C.; Fernández, E.; Bielza, P. Metabolic mechanisms of insecticide resistance in the western flower thrips, *Frankliniella occidentalis* (Pergande). *Pest. Manag. Sci.* **2005**, *61*, 1009–1015. [CrossRef] [PubMed]
33. Du, C.; Yang, B.; Wu, J.H.; Ali, S. Identification and virulence characterization of two *Akanthomyces attenuatus* isolates against *Megalurothrips usitatus* (Thysanopera: Thripidea). *Insects* **2019**, *10*, 168. [CrossRef] [PubMed]
34. Ali, S.; Huang, Z.; Ren, S.X. Production of cuticle degrading enzymes by *Isaria fumosorosea* and their evaluation as a biocontrol agent against diamondback moth. *J. Pest. Sci.* **2010**, *83*, 361–370. [CrossRef]
35. Zhao, J.; Yao, Y.N.; Wei, Y.; Huang, S.; Keyhani, N.O.; Huang, Z. Screening of *Metarhizium anisopliae* UV-induced mutants for faster growth yields a hyper-virulent isolate with greater UV and thermal tolerances. *Appl. Microbiol. Biotechnol.* **2016**, *100*, 9217–9228. [CrossRef]
36. Wu, J.; Yang, B.; Xu, J.; Cuthbertson, A.G.S.; Ali, S. Characterization and toxicity of crude toxins produced by *cordyceps fumosorosea* against *Bemisia tabaci* (Gennadius) and *Aphis craccivora* (Koch). *Toxins* **2021**, *13*, 220. [CrossRef]
37. Bradford, M.M. A rapid and sensitive method for the quantitation of microgram quantities of protein utilizing the principle of protein dye binding. *Anal. Biochem.* **1976**, *72*, 248–254. [CrossRef]
38. SAS Institute. *SAS User's Guide*; Statistics SAS Institute: Cary, NC, USA, 2000.

Article

Influence of Agronomic Factors on Mycotoxin Contamination in Maize and Changes during a 10-Day Harvest-Till-Drying Simulation Period: A Different Perspective

Bernat Borràs-Vallverdú [1], Antonio J. Ramos [1,*], Carlos Cantero-Martínez [2], Sonia Marín [1], Vicente Sanchis [1] and Jesús Fernández-Ortega [2]

[1]   AGROTECNIO-CERCA Center, Food Technology Department, University of Lleida, Rovira Roure 191, 25198 Lleida, Spain
[2]   AGROTECNIO-CERCA Center, Department of Crop and Forest Sciences, University of Lleida, Rovira Roure 191, 25198 Lleida, Spain
*   Correspondence: antonio.ramos@udl.cat

Abstract: Agronomic factors can affect mycotoxin contamination of maize, one of the most produced cereals. Maize is usually harvested at 18% moisture, but it is not microbiologically stable until it reaches 14% moisture at the drying plants. We studied how three agronomic factors (crop diversification, tillage system and nitrogen fertilization rate) can affect fungal and mycotoxin contamination (deoxynivalenol and fumonisins B_1 and B_2) in maize at harvest. In addition, changes in maize during a simulated harvest-till-drying period were studied. DON content at harvest was higher for maize under intensive tillage than using direct drilling (2695 and 474 µg kg^{-1}, respectively). We found two reasons for this: (i) soil crusting in intensive tillage plots caused the formation of pools of water that created high air humidity conditions, favouring the development of DON-producing moulds; (ii) the population of *Lumbricus terrestris*, an earthworm that would indirectly minimize fungal infection and mycotoxin production on maize kernels, is reduced in intensive tillage plots. Therefore, direct drilling is a better approach than intensive tillage for both preventing DON contamination and preserving soil quality. Concerning the simulated harvest-till-drying period, DON significantly increased between storage days 0 and 5. Water activity dropped on the 4th day, below the threshold for DON production (around 0.91). From our perspective, this study constitutes a step forward towards understanding the relationships between agronomic factors and mycotoxin contamination in maize, and towards improving food safety.

Keywords: maize; deoxynivalenol; fumonisin; tillage system; nitrogen fertilisation; crop diversification; water activity; *Fusarium*; *Lumbricus terrestris*

Key Contribution: direct drilling is a better tillage system than intensive tillage; as it not only preserves soil quality; but also helps controlling DON contamination in maize.

Citation: Borràs-Vallverdú, B.; Ramos, A.J.; Cantero-Martínez, C.; Marín, S.; Sanchis, V.; Fernández-Ortega, J. Influence of Agronomic Factors on Mycotoxin Contamination in Maize and Changes during a 10-Day Harvest-Till-Drying Simulation Period: A Different Perspective. *Toxins* **2022**, *14*, 620. https://doi.org/10.3390/toxins14090620

Received: 21 July 2022
Accepted: 30 August 2022
Published: 5 September 2022

1. Introduction

Maize is one of the most produced cereals worldwide and is used for both human consumption and animal feed. It is estimated that 1,162,352,997 tons of maize were produced in 2020 [1]. Unfortunately, maize is susceptible to toxigenic fungal contamination at all points of its supply chain (pre-harvest, harvest and post-harvest stages) [2,3]. Amongst the most prevalent and toxic fungal metabolites in maize, the mycotoxins fumonisin B_1 (FB_1), fumonisin B_2 (FB_2) and deoxynivalenol (DON) can be found [2,4]. In maize, fumonisins are primarily caused by *Fusarium verticillioides*, *Fusarium proliferatum* and *Fusarium subglutinans*, while DON is mostly caused by *Fusarium graminearum* and *Fusarium culmorum* [4–7]. Apart from being one of the major causes of economic losses in maize crops, mycotoxin contamination can have a severe impact on human and animal health.

FB$_1$ affects sphingolipid metabolism, causes oxidative stress and can cause damage to cell DNA [8]. In humans, fumonisins have been associated with a higher risk of oesophageal carcinoma [9]. In animals, FB$_1$ ingestion can cause leucoencephalomalacia (LEM) in horses, hepatocarcinogenesis in rats and pulmonary oedema in swine [10]. DON inhibits protein and DNA synthesis in eukaryotic cells, and can induce nausea, emesis, vomiting, skin inflammation, leukopenia, diarrhoea, haemorrhage in the lungs and brain, and the destruction of bone marrow [11,12]. The European Union (EU) regulates the maximum content of DON and the sum of FB$_1$ + FB$_2$ in certain foodstuffs (including maize) and provides guidance values for those and other mycotoxins in food and feed products [13–15].

Many factors can affect mycotoxin contamination in maize throughout the whole supply chain. Among them, we can find biological factors (susceptibility of the crop), environmental factors (temperature, rainfall, air relative humidity, insects/bird injuries), crop management (planting and harvest dates, tillage practices, fertilization, crop rotation, irrigation), crop harvesting (crop maturity, temperature, moisture, mechanical injury), transportation conditions, time until drying, and proper drying or storage conditions (aeration, temperature, pest/rodent control) [2,4,16].

The accepted commercial moisture for maize harvesting in NE Spain is around 18%. Sometimes, when the maize is almost ready for harvest, rain can increase the grain moisture, promoting mould proliferation and extending the period before harvesting until moisture reaches commercial standards again. In addition, in some areas drying facilities are undersized. Therefore, as all maize is harvested within an interval of a few days, it is usual to find huge amounts of maize grain outdoors waiting to be processed in the drying plants. This waiting period can sometimes be as long as 10 days. Despite the accepted commercial moisture for maize being about 18%, it has been reported that to ensure that no moulds can grow in grain nor produce mycotoxins, its maximum moisture content must be no more than 14% [17,18]. To our knowledge, there is no information about how this waiting period can influence fungal and mycotoxin contamination of the maize.

Hence, the objectives of our study were to: (i) study the impact of several agronomic factors (the crop diversification, the tillage system and the nitrogen (N) fertilization rate) on total fungal contamination, *Fusarium* spp. contamination and DON, FB$_1$ and FB$_2$ contaminations of recently harvested maize; (ii) simulate the waiting period between maize harvesting and drying for 10 days, and study the influence of waiting time and temperature on the previously mentioned variables.

2. Results and Discussion

2.1. Influence of Agronomic Factors on the Maize at Harvest Date

At harvest date (day 0) all the analyzed maize from both maturity groups, N fertilization rates and tillage systems was contaminated with DON (Table 1). On the other hand, only 12.5% of that same maize samples contained FB$_1$ and FB$_2$. Average concentrations of the contaminated samples were 826 and 196 µg toxin kg^{-1} maize for FB$_1$ and FB$_2$, respectively.

Table 1. DON contamination in maize at harvest.

FAO Maturity Group/Cropping System	Fertilization	Tillage System	Average DON Contamination (µg Toxin kg^{-1} Maize)
400/SC	0 N	DD	440
		IT	2848
	High N	DD	566
		IT	4406
700/LC	0 N	DD	654
		IT	791
	High N	DD	236
		IT	2734

SC: short cycle; LC: long cycle; 0 N: zero nitrogen rate; High N: high nitrogen rate; DD: direct drilling; IT: intensive tillage.

Multi-factor ANOVAs were carried out to study the impact of the agronomic factors on the response variables at harvest.

Neither FB_1 nor FB_2 concentrations in maize at harvest date were statistically significantly affected by any of the agronomic factors. DON content of the grains at harvest date was statistically significantly affected by the tillage system (see Table 2). Maize planted under IT had higher DON contamination (2695 µg DON kg^{-1} maize on average) than maize planted using DD (474 µg DON kg^{-1} maize on average).

Table 2. Test of between-subjects effects for DON contamination at harvest date.

	SS	df	MS	F	Sig.
Crop diversification	3.697	1	3.697	1.292	0.289
N. fert. rate	2.574	1	2.574	0.900	0.371
Tillage system	19.729	1	19.729	6.897	**0.030**
Crop diversification × N fert. rate	0.006	1	0.006	0.002	0.964
Crop diversification × Tillage system	3.260	1	3.260	1.140	0.317
N fert. Rate × Tillage system	3.598	1	3.598	1.258	0.295
Crop diversification × N fert. Rate × Tillage system	0.216	1	0.216	0.075	0.791

R Squared = 0.591 (Adjusted R Squared = 0.233). SS: sum of squares; df: degrees of freedom; MS: Mean Square; F: F-value; Sig.: significance value. Bold value is the only statistically significant factor.

It has been reported that residues of crops that were infected with *Fusarium* constitute an inoculum of the fungus for the following crop [19–22]. This inoculum tends to be particularly abundant in the case of maize [23]. Therefore, according to many authors, the removal, destruction or burial of infected crop residues is likely to reduce the *Fusarium* inoculum for the following crop, making IT a better choice than DD for controlling mycotoxin-producing fungal inoculums [19,21,22]. Mansfield, De Wolf and Kuldau (2005) reported that the DON concentration of ensiled maize was lower in maize planted using a moldboard till than in maize planted using no tillage [24]. Dill-Macky and Jones (2000) studied the DON contamination of wheat following corn, wheat or soybean, using different tillage systems [25]. DON levels were lower in wheat planted using moldboard ploughing following corn or wheat in comparison to wheat planted using no tillage following the same crops. No significant differences in DON levels in wheat were observed between the two tillage systems when the previous crop was soybean, as *F. graminearum* is not considered a pathogen of soybeans. Obst, Lepschy-Von Gleissenthall and Beck (1997) stated that the use of minimum tillage instead of mouldboard ploughing after a maize crop could result in a 10-fold increase in DON contamination of the following wheat crop [26]. Schöneberg et al. (2016) demonstrated that barley from fields with ploughed soils showed significantly less *F. graminearum* and DON content than barley from reduced tillage fields, regardless of the previous crop [27]. On the other hand, Roucou, Bergez, Méléard and Orlando (2022), who collected data from a total of 2032 maize fields located in France between 2004 and 2020, found that DON contamination in maize was not significantly different whether the crop residues of the previous year were adequately managed (mostly through soil tillage) or not [28]. Suproniené et al. (2012) studied the effect of different tillage practices (conventional tillage, reduced tillage and no tillage) on mycotoxin contamination in winter and spring wheat, but no clear relationship could be observed [29]. Furthermore, Kaukoranta, Hietaniemi, Rämö, Koivisto and Parikka (2019), who analyzed survey data from 804 spring-oat fields, found that the DON concentration of the oats was the same or higher under ploughing than under non-ploughing conditions [30]. Our results are closer to those of Kaukoranta et al. (2019), as we found a significantly higher DON contamination in maize planted under IT than in maize planted using DD.

Tillage operations can affect both the soil structure and the crop productivity [31]. Unlike no tillage, IT exposes soil to erosive agents such as wind and water. The impact of water drops induces the degradation of the soil by the breakdown of water-stable aggregates, causing soil crusting [32,33]. Soil crusting negatively affects seedling emergence, reduces water infiltration rates and water storage capacity, favors runoff, diminishes organic matter, and can cause overland flow [31,32,34]. Soils rich in silt and fine sand, such as the one in this study, are highly susceptible to soil crusting [35]. As we observed the presence of pools of water only in IT plots, most probably caused by soil crusting, we hypothesize that in these plots the pools of water created high air humidity conditions, favoring the growth of moulds throughout the whole cultivation period, including DON-producing moulds. In fact, it has been stated that under moist conditions the production of macroconidia, the production of ascopores and the ejection of ascospores of *F. graminearum* are favored [36–38]. That would help explain the higher DON contamination in maize planted under IT in comparison to maize planted using DD.

Another hypothesis that supports our results is that tillage affects soil fauna, which in turn can have an impact on *Fusarium* species. Earthworms are known for breaking down organic matter and promoting nutrient cycling along with soil microbiota, and for improving soil structure, soil porosity, soil water retention capacity, root distribution, plant growth and plant health. Frequent tillage adversely affects many earthworm species, especially those linked to the surface layers (epigeics and anecics) [39–42]. When the soil is turned over, earthworms are injured and killed, their burrows are broken, their food sources are buried, and they become exposed to harsh environmental conditions and predators. [39–41,43,44]. The common earthworm (*Lumbricus terrestris*) is one of the most important anecic earthworms and is capable of incorporating plant litter into the soil and decomposing it. Oldenburg, Kramer, Schrader and Weinert (2008) and Schrader, Kramer, Oldenburg and Weinert (2009) demonstrated that *L. terrestris* accelerates the degradation of *Fusarium* biomass and DON in the wheat straw layer, and that this earthworm is more attracted to highly *Fusarium*-infected and DON contaminated wheat straw than less infected and contaminated wheat straw [45,46]. *L. terrestris* is likely to prefer the contaminated straw as its N-content and digestibility are enhanced due to fungal colonization. Thus, *L. terrestris* most probably reduces *Fusarium* biomass in maize straw too, and consequently, minimizes *Fusarium* infection and DON contamination of maize cobs. Therefore, as the population of *L. terrestris* is smaller in IT plots, a lower DON contamination in maize planted under DD is expected than in maize planted under IT. In our case, we did not sample earthworms during the experiment. However, Santiveri Morata, Cantero-Martínez, Ojeda Domínguez and Angás Pueyo (2004) studied the population of earthworms in the same field where this study was performed, and found that under DD the population of worms was higher than in more aggressive tillage systems [47].

The moisture content of the grains at harvest date was statistically significantly affected by the crop diversification and the tillage system. Moisture was higher in SC maize (21.33% on average) than in LC maize (16.93% on average), and higher in maize under IT (20.45% on average) than in maize planted using DD (17.82% on average). Likewise, the a_w of the grains was significantly affected by the crop diversification, being greater in SC maize than in LC maize (0.927 and 0.897 on average, respectively). It should be noted, though, that the differences in moisture and a_w between different maturity groups could easily be modified by the harvesting dates.

It has been reported that no tillage is associated with soil with a higher water holding capacity and higher soil moisture in surface soil layers in comparison with IT [48,49]. Therefore, one might think that the maize kernels obtained from DD-planted maize would have a higher moisture than maize kernels obtained from maize under IT, but that was not the case in our study.

The log of total fungal contamination was significantly affected by the crop diversification (*p*-value = 0.046), being higher in LC maize (5.28 on average) than in SC maize (4.81 on average).

No effect of the agronomic factors was observed in the log of *Fusarium* spp., FB_1 or FB_2 contaminations. In this context, Ono et al. (2011) observed no significant differences in the *Fusarium* sp. counts and the fumonisin concentrations between non-tilled and conventional-tilled maize [50]. Similarly, Ariño et al. (2009) found no significant differences in the fumonisin contents of maize planted using minimum tillage and ploughing [51]. Even so, it is necessary to emphasize that the low incidence of FB_1 and FB_2 contamination on maize at harvest (12.5%) makes it rather difficult to observe differences in the concentration of these toxins due to agronomic factors.

Regarding how N fertilization can affect fumonisin contamination in maize, previous research has shown contrasting results. A shared vision is that a balanced fertilization is the best approach to minimize fumonisin concentrations, as stress due to N deficiency or high N rates can significantly raise fumonisin levels [50,52–56]. In our study, no differences in fumonisin contamination were observed between 0 N and High N fertilization rates.

2.2. Correlations between the Studied Variables at the Harvest Date

Principal Component Analysis was performed in search of correlations between response variables at harvest (see correlation matrix heatmap in Figure 1). The variables studied were moisture, a_w, the log of total fungal contamination, the log of *Fusarium* spp. contamination, and the different mycotoxin contaminations (DON, FB_1 and FB_2). Following the criteria of choosing the principal components with eigenvalues > 1, three principal components were taken, which accounted for 81.32% of the total variance.

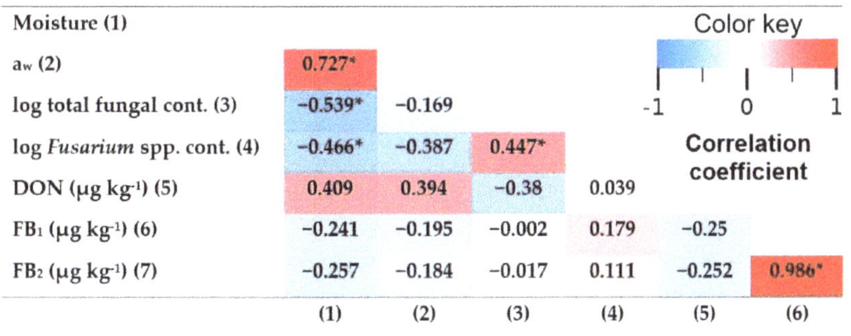

	(1)	(2)	(3)	(4)	(5)	(6)
Moisture (1)						
a_w (2)	0.727*					
log total fungal cont. (3)	−0.539*	−0.169				
log *Fusarium* spp. cont. (4)	−0.466*	−0.387	0.447*			
DON (µg kg⁻¹) (5)	0.409	0.394	−0.38	0.039		
FB₁ (µg kg⁻¹) (6)	−0.241	−0.195	−0.002	0.179	−0.25	
FB₂ (µg kg⁻¹) (7)	−0.257	−0.184	−0.017	0.111	−0.252	0.986*

Figure 1. Correlation matrix heatmap based on the correlation coefficients from the PCA at harvest date. A darker blue color indicates a stronger negative correlation, while a darker red color indicates a stronger positive correlation. * indicates a significant correlation (*p*-value < 0.05).

Few variables were significantly correlated. FB_1 contamination was significantly positively correlated with FB_2 contamination (r = 0.986, *p*-value < 0.001). That is in accordance with the results of Carbas et al. (2021) and Cao et al. (2013), who also found significant positive correlations between FB_1 and FB_2 contaminations (r = 0.96 and r = 0.99, respectively) [57,58].

There was a significant positive correlation between the log of total fungal contamination and the log of *Fusarium* spp. contamination, indicating that *Fusarium* spp. is of considerable relevance to total fungal contamination.

Moisture was significantly positively correlated with a_w (r = 0.727, *p*-value = 0.001), and significantly negatively correlated with the log of total fungal contamination (r = 0.539, *p*-value = 0.016) and with the log of *Fusarium* spp. contamination (r = −0.466, *p*-value = 0.035). Cao et al. (2013) also described a significantly negative correlation between moisture and *Fusarium* spp. contamination (r = −0.68, *p*-value < 0.05).

Fusarium spp. contamination at harvest date was not significantly correlated with the concentration of any of the studied mycotoxins (DON, FB_1 and FB_2) in the same period. This could be explained by there being non-DON/FB_1/FB_2-producing *Fusarium* spp. strains colonizing our maize, and/or because a higher count of DON/FB_1/FB_2-producing *Fusarium* spp. at harvest date does not necessarily imply a higher concentration of these mycotoxins. Factors such as a_w, temperature and relative humidity can affect mycotoxin production [59,60]. Similarly, Lanza et al. (2017) found no association either between fumonisin levels and the frequency of *Fusarium* spp. in maize kernels [61]. On the other hand, Schöneberg et al. (2016) found that *F. graminearum* was positively correlated with DON content in barley (r = 0.72, *p*-value < 0.001) [27].

No significant correlations were observed between DON and FB_1 or FB_2 concentrations. That is consistent with the bibliography, as it has been described that in maize DON is produced primarily by *F. graminearum* and *F. culmorum*, while FB_1 and FB_2 are mainly produced by *F. verticillioides*, *F. proliferatum* and *F. subglutinans* [4–7].

DON was positively correlated with moisture and a_w, but the correlations were not significant (*p*-values of 0.058 and 0.066, respectively).

2.3. Effect of Time and Temperature on Maize Moisture, a_w, Microbial Counts and Mycotoxin Contamination after Harvest

Multi-factor ANOVAs were carried out to determine the effect of time and temperature (15 or 25 °C) on the studied variables. On one side, moisture, a_w and microbial counts were studied on days 0, 4, 7 and 10. On the other side, mycotoxin contamination was studied on days 0, 5 and 10. All the data are available in a spreadsheet in the Supplementary Materials.

No significant effect of time nor temperature was observed on the moisture, the total fungal contamination or the *Fusarium* spp. contamination during the 10 days of the experiment. By contrast, the variable time significantly affected the evolution of a_w (*p*-value = 0.001), which dropped on day 4 for both temperatures (Figure 2). Statistically significant differences were observed between a_w on day 0 and a_w on days 4, 7 and 10.

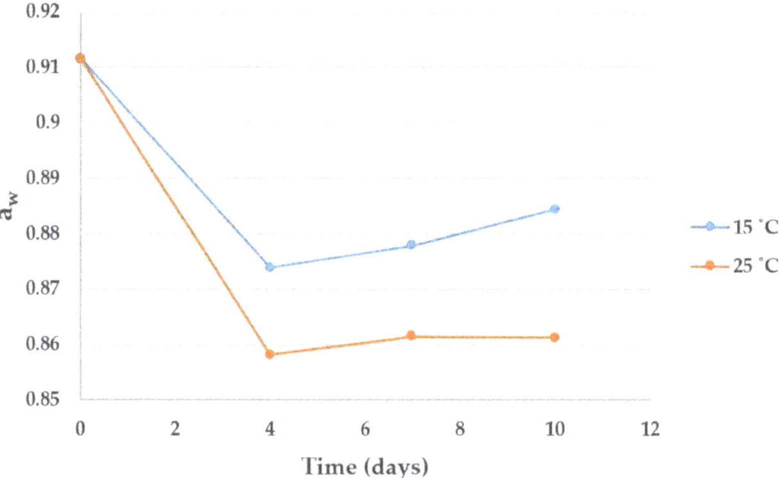

Figure 2. Influence of time and temperature on the evolution of a_w.

DON, FB_1 and FB_2 contaminations were not affected by time or temperature, although in the case of DON time was close to being significant (*p*-value = 0.078). Thus, statistically significant differences were observed in DON concentrations between days 0 and 5 (*p*-value = 0.049) but not between days 0 and 10 (*p*-value = 0.051) or days 5 and 10 (*p*-value = 0.989) (see Table 3). Regarding FB_1 and FB_2 contamination, the tendency was the same as that at harvest: a low prevalence of these toxins. On days 5 and 10, only 15.63 and 18.75% of samples contained at least one of the studied fumonisins. The average contamination of contaminated samples on days 5 and 10 was 1938 and 1709 µg toxin kg^{-1} maize for FB_1, and 1068 and 1279 µg toxin kg^{-1} maize for FB_2.

Table 3. Influence of time and temperature on the evolution of DON concentrations (μg DON kg^{-1} maize).

Temperature	DON Concentration (μg DON kg^{-1} Maize)		
	Day 0	Day 5	Day 10
15 °C	1584 ± 1932	2367 ± 2983	2649 ± 2349
25 °C	1584 ± 1932	3771 ± 3597	3469 ± 4300

Presented values correspond to mean and standard deviation.

As an increase in DON concentration was observed in the 0–5 days period, and *Fusarium* spp. counts remained stable during the whole 10 days period, the absence of DON production during the 5–10 days period could be attributed to the drop in a_w during the first 4 days. If a_w levels had remained constant since harvest, DON contamination most likely would have increased continuously. Considering these results, we could say that under the tested temperatures (15 and 25 °C), there are DON-producing *Fusarium* spp. species in maize that can produce DON at an approximate a_w of at least 0.91, while at an a_w of 0.88 they can no longer produce this toxin.

Our results are in line with those obtained by Comerio, Fernández Pinto and Vaamonde (1999), who studied the DON production of *F. graminearum* in wheat at different a_w [62]. They found that at an a_w = 0.925 DON was produced, but not at a_w = 0.900; therefore, the limiting a_w for DON production under those conditions was close to 0.900. Other studies have suggested slightly higher values under similar conditions. Ramirez, Chulze and Magan (2006) studied the DON production of *F. graminearum* on wheat and found mycotoxin production at a_w = 0.95 at the temperatures of 15, 25 and 30 °C, but they did not find DON production at a_w = 0.93 under any temperature [63]. Schmidt-Heydt, Parra, Geisen and Magan (2011) found that *F. culmorum* and *F. graminearum* could produce DON at an a_w = 0.93 at 25 °C in YES medium after a 9 day incubation, but not at a_w = 0.90 under any of the tested conditions [64].

3. Conclusions

At harvest, all maize samples were contaminated with DON (1584 ± 1578 μg DON kg^{-1} maize), while only 12.5% of the maize samples were contaminated with FB$_1$ and FB$_2$ (average contaminations of contaminated samples were 826 and 196 μg toxin kg^{-1} maize, respectively). No effect of the crop diversification or the N fertilization rate was observed on the maize DON contamination. The only agronomic factor that significantly affected the DON content of grains was the tillage system. Maize planted under IT presented a greater DON contamination (2695 μg DON kg^{-1} maize on average) than maize planted using DD (474 μg DON kg^{-1} maize on average). Two main reasons support these results. The first reason is that in IT plots the degradation of the soil resulting from the continuous tillage caused soil crusting, which induced the formation of pools of water, creating high air humidity conditions, which favored the growth of DON-producing moulds. The second reason is that the frequent tillage in IT plots causes a decrease in the population of *L. terrestris*. This earthworm is likely to reduce *Fusarium* infection and DON contamination in maize straw. Consequently, maize cobs under DD are expected to be less infected and contaminated. Hence, DD would be a better approach than IT not only in terms of controlling DON contamination, but also from the agronomic point of view. More studies that employ long-term IT and DD plots are needed to assess precisely how the tillage system can influence the mycotoxin contamination of grains.

No significant correlations were found between the log of *Fusarium* spp. contamination at harvest date and the concentration of any of the studied mycotoxins in the same period.

During the 10-day storage, no effect of time or temperature was observed on the moisture, the total fungal contamination, the *Fusarium* spp. contamination or the FB$_1$ and FB$_2$ contaminations. Time affected the evolution of a_w, which fell on day 4 for both temperatures. DON concentration on day 5 was significantly higher than on day 0, but there were no significant differences between days 5 and 10. Therefore, it is predictable that continued DON production was held back by the a_w drop in the first 4 days of storage, meaning the minimum a_w for the DON-producing species colonizing our maize to produce this toxin is around 0.91.

To our knowledge, this is the first study that relates soil crusting and the consequent formation of pools of water in maize plots under IT with a higher DON grain contamination in comparison with maize plots under DD, and is also the first work to question how the harvest-till-drying period of maize can affect fungal and mycotoxin contamination.

4. Materials and Methods

4.1. Climate and Soil Characteristics, Experimental Design and Crop Management

Maize was planted in an experimental field in Agramunt, NE Spain (41°48′ N, 1°07′ E, 330 m asl). The soil in this area is classified as xerofluvent typic [65]. Many soil characteristics were measured: the average pH of the soils was (H_2O, 1:2.5) 8.5; the electrical conductivity (1:5) was 0.15 dSm^{-1}; the soil organic carbon (SOC) concentration (0–30 cm) was 8.6 g kg^{-1}; the water available holding capacity (between -33 kPa and -1500 kPa) was 10% (v/v). The climate of the area is semiarid Mediterranean with a continental trend. Climate was monitored with a weather station placed in the experimental field. During the last 30 years, the mean annual precipitation was 442 mm, the mean annual temperature was 14.6 °C, and the mean annual potential evapotranspiration (PET) was 855 mm. The winter is cold, with some days below 0 °C in January. For that, soil temperature does not reach 8 °C until the beginning of April, when the planting date for maize starts. Additionally, the climate imposes hot summers, reaching temperatures over 35 °C in July and August.

The experimental design was a split-plot with 3 blocks. The plots were 50 m × 3 m = 150 m^2 and 4 rows of maize were planted in each plot (rows spaced 73 cm apart). Three agronomic factors were evaluated: the crop diversification, the tillage system and the N fertilization rate. For the crop diversification, a monocropping long-cycle maize (LC maize) (FAO 700 maturity group, Pioneer's P1570 hybrid) was compared against a legume–maize double cropping, using short-cycle maize (SC maize) (FAO 400 maturity group, Pioneer's P0312 hybrid) as the main crop and vetch (*Vicia sativa* L., var. Prontivesa) as the secondary crop.

In the case of the tillage system, intensive tillage (IT) and direct drilling (DD) were studied. IT consisted of subsolate (35 cm depth), disc harrow and rototiller, while DD consisted of the application of herbicide (1.5 L ha^{-1} of 36% glyphosate [*N*-(phosphonomethyl)-glycine]) and sowing directly the seeds into the soil. In reference to the N fertilization rate, a zero N rate (0 N) and high N rate (High N) were evaluated. The rate of mineral fertilization applied was 400 kg N ha^{-1} for LC maize, while it was reduced to 300 kg N ha^{-1} in SC maize because of the possible fixation of the preceding legume crop. N fertilization was distributed between 2 top-dressing fertilizations with ammonium nitrate (34.5% N), with a rate of 150 kg N ha^{-1} in each one at stages V3–V5 (May in LC maize and June in SC maize) and V7–V8 respectively (June in LC maize and July in SC maize). In addition, for LC maize, a 100 kg N ha^{-1} pre-emergence fertilization was carried out during April with urea (46% N). The experiment was carried out over 3 years (2019, 2020 and 2021), although the present study was carried out with the third year's harvest. LC and SC maize were seeded in April and June, respectively. Accordingly, its flowering took place in July and August, respectively. Vetch was sown in December. In both maturity groups, the planting rate was 90,000 seeds ha^{-1}, with a row spacing of 73 cm. In the case of vetch, the planting density was 267 plants m^{-2}. All maize plots received equally a pre-emergence herbicide treatment with 7 L ha^{-1} of Primextra Gold (Terbuthylazine 18.75% + *S*-Metolachlor 31.25% (SE) w/v). For each tillage system and plant species, the harvest residue was treated differently. In the case of maize and IT, it was integrated into the soil by tillage, whereas in DD, it was chopped and spread on the soil surface. Vetch was harvested for forage at a cutting height of 5 cm, so all the biomass was exported from the plots. The irrigation rate was determined using Dastane's methods [66] for calculating crop water requirements on a weekly basis. Irrigation was carried out by sprinkling, starting in March and ending in October. The amount of irrigation used and mean meteorological conditions in the experimental field, obtained from an on-site weather station, are shown in Figure 3.

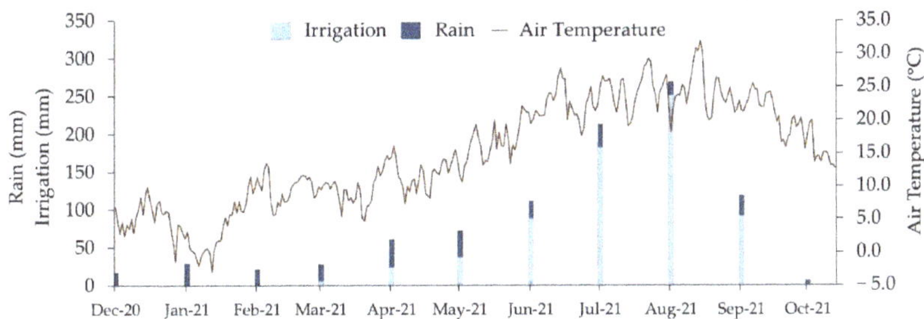

Figure 3. Irrigation and meteorological conditions in the experimental field.

4.2. Maize Harvesting and Storage

Cob samples of a total of 16 different plots were taken (2 cultivars × 2 tillage systems × 2 N fertilization rates × 2 blocks). Both cultivars were harvested when the maize was close to the commercial moisture (18 %). That was the 21st and 26th of October 2021 for LC and SC maize, respectively. On the harvest day, around 1.5 kg of maize (approximately 8–10 maize cobs) was sampled from each plot. The different maize cobs were collected throughout the entire plot, being representative of the area of study. As not to alter the microbiota of the samples, the cobs were picked up using different sterile nitrile gloves for each plot. The maize from each plot was deposited and transported in a different sterile plastic bag. In the laboratory, cobs were shelled under sterile conditions in a laminar flow cabinet. The kernels from each plot were split into two different sterile plastic bags, which were kept at different temperatures: 15 or 25 °C, for 10 days. Those specific temperatures were chosen to simulate the average maximum and minimum daily temperatures in the area at the time of harvest.

4.3. Laboratory Determinations

Different determinations were performed on the harvest day (day 0) and the following days for the maize from each plot. Moisture (%), water activity (a_w), total fungal contamination (CFU g^{-1} maize) and *Fusarium* spp. contamination (CFU g^{-1} maize) were determined on days 0, 4, 7 and 10. DON, FB_1 and FB_2 contamination were determined on days 0, 5 and 10.

4.3.1. Moisture

Approximately 15 g of maize kernels were precisely weighed into pre-weighed glass jars. The jars were put in an oven (JP Selecta 210, JP Selecta S.A., Abrera, Spain) at 105 °C for 16 h, and after that period were weighed again. The moisture was calculated according to Equation (1). Three replicates were carried out for each plot, storage time and storage temperature. Average moisture and standard deviation were calculated.

$$Moisture\ (\%) = \frac{W_0 - W_f}{W_0 - W_j} * 100 \tag{1}$$

where W_0 is the weight of the glass jar and the maize before drying, W_f is the weight of the glass jar and the maize after drying, and W_j is the weight of the glass jar.

4.3.2. Water Activity (a_w)

The a_w of whole maize kernels for each plot, storage time and storage temperature was measured using the AquaLab Series 3 TE (AquaLab S.L., Sabadell, Spain). A sample of about 3 g was introduced into the water activity meter, and a_w was properly read.

4.3.3. Total Fungal Contamination and *Fusarium* spp. Contamination

One maize sample from each plot, storage time and storage temperature was analyzed for total fungal contamination and *Fusarium* spp. contamination. Approximately 20 g of kernels was ground using a disinfected IKA A11 (IKA®-Werke GmbH & Co. KG, Staufen, Germany) mill for 30 s. Ten grams of the resulting flour was weighed in a sterile Stomacher bag with a lateral filter. Then, 90 mL of sterile saline peptone water was added to the bag (10^{-1} dilution). The flour and the saline peptone water were mixed in a laboratory blender (Stomacher 400, Seward Ltd., Worthing, UK) for 120 s at normal speed. A series of dilutions were prepared based on the filtered extract using saline peptone water (up to the 10^{-6} dilution). Then, 0.1 mL of each dilution was plated into Petri plates containing Chloramphenicol Glucose Agar (CGA) (for total fungal contamination) or Malachite Green Agar 2.5 (MGA) (a selective medium for *Fusarium* spp.).

The inoculum was spread across the Petri plates with a Digralsky spreader, and the plates were incubated upside down at 25 °C. Plate readings were performed after 3 days of incubation for CGA plates and 4 days of incubation for MGA plates.

4.3.4. DON, FB_1 and FB_2 Contamination

Extraction of DON, FB_1 and FB_2

One sample from each plot, storage time and storage temperature was analyzed for its DON, FB_1 and FB_2 content. An amount of 17 g of each sample were ground in a IKA A11 mill for 30 s. Seven grams of ground maize were transferred into a 50 mL Falcon tube for DON analysis, and another 7 g of ground maize were put into another 50 mL Falcon tube for FB_1 and FB_2 analysis.

DON Extraction and Sample Preparation

DON extraction and analysis were based on the study of Borràs-Vallverdú, Ramos, Marín, Sanchis and Rodríguez-Bencomo (2020) [67]. An amount of 1.4 g of NaCl and 40 mL of Milli-Q water were added to the Falcon tube with the ground maize. The mixture was vortexed for 30 s and ultrasound-treated with the Bransonic M2800H-E (Branson Ultrasonic SA, Carouge, Switzerland) at maximum power for 15 min. After that, the Falcon tubes were centrifuged in a Hettich 320R centrifuge (Andreas Hettich GmbH & Co. KG, Tuttlingen, Germany) at $8965 \times g$ for 10 min at 20 °C. The supernatant was vacuum filtered using 90 mm glass microfiber filters (Whatman, Buckinghamshire, UK). DonPrep immunoaffinity columns (Biopharm AG, Darmstadt, Germany) were prepared by adding 10 mL of Milli-Q water. Then, 8 mL of the filtered supernatant was collected and passed through the immunoaffinity column. After that, 1.5 mL of methanol was added to elute the toxin. Backflushing was done three times, and then another 0.5 mL of methanol was passed through the column. The 2 mL of collected methanolic extract was evaporated at 40 °C (Stuart SBH200D/3 block heater, Cole-Parmer©, Staffordshire, UK) under a gentle stream of N_2. The residue was re-suspended in 0.8 mL of MeOH:H_2O 10:90 (*v:v*), vortexed, filtered through 0.22 μm PTFE filters and analyzed by HPLC-DAD according to the following section.

DON HPLC-DAD Analysis

HPLC-DAD determination of DON was performed using an Agilent Technologies 1260 Infinity HPLC system (Santa Clara, CA, USA) coupled with an Agilent 1260 Infinity II Diode Array Detector (DAD). A Phenomenex® Gemini C18 column (Torrance, CA, USA) was used (150 × 4.6 mm, 5 μm particle size, 110 Å pore size). Absorbance reading was performed at 220 nm. Three mobile phases were prepared: phase A (methanol:water 10:90, *v:v*), phase B (acetonitrile:water 20:80, *v:v*) and phase C (100% methanol). The gradient applied was as follows: 0 min 100% A; 10 min 60% A and 40% B; 13 min 60% A and 40% B; 15 min 100% C; 25 min 100% C; 29 min 100% A until 40 min (for re-equilibrating the column). The flow rate was set at 1 mL/min. The column temperature was 40 °C, and the injection volume was 50 µL. DON retention time was 10.0 min. Quantification was carried out by using DON calibration curves prepared in methanol:water 10:90, *v:v*.

LOD and LOQ, considered as three and ten times the signal of the blank, respectively, were 12.6 and 42.0 µg kg^{-1}. Recovery was calculated using artificially DON-contaminated maize, extracting and analysing the mycotoxins as previously stated. Recovery was studied per triplicate at three different DON concentrations: 2.286, 1.143 and 0.571 µg DON kg^{-1} maize. The respective average recoveries and standard deviations were 81.7 ± 9.5, 87.4 ± 13.3 and 91.3 ± 14.5%.

FB$_1$ and FB$_2$ Extraction and Sample Preparation

Fumonisin extraction and analysis were based on the study of Belajova and Rauova (2010) [68]. An amount of 1.4 g of NaCl and 35 mL of H$_2$O:ACN:MeOH 50:25:25 (*v:v:v*) were added to the Falcon tube with the ground maize. The mixture was vortexed for 30 s and ultrasound-treated with the Bransonic M2800H-E (Branson Ultrasonic SA, Carouge, Switzerland) at maximum power for 15 min. After that, the Falcon tubes were centrifuged in a Hettich 320R centrifuge (Andreas Hettich GmbH & Co. KG, Tuttlingen, Germany) at 8965× *g* for 10 min at 20 °C. The supernatant was vacuum filtered using 90 mm glass microfiber filters (Whatman, Buckinghamshire, UK). The solution to be analyzed was prepared by mixing 3.5 mL of the filtered supernatant with 46.5 mL of PBS in another 50 mL Falcon tube. The whole content of the Falcon tube was passed through a Fumoniprep immunoaffinity column (Biopharm AG, Darmstadt, Germany). After that, 1.5 mL of methanol was added to collect the toxin. Backflushing was done three times, and then 1.5 mL of Milli-Q water was passed through the column. The 3 mL of collected solution was evaporated at 40 °C (Stuart SBH200D/3 block heater, Cole-Parmer©, Staffordshire, UK) under a gentle stream of N$_2$. The residue was re-suspended in 0.8 mL of MeOH:H$_2$O 50:50 (*v:v*), vortexed, filtered through 0.22 µm PTFE filters and analyzed by HPLC-FLD according to the following section.

FB$_1$ and FB$_2$ HPLC-FLD Analysis

HPLC-FLD determination of FB$_1$ and FB$_2$ was performed using an Agilent Technologies 1260 Infinity HPLC system (Santa Clara, CA, USA) coupled with an Agilent 1260 Infinity Fluorescence Detector (FLD). A Phenomenex® Kinetex PFP column (Torrance, CA, USA) was used (150 × 4.6 mm, 5 µm particle size, 110 Å pore size). Excitation and emission were performed at 335 and 460 nm, respectively. Three mobile phases were prepared: phase A (acetonitrile), phase B (methanol) and phase C (0.1% acetic acid). The gradient applied was as follows: 0 min 15% A and 85% C; 10 min 5% A, 61% B and 34% C; 14 min 5% A, 61% B and 34% C; 16 min 5% A, 72% B and 23% C; 20 min 15% A and 85% C (for re-equilibrating the column). The flow rate was set at 1.2 mL/min. The column temperature was 40 °C, and the injection volume was 50 µL. FB$_1$ and FB$_2$ retention times were 15 and 17.8 min, respectively. Quantification was carried out by using FB$_1$ and FB$_2$ calibration curves prepared in methanol:water 50:50, *v:v*.

Prior to injection, samples were derivatized. The derivatization mixture (DM) for the analysis of fumonisins was prepared as follows: 40 mg of ortho-phthaldialdehyde was dissolved in 1 mL of methanol and diluted in 10 mL of 0.1 M disodium tetraborate. Then, 50 µL of 2-mercaptoethanol was added and the mixture was vortexed. The prepared mixture was stored in an amber glass vial at 4 °C for a maximum of 7 days. The injector was programmed to draw 37.5 µL of DM and 12.5 µL of the sample to be analyzed, and then we mixed them for 0.3 min before injection.

LOD and LOQ, considered as three and ten times the signal of the blank, were 10.0 and 33.3 µg kg^{-1} for FB$_1$ and 16.0 and 53.3 µg kg^{-1} for FB$_2$, respectively. Recovery was calculated using artificially fumonisin-contaminated maize, extracting and analysing the mycotoxins as previously stated. Recovery was studied per triplicate at three different fumonisin concentrations: 0.855 + 0.855, 0.57 + 0.57 and 0.285 + 0.285 (µg FB$_1$ + µg FB$_2$) kg^{-1} maize. For FB$_1$, the respective average recoveries and standard deviations were 82.1 ± 8.7,

86.8 $\pm$ 9.2 and 77.0 $\pm$ 9.5%. For FB$_2$, those values were 102.6 $\pm$ 21.5, 101.6 $\pm$ 27.7 and 88.6 $\pm$ 27.6%.

4.4. Reagents and Chemicals

DON was from Romer Labs (Tulln, Austria). FB$_1$ and FB$_2$ were from Sigma (St. Louis, MO, USA), ortho-phthaldialdehyde was from Merck (Darmstadt, Germany) and 2-mercaptoethanol was from Scharlau (Sentmenat, Spain). Methanol HPLC grade, acetonitrile HPLC gradient grade and NaCl were from Fisher Scientific UK Limited (Loughborough, UK).

CGA was from Biokar (Barcelona, Spain). MGA was prepared in the laboratory according to Castellá et al. (1997) [69]. Peptone was from Biokar (Barcelona, Spain), KH$_2$PO$_4$ and chloramphenicol were from Scharlau (Sentmenat, Spain) and MgSO$_4\cdot$7 H$_2$O was from Quality chemicals (Esparreguera, Spain). Malachite green (C$_{48}$H$_{50}$N$_4$O$_4\cdot$2C$_2$H$_2$O$_4$) was from Probus (Badalona, Spain) and agar was from Condalab (Torrejón de Ardoz, Spain).

4.5. Statistics

Statistical analyses were carried out using the SPSS program for Windows (version 22) (IBM Corp., Armonk, New York, NY, USA; https://www.ibm.com/es-es/analytics/spss-statistics-software, access on 28 August 2022). The significance level was established at 0.05. Descriptive statistics, Principal Compounds Analysis and multiple-factor ANOVAs were performed. LSD tests were used to evaluate significantly statistical differences among groups in a variable. Graphics were drawn using Microsoft Excel 2013.

Supplementary Materials: The following supporting information can be downloaded at: https://www.mdpi.com/article/10.3390/toxins14090620/s1. Supplementary Document: Effect of Time and Temperature on Maize Moisture, aw, Microbial Counts and Mycotoxin Contamination after Harvest.

Author Contributions: Conceptualization, A.J.R., C.C.-M. and S.M.; methodology, B.B.-V.; validation, B.B.-V.; formal analysis, B.B.-V.; investigation, B.B.-V.; resources, A.J.R., C.C.-M., S.M. and V.S.; writing—original draft preparation, B.B.-V. and J.F.-O.; writing—review and editing, A.J.R., C.C.-M., S.M. and V.S.; visualization, B.B.-V. and J.F.-O.; project administration, A.J.R., C.C.-M. and S.M.; funding acquisition, A.J.R., C.C.-M., S.M. and V.S. All authors have read and agreed to the published version of the manuscript.

Funding: This work is part of the R+D+I project PID2020-114836RB-I00, financed by MCIN/AEI/10.13039/501100011033. B.B.-V. and J.F.-O. have been funded by the FD pre-doctoral fellowship (PRE2018-085278 and PRE2018-084610 respectively) of the Spanish Ministry of Science, Innovation and Universities.

Data Availability Statement: The data presented in this study are available in this article and Supplementary Materials.

Conflicts of Interest: The authors declare no conflict of interest.

References

1. FAO FAOSTAT. Production. Crops and Livestock Products. Available online: https://www.fao.org/faostat/en/#data/QCL (accessed on 10 March 2022).
2. Munkvold, G.P. Cultural and genetic approaches to managing mycotoxins in Maize. *Annu. Rev. Phytopathol.* **2003**, *41*, 99–116. [CrossRef] [PubMed]
3. Tian, Y.; Tan, Y.; Liu, N.; Liao, Y.; Sun, C.; Wang, S.; Wu, A. Functional agents to biologically control deoxynivalenol contamination in cereal grains. *Front. Microbiol.* **2016**, *7*, 00395. [CrossRef]
4. Neme, K.; Mohammed, A. Mycotoxin occurrence in grains and the role of postharvest management as a mitigation strategies. A Review. *Food Control* **2017**, *78*, 412–425. [CrossRef]
5. Folcher, L.; Delos, M.; Marengue, E.; Jarry, M.; Weissenberger, A.; Eychenne, N.; Regnault-Roger, C. Lower mycotoxin levels in Bt Maize grain. *Agron. Sustain. Dev.* **2010**, *30*, 711–719. [CrossRef]
6. Lanza, F.E.; Zambolim, L.; Veras da Costa, R.; Vieira Queiroz, V.A.; Cota, L.V.; Dionísia da Silva, D.; Coelho de Souza, A.G.; Fontes Figueiredo, J.E. Prevalence of fumonisin-producing *Fusarium* species in Brazilian corn grains. *Crop Prot.* **2014**, *65*, 232–237. [CrossRef]
7. Logrieco, A.; Bottalico, A.; Mulé, G.; Moretti, A.; Perrone, G. Epidemiology of toxigenic fungi and their associated mycotoxins for some Mediterranean crops. *Eur. J. Plant Pathol.* **2003**, *109*, 645–667. [CrossRef]

8.	Qu, L.; Wang, L.; Ji, H.; Fang, Y.; Lei, P.; Zhang, X.; Jin, L.; Sun, D.; Dong, H. Toxic mechanism and biological detoxification of fumonisins. *Toxins* **2022**, *14*, 182.

9.	Chu, F.S.; Li, G.Y. Simultaneous occurrence of fumonisin B_1 and other mycotoxins in moldy corn collected from the People's Republic of China in regions with high incidences of esophageal cancer. *Appl. Environ. Microbiol.* **1994**, *60*, 847–852. [CrossRef]

10.	Thiel, P.G.; Marasas, W.F.O.; Sydenham, E.W.; Shephard, G.S.; Gelderblom, W.C.A. The implications of naturally occurring levels of fumonisins in corn for human and animal health. *Mycopathologia* **1992**, *117*, 3–9. [CrossRef]

11.	Ueno, Y. Mode of action of trichothecenes. *Pure Appl. Chem.* **1977**, *49*, 1737–1745.

12.	Pestka, J.J. Deoxynivalenol: Toxicity, Mechanisms and Animal Health Risks. *Anim. Feed Sci. Technol.* **2007**, *137*, 283–298. [CrossRef]

13.	Commission of the European Communities. Regulation (EC) No 1881/2006 of 19 December 2006 setting maximum levels for certain contaminants in foodstuffs. *Off. J. Eur. Union* **2006**, *L364*, 5–24.

14.	Commission of the European Communities. Commission Recommention of 17 August 2006 on the prevention and reduction of *Fusarium* toxins in cereals and cereal products. *Off. J. Eur. Union* **2006**, *L234*, 35–40.

15.	Commission of the European Communities. Commission recommendation of 17 August 2006 on the presence of deoxynivalenol, zearalenone, ochratoxin A, T-2 and HT-2 and fumonisins in products intended for animal feeding. *Off. J. Eur. Union* **2006**, *L229*, 7–9.

16.	Liu, Z.; Zhang, G.; Zhang, Y.; Jin, Q.; Zhao, J., Li, J. Factors controlling mycotoxin contamination in Maize and food in the Hebei province, China. *Agron. Sustain. Dev.* **2016**, *36*, 39. [CrossRef]

17.	Richard, J.L. Some major mycotoxins and their mycotoxicoses—An overview. *Int. J. Food Microbiol.* **2007**, *119*, 3–10. [CrossRef]

18.	Channaiah, L.H.; Maier, D.E. Best Stored Maize Management Practices for the Prevention of Mycotoxin Contamination. In *Mycotoxin Reduction in Grain Chains*; Wiley: Hoboken, NJ, USA, 2014; pp. 78–88.

19.	Cotten, T.K.; Munkvold, G.P. Survival of *Fusarium moniliforme, F. proliferatum,* and *F. subglutinans* in Maize stalk residue. *Phytopathology* **1998**, *88*, 550–555. [CrossRef]

20.	Council for Agricultural Science and Technology. *Mycotoxins—Risks in Plant, Animal and Human Systems*; CAST: Ames, IA, USA, 2003; ISBN 1-887383-22-0.

21.	Edwards, S.G. Influence of agricultural practices on *Fusarium* infection of cereals and subsequent contamination of grain by trichothecene mycotoxins. *Toxicol. Lett.* **2004**, *153*, 29–35. [CrossRef]

22.	Maiorano, A.; Blandino, M.; Reyneri, A.; Vanara, F. Effects of Maize residues on the *Fusarium* spp. Infection and deoxynivalenol (DON) contamination of wheat grain. *Crop Prot.* **2008**, *27*, 182–188. [CrossRef]

23.	Rossi, V.; Scandolara, A.; Battilani, P. Effect of environmental conditions on spore production by *Fusarium verticillioides*, the causal agent of maize ear rot. *Eur. J. Plant Pathol.* **2008**, *123*, 159–169. [CrossRef]

24.	Mansfield, M.A.; De Wolf, E.D.; Kuldau, G.A. Relationships between weather conditions, agronomic practices, and fermentation characteristics with deoxynivalenol content in fresh and ensiled Maize. *Plant Dis.* **2005**, *89*, 1151–1157. [CrossRef]

25.	Dill-Macky, R.; Jones, R.K. The effect of previous crop residues and tillage on fusarium head blight of wheat. *Plant Dis.* **2000**, *84*, 71–76. [CrossRef]

26.	Obst, A.; Lepschy-Von Gleissenthall, J.; Beck, R. On The etiology of *Fusarium* head blight of wheat in South Germany—Preceding crops, weather conditions for inoculum production and head infection, proneness of the crop to infection and mycotoxin production. *Cereal Res. Commun.* **1997**, *25*, 699–703. [CrossRef]

27.	Schöneberg, T.; Martin, C.; Wettstein, F.E.; Bucheli, T.D.; Mascher, F.; Bertossa, M.; Musa, T.; Keller, B.; Vogelgsang, S. *Fusarium* and mycotoxin spectra in Swiss barley are affected by various cropping techniques. *Food Addit. Contam. Part A* **2016**, *33*, 1608–1619. [CrossRef]

28.	Roucou, A.; Bergez, C.; Méléard, B.; Orlando, B. An agro-climatic approach to developing a national prevention tool for deoxynivalenol in French Maize-growing areas. *Toxins* **2022**, *14*, 74. [CrossRef]

29.	Supronienė, S.; Mankevičienė, A.; Kadžienė, G.; Kačergius, A.; Feiza, V.; Feizienė, D.; Semaškienė, R.; Dabkevičius, Z.; Tamošiūnas, K. The impact of tillage and fertilization on *Fusarium* infection and mycotoxin production in wheat grains. *Žemdirbystė* **2012**, *99*, 265–272.

30.	Kaukoranta, T.; Hietaniemi, V.; Rämö, S.; Koivisto, T.; Parikka, P. Contrasting responses of T-2, HT-2 and DON mycotoxins and *Fusarium* species in oat to climate, weather, tillage and cereal intensity. *Eur. J. Plant Pathol.* **2019**, *155*, 93–110. [CrossRef]

31.	Ramos, M.C.; Pareja-Sánchez, E.; Plaza-Bonilla, D.; Cantero-Martínez, C.; Lampurlanés, J. Soil sealing and soil water content under no-tillage and conventional tillage in irrigated corn: Effects on grain yield. *Hydrol. Process.* **2019**, *33*, 2095–2109. [CrossRef]

32.	Pareja-Sánchez, E.; Plaza-Bonilla, D.; Ramos, M.C.; Lampurlanés, J.; Álvaro-Fuentes, J.; Cantero-Martínez, C. Long-term no-till as a means to maintain soil surface structure in an agroecosystem transformed into irrigation. *Soil Tillage Res.* **2017**, *174*, 221–230. [CrossRef]

33.	Arjmand Sajjadi, S.; Mahmoodabadi, M. Aggregate breakdown and surface seal development influenced by rain intensity, slope gradient and soil particle size. *Solid Earth* **2015**, *6*, 311–321. [CrossRef]

34.	Awadhwal, N.K.; Thierstein, G.E. Soil crust and its impact on crop establishment: A review. *Soil Tillage Res.* **1985**, *5*, 289–302. [CrossRef]

35.	Ramos, M.C.; Nacci, S.; Pla, I. Soil sealing and its influence on erosion rates for some soils in the Mediterranean area. *Soil Sci.* **2000**, *165*, 398–403. [CrossRef]

36. Manstretta, V.; Rossi, V. Effects of weather variables on ascospore discharge from *Fusarium graminearum* perithecia. *PLoS ONE* **2015**, *10*, e0138860. [CrossRef]

37. Sutton, J.C. Epidemiology of wheat head blight and Maize ear rot Caused by *Fusarium graminearum*. *Can. J. Plant Pathol.* **1982**, *4*, 195–209. [CrossRef]

38. Parry, D.W.; Jenkinson, P.; McLeod, L. *Fusarium* ear blight (Scab) in Small Grain Cereals—A review. *Plant Pathol.* **1995**, *44*, 207–238. [CrossRef]

39. Nieminen, M.; Ketoja, E.; Mikola, J.; Terhivuo, J.; Sirén, T.; Nuutinen, V. Local land use effects and regional environmental limits on earthworm communities in Finnish arable landscapes. *Ecol. Appl.* **2011**, *21*, 3162–3177. [CrossRef]

40. Briones, M.J.I.; Schmidt, O. Conventional tillage decreases the abundance and biomass of earthworms and alters their community structure in a global meta-analysis. *Glob. Chang. Biol.* **2017**, *23*, 4396–4419. [CrossRef]

41. Ojha, R.B.; Devkota, D. Earthworms: "Soil and ecosystem engineers"—A review. *World J. Agric. Res.* **2014**, *2*, 257–260. [CrossRef]

42. Imaz, M.J.; Virto, I.; Bescansa, P.; Enrique, A.; Fernandez-Ugalde, O.; Karlen, D.L. Soil quality indicator response to tillage and residue management on semi-arid Mediterranean cropland. *Soil Tillage Res.* **2010**, *107*, 17–25. [CrossRef]

43. Chan, K.Y. An overview of some tillage impacts on earthworm population abundance and diversity—Implications for functioning in soils. *Soil Tillage Res.* **2001**, *57*, 179–191. [CrossRef]

44. Kladivko, E.J. Tillage systems and soil ecology. *Soil Tillage Res.* **2001**, *61*, 61–76. [CrossRef]

45. Oldenburg, E.; Kramer, S.; Schrader, S.; Weinert, J. Impact of the Earthworm *Lumbricus terrestris* on the degradation of fusarium-infected and deoxynivalenol-contaminated wheat straw. *Soil Biol. Biochem.* **2008**, *40*, 3049–3053. [CrossRef]

46. Schrader, S.; Kramer, S.; Oldenburg, E.; Weinert, J. Uptake of deoxynivalenol by earthworms from *Fusarium*-infected wheat straw. *Mycotoxin Res.* **2009**, *25*, 53–58. [CrossRef] [PubMed]

47. Santiveri Morata, F.; Cantero-Martínez, C.; Ojeda Domínguez, L.; Angás Pueyo, P. Técnicas de laboreo del suelo en zonas de secano semi-árido. *Agric. Rev. Agropecu. Ganad.* **2004**, *866*, 724–729.

48. Bockus, W.W.; Shroyer, J.P. The impact of reduced tillage on soilborne plant pathogens. *Annu. Rev. Phytopathol.* **1998**, *36*, 485–500. [CrossRef]

49. Dwyer, L.M.; Ma, B.; Stewart, D.W.; Hayhoe, H.; Balchin, D.; Culley, J.L.B.; Mcgovern, M. Root mass distribution under conventional and conservation tillage. *Can. J. Soil Sci.* **1996**, *76*, 23–28.

50. Ono, E.; Moreno, E.; Ono, M.; Rossi, C.; Saito, G.; Vizoni, É.; Sugiura, Y.; Hirooka, E. Effect of cropping systems and crop successions on fumonisin levels in corn from Northern Paraná State, Brazil. *Eur. J. Plant Pathol.* **2011**, *131*, 653–660. [CrossRef]

51. Ariño, A.; Herrera, M.; Juan, T.; Estopañán, G.; Carramiñana, J.; García, C.; Herrera, A. Influence of agricultural practices on the contamination of Maize by fumonisin mycotoxins. *J. Food Prot.* **2009**, *72*, 898–902. [CrossRef]

52. Marocco, A.; Gavazzi, C.; Pietri, A.; Tabaglio, V. On Fumonisin incidence in monoculture Maize under no-till, conventional tillage and two nitrogen fertilisation levels. *J. Sci. Food Agric.* **2008**, *88*, 1217–1221. [CrossRef]

53. Marocco, A.; Tabaglio, V.; Pietri, A.; Gavazzi, C. Monoculture Maize (*Zea mays L.*) cropped under conventional tillage, no-tillage and n fertilization: (II) Fumonisin incidence on kernels. *Ital. J. Agron.* **2009**, *4*, 69–75. [CrossRef]

54. Battilani, P.; Pietri, A.; Barbano, C.; Scandolara, A.; Bertuzzi, T.; Marocco, A. Logistic regression modeling of cropping systems to predict fumonisin contamination in Maize. *J. Agric. Food Chem.* **2008**, *56*, 10433–10438. [CrossRef]

55. Madege, R.R.; Audenaert, K.; Kimanya, M.; Tiisekwa, B.; De Meulenaer, B.; Bekaert, B.; Landschoot, S.; Haesaert, G. Control of *Fusarium verticillioides* (Sacc.) nirenberg and fumonisins by using a combination of crop protection products and fertilization. *Toxins* **2018**, *10*, 67. [CrossRef] [PubMed]

56. Blandino, M.; Vanara, F.; Reyneri, A. Influence of nitrogen fertilization on mycotoxin contamination of Maize kernels. *Crop Prot.* **2008**, *27*, 222–230. [CrossRef]

57. Carbas, B.; Simões, D.; Soares, A.; Freitas, A.; Ferreira, B.; Carvalho, A.R.F.; Silva, A.S.; Pinto, T.; Diogo, E.; Andrade, E.; et al. Occurrence of *Fusarium* spp. in Maize grain harvested in Portugal and accumulation of related mycotoxins during storage. *Foods* **2021**, *10*, 375. [CrossRef]

58. Cao, A.; Santiago, R.; Ramos, A.J.; Marín, S.; Reid, L.M.; Butrón, A. Environmental factors related to fungal infection and fumonisin accumulation during the development and drying of white Maize kernels. *Int. J. Food Microbiol.* **2013**, *164*, 15–22. [CrossRef] [PubMed]

59. Ferrigo, D.; Raiola, A.; Causin, R. *Fusarium* Toxins in cereals: Occurrence, legislation, factors promoting the appearance and their management. *Molecules* **2016**, *21*, 627. [CrossRef]

60. Daou, R.; Joubrane, K.; Maroun, R.G.; Khabbaz, L.R.; Ismail, A.; El Khoury, A. Mycotoxins: Factors influencing production and control strategies. *AIMS Agric. Food* **2021**, *6*, 416–447. [CrossRef]

61. Lanza, F.E.; Zambolim, L.; Costa, R.V.; Figueiredo, J.E.F.; Silva, D.D.; Queiroz, V.A.V.; Guimarães, E.A.; Cota, L.V. Symptomatological aspects associated with fungal incidence and fumonisin levels in corn kernels. *Trop. Plant Pathol.* **2017**, *42*, 304–308. [CrossRef]

62. Comerio, R.M.; Fernández Pinto, V.E.; Vaamonde, G. Influence of water activity on deoxynivalenol accumulation in wheat. *Mycotoxin Res.* **1999**, *15*, 24–32. [CrossRef]

63. Ramirez, M.L.; Chulze, S.; Magan, N. Temperature and water activity effects on growth and temporal deoxynivalenol production by two argentinean strains of *Fusarium graminearum* on irradiated wheat grain. *Int. J. Food Microbiol.* **2006**, *106*, 291–296. [CrossRef] [PubMed]

64. Schmidt-Heydt, M.; Parra, R.; Geisen, R.; Magan, N. Modelling the relationship between environmental factors, transcriptional genes and deoxynivalenol mycotoxin production by strains of two *Fusarium* Species. *J. R. Soc. Interface* **2011**, *8*, 117–126. [CrossRef] [PubMed]
65. Soil Science Division Staff. *Soil Survey Manual*; Ditzler, C., Scheffe, K., Monger, H.C., Eds.; USDA Handbook No 18; Government Printing Office: Washington, DC, USA, 2017.
66. Dastane, N.G. *Effective Rainfall in Irrigated Agriculture*; Irrigation and Drainage Paper; FAO: Rome, Italy, 1978; Volume 25, ISBN 925100272X.
67. Borràs-Vallverdú, B.; Ramos, A.J.; Marín, S.; Sanchis, V.; Rodríguez-Bencomo, J.J. Deoxynivalenol degradation in wheat kernels by exposition to ammonia vapours: A tentative strategy for detoxification. *Food Control* **2020**, *118*, 107444. [CrossRef]
68. Belajova, E.; Rauova, D. Single laboratory-validated HPLC methods for determination of ochratoxin A, fumonisin B_1 and B_2, zearalenone and deoxynivalenol in cereals and cereal-based foods. *J. Food Nutr. Res.* **2010**, *49*, 57–68.
69. Castellá, G.; Bragulat, M.R.; Rubiales, M.V.; Cabañes, F.J. Malachite green agar, a new Selective medium for *Fusarium* spp. *Mycopathologia* **1997**, *137*, 173–178. [CrossRef]

Article

The Neurotoxic Effect of Ochratoxin-A on the Hippocampal Neurogenic Niche of Adult Mouse Brain

Eva Mateo [1,†], Rik Paulus Bernardus Tonino [2,†], Antolin Canto [3,†], Antonio Monroy Noyola [4], Maria Miranda [3], Jose Miguel Soria [3,*] and María Angeles Garcia Esparza [5,*]

1 Department of Microbiology and Ecology, School of Medicine and Dentistry, University of Valencia, 46001 Valencia, Spain
2 Haematology Department, Leiden University, 1043 AJ Leiden, The Netherlands
3 Department of Biomedical Sciences, Cardenal Herrera University-CEU Universities, 46001 Valencia, Spain
4 Neuroprotection Laboratory, Faculty of Pharmacy, Autonomous University of the State of Morelos, Cuernavaca 98100, Morelos, Mexico
5 Department of Pharmacy, Cardenal Herrera University-CEU Universities, 46001 Valencia, Spain
* Correspondence: jose.soria@uchceu.es (J.M.S.); maria.garcia2@uchceu.es (M.A.G.E.)
† These authors contributed equally to this work.

Abstract: Ochratoxin A (OTA) is a common secondary metabolite of *Aspergillus ochraceus, A. carbonarius*, and *Penicillium verrucosum*. This mycotoxin is largely present as a contaminant in several cereal crops and human foodstuffs, including grapes, corn, nuts, and figs, among others. Preclinical studies have reported the involvement of OTA in metabolic, physiologic, and immunologic disturbances as well as in carcinogenesis. More recently, it has also been suggested that OTA may impair hippocampal neurogenesis in vivo and that this might be associated with learning and memory deficits. Furthermore, aside from its widely proven toxicity in tissues other than the brain, there is reason to believe that OTA contributes to neurodegenerative disorders. Thus, in this present in vivo study, we investigated this possibility by intraperitoneally (i.p.) administering 3.5 mg OTA/kg body weight to adult male mice to assess whether chronic exposure to this mycotoxin negatively affects cell viability in the dentate gyrus of the hippocampus. Immunohistochemistry assays showed that doses of 3.5 mg/kg caused a significant and dose-dependent reduction in repetitive cell division and branching (from 12% to 62%). Moreover, the number of countable astrocytes ($p < 0.001$), young neurons ($p < 0.001$), and mature neurons ($p < 0.001$) negatively correlated with the number of i.p. OTA injections administered (one, two, three, or six repeated doses). Our results show that OTA induced adverse effects in the hippocampus cells of adult mice brain tissue when administered in cumulative doses.

Keywords: ochratoxin A; brain; hippocampus; neurogenic niche; neurotoxicity; cell morphology

Key Contribution: Our results suggest that OTA may impair hippocampal neurogenesis in vivo and induce adverse effects in the hippocampus cells of adult mice brain tissue when administered in cumulative doses.

Citation: Mateo, E.; Tonino, R.P.B.; Canto, A.; Monroy Noyola, A.; Miranda, M.; Soria, J.M.; Garcia Esparza, M.A. The Neurotoxic Effect of Ochratoxin-A on the Hippocampal Neurogenic Niche of Adult Mouse Brain. *Toxins* **2022**, *14*, 624. https://doi.org/10.3390/toxins14090624

Received: 1 July 2022
Accepted: 31 August 2022
Published: 6 September 2022

Publisher's Note: MDPI stays neutral with regard to jurisdictional claims in published maps and institutional affiliations.

1. Introduction

The mycotoxin ochratoxin A (OTA), or 7-carboxy 5-chloro-8-hydroxy-3, 4 dihydro-3-R-methylcoumarin-7-L-β-phenylalanine (Figure 1), is a common metabolite produced by *Aspergillus ochraceus, A. carbonarius*, and *Penicillium verrucosum* [1].

OTA is rapidly absorbed and distributed but slowly eliminated and excreted, leading to potential accumulation in the body, which is due mainly to its binding to plasma proteins and a low metabolism rate. Plasma half-lives range from several days in rodents and pigs to several weeks in nonhuman primates and humans [2,3].

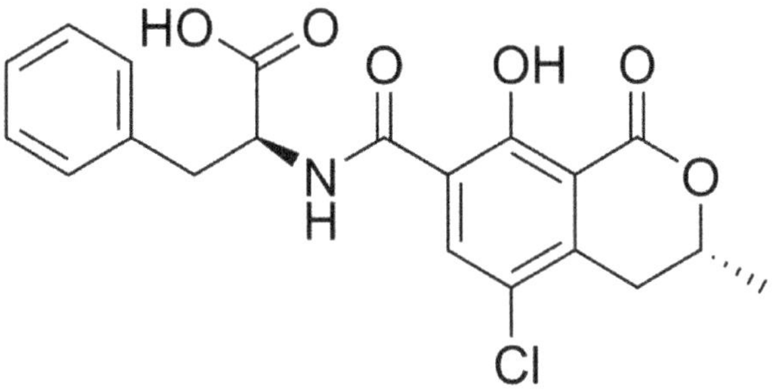

Figure 1. The chemical structure of ochratoxin A.

OTA has been found in barley, oats, rye, wheat, cereals, grains, beans, corn, spices, and products such as coffee, grape juice, wine, beer, and bread [4,5]. In vitro and in vivo studies have shown that this toxin induces immune toxicity, hepatotoxicity, nephrotoxicity, and reproductive and developmental toxicity [6].

Thus, previous studies have described that severe health hazards are associated with mycotoxin exposure, their molecular signaling pathways and processes of toxicity, and their genotoxic and cytotoxic effects on humans and animals [7]. For humans, however, hazard identification has been more difficult. Several adverse human health effects, including the kidney diseases Balkan endemic nephropathy (BEN) and chronic interstitial nephropathy (CIN), have been associated with exposure to OTA; however, these associations have thus far been less conclusive than those for OTA-associated adverse effects in laboratory animal studies [8].

Furthermore, because it causes damage at the molecular level (including causing single-strand DNA breaks, covalent DNA adduction, and DNA oxidation), it has been reported as a possible xenobiotic carcinogen in humans [9–11] and as a teratogenic agent in several laboratory and farm animals, including rats, mice, hamsters, quails, and chickens [12].

Several experimental studies have suggested that OTA also has a neurotoxic effect in humans. Indeed, OTA, as well as its non-toxic corresponding metabolite, ochratoxin α (OTα) [13], has been identified and quantified in human blood and urine [14–17]. In turn, studies in rat and porcine brain capillary endothelial cells have shown its permeability, allowing it to reach the brain [17,18]. However, the cellular mechanisms responsible for the neurotoxicity of OTA have not been clearly elucidated, although oxidative stress, DNA damage, and mitochondrial dysfunction are plausible possibilities [19–22]. These toxic mechanisms are implicated in neurodegenerative disorders such as Alzheimer's and Parkinson's disease, and thus, OTA is considered an environmental factor that could contribute to the development of oxidative stress and neurodegeneration [21].

Sava et al. [23] suggested that OTA impairs hippocampal neurogenesis in vivo and that this might be associated with learning [24] and memory deficits [23,25]. The neurotoxicity of OTA has been shown to be most pronounced in the ventral mesencephalon, hippocampus, and striatum brain regions [26], even when its bioconcentration was significantly lower than in other regions of the brain. In addition, the half-life of OTA in the hippocampus was found to be only 42.5 h [27]. In turn, the hippocampus (a primary site of neurodegeneration in Alzheimer's disease) exhibited relatively low OTA levels with concurrently pronounced OTA neurotoxicity [26]. In this context, it has been hypothesized that low-level exposure to OTA may exert delayed neurotoxic effects which could, in turn, contribute to the development of neurodegenerative disorders.

After the tolerable daily intake dose of OTA was set to 3 ng/kg of body weight, the EU Scientific Committee on Food recommended a reduction in human exposure as low as reasonably possible due to concerns regarding the potential genotoxicity and carcinogenicity of this mycotoxin [1,3]. Extensive data have been published on the toxic effects of OTA, but very little has been published regarding long-term in vivo hippocampal exposure to this mycotoxin. In this context, it would be useful to elucidate the neurotoxicity effects in long-term research. Short-term in vitro studies with high doses have shown that neural stem/progenitor cells are vulnerable to OTA [28], but data on its long-term neurotoxic effect in vivo are still lacking. Such long-term studies might show whether physiological levels of OTA exert neurotoxicity through oxidative stress. Thus, the aim of this study was to perform an 18-day in vivo toxicity assay to investigate the long-term effects of OTA exposure on neurogenesis in the hippocampal dental gyrus of adult mouse brains.

2. Results

2.1. Ochratoxin A Reduces the Number of Astrocytes in the Dentate Gyrus

Twenty-one animals were treated with increasing accumulative doses of 3.5 mg of OTA (n = 17) or vehicle (n = 4) per kilogram of body weight and were sacrificed 3 days after the last scheduled injection. Immunofluorescence staining for GFAP was performed to determine whether the OTA had influenced the number of astrocytes present in the hippocampus (Figure 2). Astrocytes (type B cells) are precursors to progenitor cells (type C cells), which in turn are precursors of migrating neuroblasts (type A cells). Furthermore, astrocyte activity is crucial for the correct formation and function of the blood−brain barrier, which are vital for providing the appropriate environment for proper neuronal functioning and protecting the central nervous system from injury and disease.

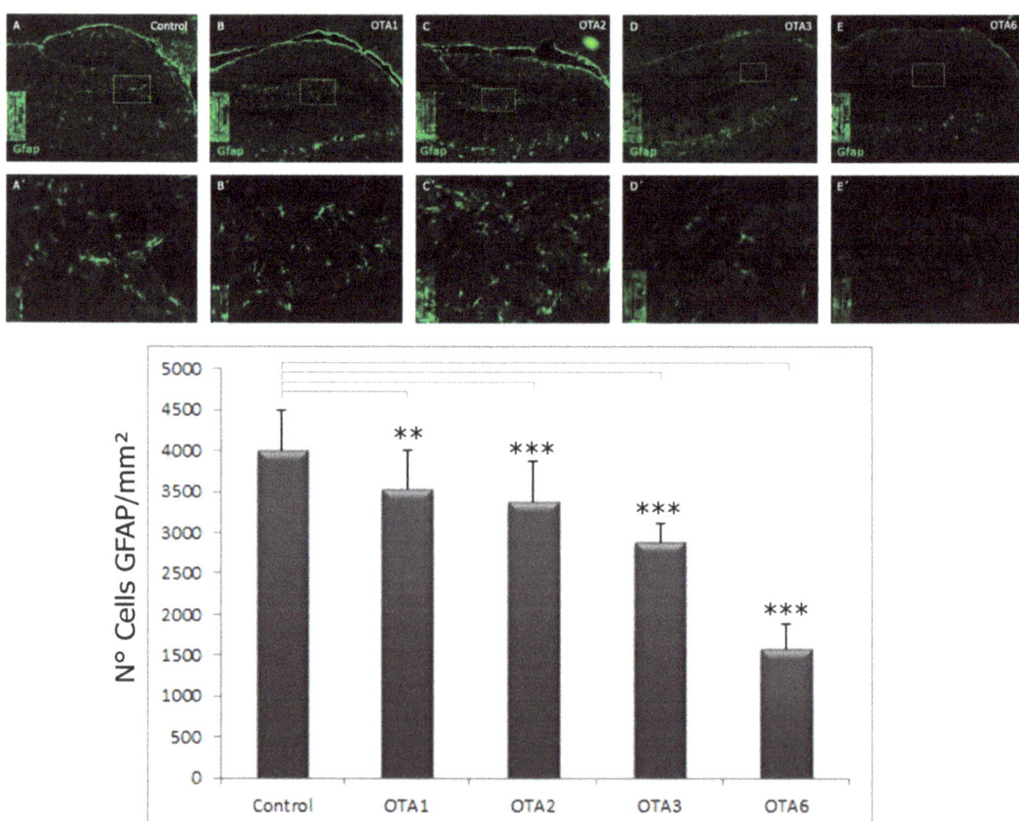

Figure 2. Photomicrographs of GFAP-labeled cells in the control, OTA1, OTA2, OTA3, and OTA6 groups (which received one to six injections of ochratoxin A, respectively), taken at 20× magnification. Compared to the control, the number of GFAP-positive cells decreased as the number of ochratoxin A treatments increased. (Control n = 4, OTA1 n = 4, OTA2 n = 4, OTA3 n = 4, OTA6 n = 5). ** = $p < 0.01$, and *** = $p < 0.001$.

As shown in Figure 2, compared to the controls, we found a significant decrease in GFAP cell expression in the dentate gyrus after treatment with OTA. The correlation between the number of doses and the decrease in GFAP-labeled cells indicated that this effect was dose-dependent. After six cumulative doses, there was a dramatic decrease (by more than 60%) in countable cells compared to the controls, whereas one injection of OTA only decreased the number of countable cells by 11.8% (control = 3996 cells/mm^2; OTA1 = 3523 cells/mm^2; OTA6 = 1591 cells/mm^2). Compared to the controls, there were significantly fewer countable cells in the OTA1 ($p < 0.01$) and OTA2, OTA3, and OTA6 ($p < 0.001$). All groups differed significantly from one another ($p < 0.01$). No significant differences were found between OTA1 and OTA2.

Furthermore, we observed an apparent morphological change in the cell structure, as shown in Figure 3. Astrocytic processes seemed less profound in treated animals, and as the number of doses increased, the astrocytes became difficult to morphologically identify. The statistical significance of this change compared to the control was calculated using one-way ANOVA with a post-hoc LSD test. Figure 4 shows the results obtained in the quantitative morphological astrocyte study. The astrocytes from the control animals were significantly longer and more ramified than those from animals that received three or six OTA doses. The administration of two doses of OTA did not affect the number of branches per cell or the branch lengths. These results agree with those obtained for GFAP expression (Figure 2) and may confirm the possible neurotoxic effects of OTA.

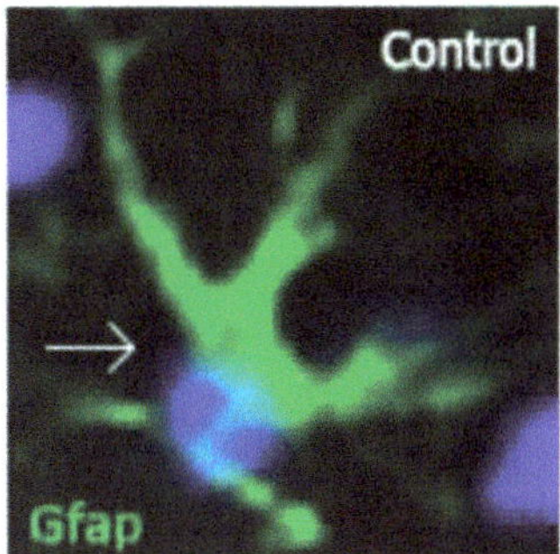

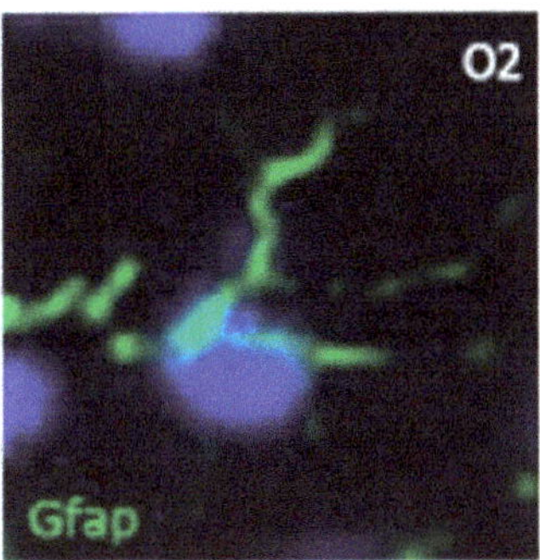

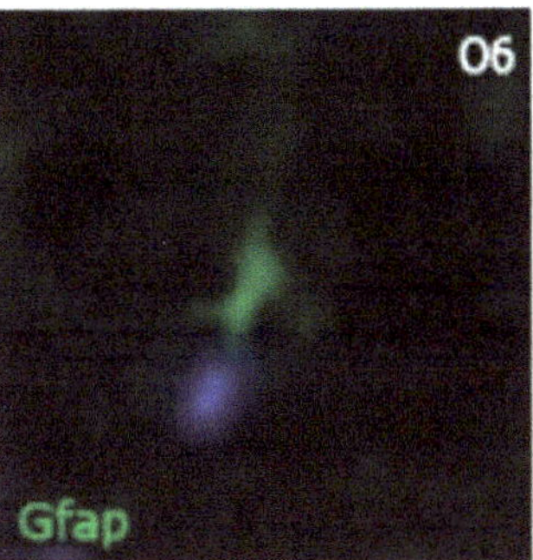

Figure 3. Individual GFAP-labeled cells in the control group and after two (O2) or six (O6) doses of ochratoxin A. Note how the morphology of these cells changed as the number of doses increased, with the cellular body appearing to decrease in volume and the cytoplasmic processes retracting. These photomicrographs were obtained at a 40× magnification.

2.2. Ochratoxin A Reduces the Number of Young Neurons in the Dentate Gyrus

Once we determined that OTA negatively affected astrocytes, we examined the vulnerability of young neurons (type A cells) to this mycotoxin. Counting the DCX-stained cells in the DG revealed a significant difference in all the groups compared to the control ($p < 0.001$; Table 1 and Figure 5). However, the differences were less clear for astrocytes. After one and six doses of OTA, 16% and 38.7% fewer cells were counted, respectively (control = 3825, OTA1 = 3217, and OTA6 = 234 cells/mm^2). Again, no significant difference was detected when comparing OTA1 and OTA2. Nonetheless, OTA3 and OTA6 significantly differed from OTA1 and OTA2, while OTA6 had significantly fewer DCX-positive cells ($p < 0.05$) compared to OTA3. In terms of astrocytes, small morphological changes, albeit fewer evident alterations, were observed for the young neurons (Figure 6). In addition, there was also an interesting difference in the fluorescence, with the overall intensity and single-cell level fluorescence being reduced. We calculated the statistical significance compared to the control using one-way ANOVA, applying LSD as the post-hoc test. These data are expressed as the mean ± standard error.

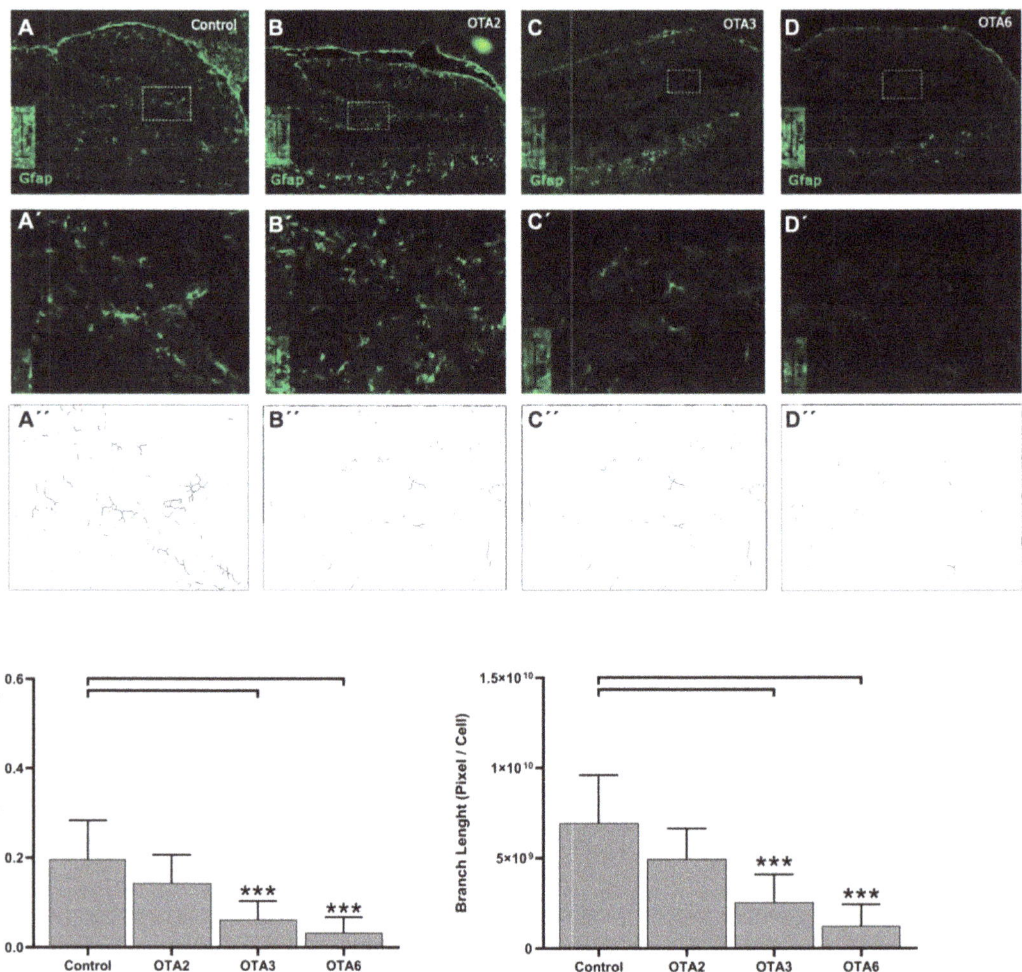

Figure 4. Photomicrographs of GFAP-labeled cells in the control, OTA2, OTA3, and OTA6 groups (which received two, three, or six ochratoxin A injections, respectively) at a $20\times$ magnification Compared to the control: *** = $p < 0.001$. An example of the branch quantification and the branches per cell and branch lengths for each photo are also shown. (Control n = 4, OTA2 n = 4, OTA3 n = 4, OTA6 n = 5).

Table 1. Number of DCX and DAPI co-labeled cells per square millimeter in the controls and OTA1, OTA2, OTA3, and OTA6 groups, which received one to six doses of ochratoxin-A, respectively. Statistical analysis was one-way ANOVA with a post-hoc LSD test. (Control n = 4, OTA1 n = 4, OTA2 n = 4, OTA3 n = 4, OTA6 n = 5).

	Control	OTA1	OTA2	OTA3	OTA6
N = cells/mm^2	4	4	4	4	5
	3825 + 182	3217 + 1445	2999 + 175	2686 + 99	2343 + 154
Significance compared to the control	-	<0.001	<0.001	<0.001	<0.001

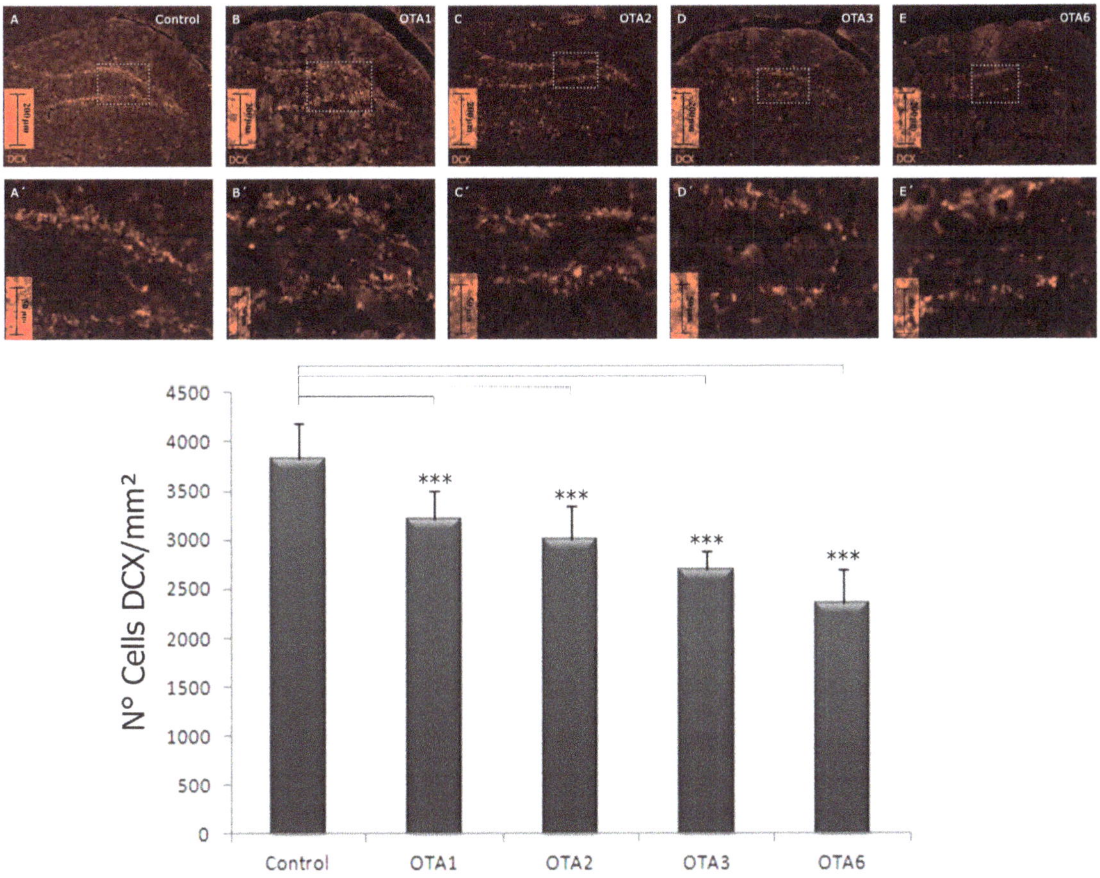

Figure 5. Photomicrographs of DCX-labeled cells in the dentate gyrus of the control, OTA1, OTA2, OTA3, and OTA6 groups (which received one to six injections with ochratoxin A, respectively) at a magnification of 20×. Compared to the control, the number of DCX-positive cells decreased as the number of ochratoxin A treatments increased: *** = $p < 0.001$. Games–Howell statistical analysis. (Control n = 4, OTA1 n = 4, OTA2 n = 4, OTA3 n = 4, OTA6 n = 5).

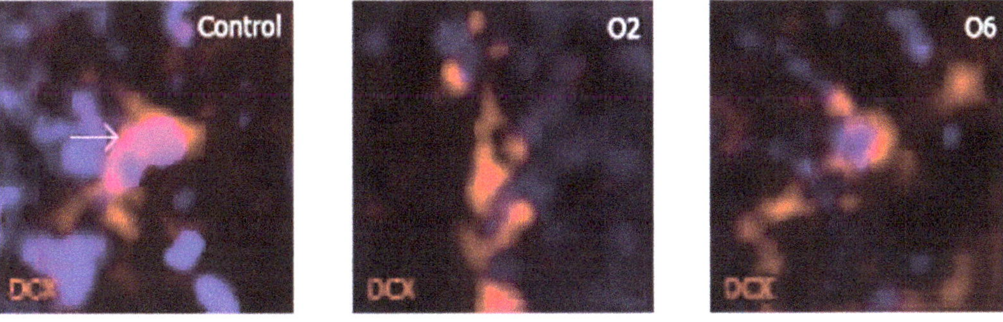

Figure 6. Photomicrographs of individual DCX-labeled young neurons demonstrating small morphological changes as the doses of ochratoxin A increased. The sizes of the cells appear to have decreased compared to the control group, although these changes do not appear to be as prominent as with the GFAP-labeled cells.

2.3. Ochratoxin A Reduces the Number of Mature Neurons in the Dentate Gyrus

Given the finding that OTA reduces young neurons and astrocytes, which are both types of cells that can proliferate, we examined whether mature neurons were also influenced by OTA, with significant differences found in all the groups ($p < 0.001$). Notably, the decrease in the countable mature neurons was more striking than the decrease in young neurons and astrocytes: 31.6% and 62.4% of the labeled cells had perished after 1 and 6 doses of OTA, respectively (control = 7990, OTA1 = 5466, and OTA6 = 3006 cells/mm^2; Table 2 and Figure 7).

Table 2. Numbers of MAP2 and DAPI co-labeled cells per square millimeter in the controls and OTA1, OTA2, OTA3, and OTA6 groups, which received one to six doses of ochratoxin-A, respectively. Statistical analysis was one-way ANOVA with a post-hoc Games−Howell test. (Control n = 4, OTA1 n = 4, OTA2 n = 4, OTA3 n = 4, OTA6 n = 5).

	Control	OTA1	OTA2	OTA3	OTA6
N = cells/mm^2	4	4	4	4	5
	7990 ± 426	5466 ± 488	5247 ± 297	4095 ± 617	3006 ± 694
Significance compared to the control	-	<0.001	<0.001	<0.001	<0.001

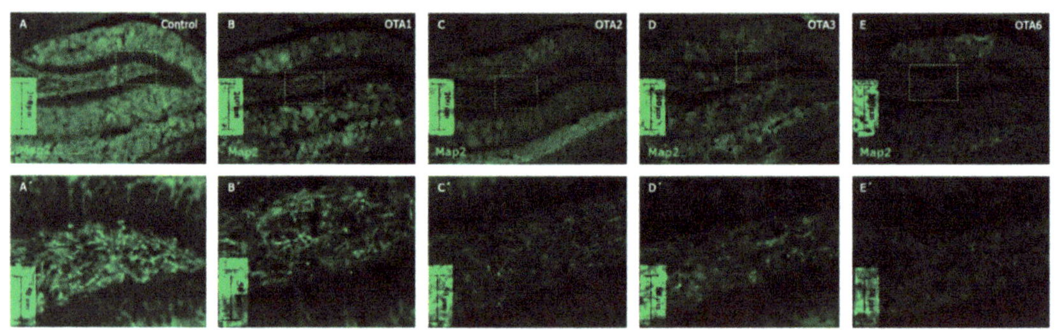

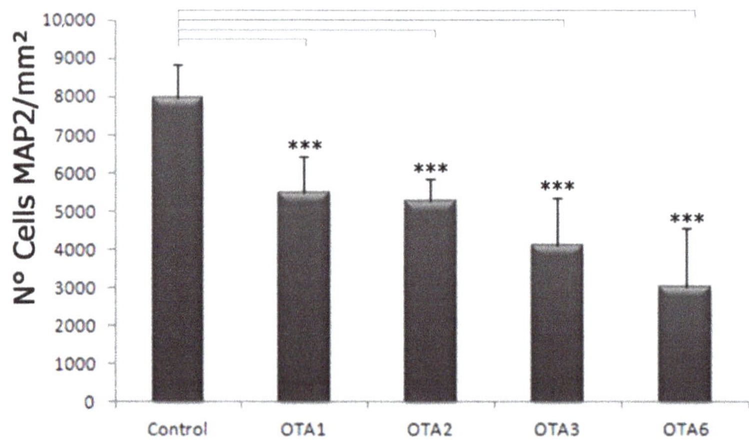

Figure 7. Photomicrographs of MAP2-labeled cells in the dentate gyrus of the control, O1, O2, O3, and O6 groups taken at a 20× magnification. Compared to the control, the number of MAP2-positive cells decreased as the number of ochratoxin A treatments increased. *** = $p < 0.001$). Statistical analysis was one-way ANOVA with a post-hoc LSD test. (Control n = 4, OTA1 n = 4, OTA2 n = 4, OTA3 n = 4, OTA6 n = 5).

Interestingly, we did not find any statistically significant differences when comparing OTA1 to OTA2 and OTA3 to OTA6. This might indicate that the acute mature neuron response to OTA is more prominent than the chronic effect or that there is a group of neurons that perish at low concentrations of OTA while others are not that vulnerable to this mycotoxin. Another interesting finding was that in the case of MAP2 labeling, apart from the changed morphology (Figure 7), the fluorescence intensity was also lower than for DCX or GFAP labeling, as shown in Figure 8. Whereas in the control the fluorescence intensity was high, in OTA6, even the best-labeled cells were hard to identify. The statistical significance of the fluorescence intensity compared to the control group was calculated using a Games–Howell test (no-homogeneous variance post-hoc test), and the data are expressed as the mean ± standard error.

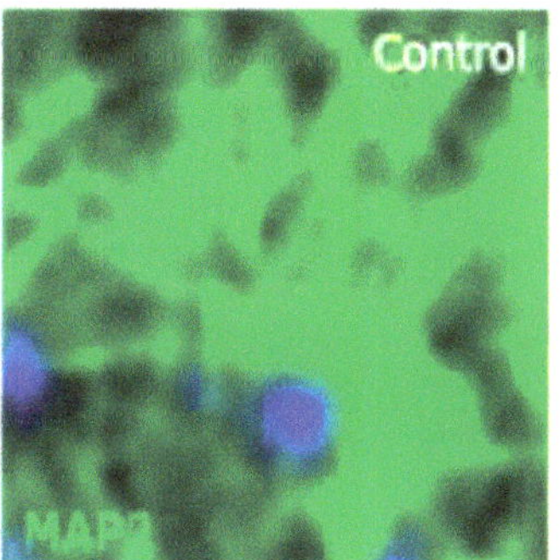

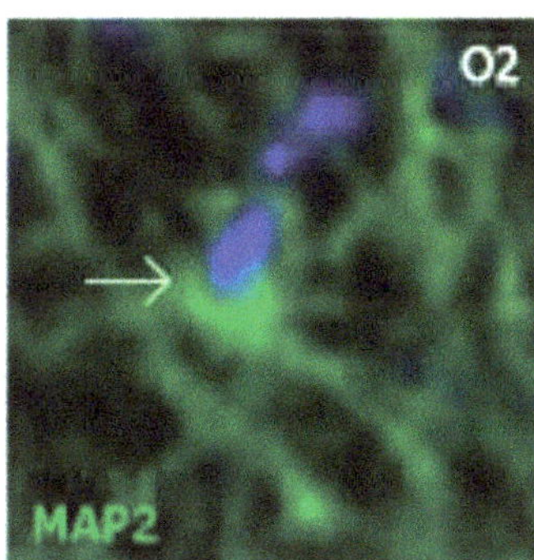

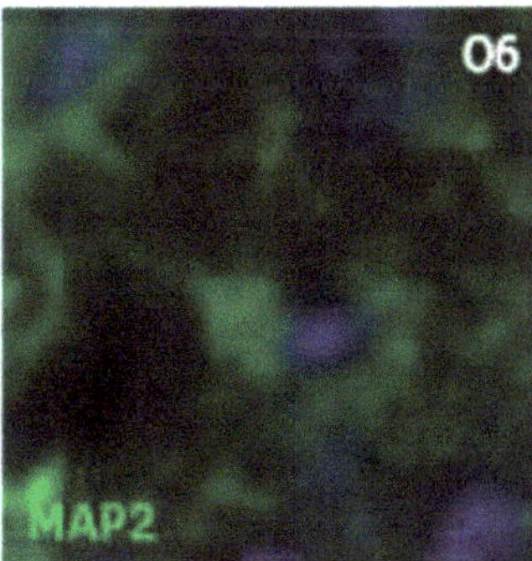

Figure 8. Photomicrographs of individual MAP2-labeled mature neurons and DAPI-labeled nuclei demonstrating their changing morphology as the number of doses of ochratoxin A increased. Whereas many dendrites and cells were present in the control, no comparable structures were present in the O6 group (6 doses of ochratoxin A). Very few dendrites could be found, and the cell body fluorescence was strongly decreased in the O6 group. The images were taken at a 40× magnification and digitally magnified to 400×. The arrow indicates a Map2/DAPI co-labeled cell.

3. Discussion

The mycotoxin OTA, which is found in many foodstuffs [4,5], produces a toxic effect in various organisms. It has been described as carcinogenic, nephro-, immuno-, and genotoxic, is related to increased oxidative stress [6], and is considered neurotoxic [25–27,29,30]. For humans, however, hazard identification has been more difficult. Several adverse human health effects, including the kidney diseases Balkan endemic nephropathy (BEN) and chronic interstitial nephropathy (CIN), have been associated with exposure to OTA; however, these associations have thus far been less conclusive than those for OTA-associated adverse effects in laboratory animal studies [8].

The relationship between OTA and neurotoxicity has been established in vitro [19,31,32]. However, very little has been published about the neurotoxic effects of OTA in in vivo pre-clinical studies.

In this current study, we demonstrated how OTA can have a significant detrimental impact on the hippocampus of adult mice. In vivo OTA significantly affected astrocytes (neuronal stem cells) and young and mature neurons after one dose (3.5 mg/kg body weight). Fewer GFAP/DAPI co-labeled cells were countable per square millimeter, with a decrease of 11.8% after one dose of OTA and up to 61.3% after six doses. Furthermore, DCX/DAPI co-labeled cells decreased by 16% and 38.7% after one and six doses, respectively. In the case of Map2/DAPI co-labeled cells, there were 31.6% and 62.4% less after one and six doses, respectively. Since consecutive doses caused more damage, we now know that chronic exposure to OTA increasingly affects neural precursors as well as differentiated neurons.

It appears that young neurons were less vulnerable than stem cells and mature neurons, given that 61.3% of the young neurons survived the six doses of OTA compared to 37.6% and 39.8% for mature neurons and astrocytes, respectively. Together with the subventricular zone, the subgranular zone (SGZ) of the dentate gyrus serves as a source of neural stem cells in the process of adult hippocampal neurogenesis [33]. Unlike the neurons of the subventricular zone, the newly generated neurons in the SGZ do not migrate out of the dentate gyrus. Hence, the effect of OTA on neurogenesis in the hippocampus can be measured by focusing only on the dentate gyrus. Thus, to the best of our knowledge, this is the first in vivo demonstration that the dentate gyrus was negatively affected by chronic exposure to OTA.

Type A, B, and C cells are involved in neurogeneration. Type B cells (astrocytes) generate type C cells (proliferative precursors), which in turn give rise to type A cells, or migrating neuroblasts [34–36]. Thus, any negative effect on type B cells might lead to a decrease in type C cells and, consequently, deprive the type A cell population. Zurich et al. [37] described that OTA affects the neuroprotective capacity of glial cells by inhibiting GFAP with a consequent decrease in GFAP-positive cells, indicating a deprivation of type B and C cells. Because type A cells are derived from type C cells, the reduction of DCX-positive cells found in our study correlates with the findings of Zurich et al. [37].

Given that MAP2-labeled mature neurons were also affected, in our work, an inverse linear correlation between the OTA doses and the labeled cell count was observed for every cell marker we employed. Indeed, in previous in vitro and in vivo research, dose-dependent relationships were also found in other neurogenic zones when using OTA [19,22,25,32]. Based on this knowledge, we decided to look for a similar correlation when cumulative OTA doses were repetitively administered over time; our results show a dose-dependent relationship between the administration of OTA and cell survival.

An interesting observation was that there was no significant difference between the OTA1 and OTA2 groups for any of the three immunolabels. We hypothesized that the accumulated damage might not just be dose-dependent but could also act via a more complex system involving the blood−brain barrier. As reported by Sava et al. [27], i.p. treatment with one dose of 3.5 mg OTA/kg body weight led to significant changes in hippocampal neurogenesis, a result that also correlates with our findings after one dose of OTA. Surprisingly, our results also show that a second dose of OTA did not impact the neurons as much as we expected, probably because of the ability of the brain to recover and the half-life of 42.5 h for OTA in the hippocampus. This suggests that OTA does not accumulate in the dentate gyrus.

Alternatively, the difference in morphology we observed in OTA2 compared to OTA1 could indicate impaired astrocyte function after exposure to OTA. Because astrocytes are important for the function of the blood−brain barrier [38,39], impaired astrocyte function might lead to severe deregulation of its main functions in the support of the blood−brain barrier. These functions include the removal of toxic residues and controlling glycogen accumulation [40], which could possibly cause less efficient expulsion of OTA from the brain.

A reduction in the number and length of astrocyte branches was observed after brain hypoxia/ischemia insults, and 72 h after a hypoxia/ischemia insult in aged rats, the brain astrocytes had fewer and shorter branches than those in control animals [41]. Another work using the same treatment also confirmed a reduction in the number of branches in aged astrocytes [42]. These authors suggested that these astrocyte alterations may be related to the impairment of these cells in providing support to neurons. In turn, this may decrease the strength of the synapses and alter the metabolism of several neurotransmitters.

In this current study, we demonstrated an OTA-related reduction in the length and number of astrocyte branches. Importantly, these adverse effects were only statistically significant after the administration of three or six doses of OTA but not after receiving only two OTA doses. OTA likely has a longer half-life in the hippocampus, and thus, impaired astrocyte function could allow it to accumulate to higher concentrations, thereby increasing its neurotoxicity. Taken together, this could perhaps explain why a second dose of OTA did not lead to any significant differences compared to a single dose, even though altering the underlying processes would enable subsequent doses to cause measurable effects.

In this report, we did not study the selectivity of OTA toxic effects on astrocytes and neurons [32]. However, we did find that the decrease in the number of young neurons caused by OTA was not as severe as for astrocytes or mature neurons, especially after six doses. This could indicate that type B and C cells and mature neurons are more vulnerable to OTA than type A cells. In the longer term, we might expect the number of young neurons to decrease more because they are derived from type C cells, which seem to be more directly affected by OTA. Of course, based on our work, we cannot exclude the possibility that changes in hippocampal volume after OTA treatment might have contributed to the reduction we observed in cell numbers per square millimeter.

Nonetheless, our findings demonstrate an adverse effect of chronic levels of OTA on the cell survival of astrocytes and neurons in the dentate gyrus, leading to a progressive reduction of neurogenic capacity. Whether these changes are long-lasting and how they affect hippocampal function merits further investigation. In this study, we demonstrated that, at a cellular level, the hippocampus can be severely damaged by OTA. In this study, experiments with hematoxylin and eosin were performed, and no significant macro-structural changes were observed; however, new studies should be performed using electron microscopy in order to study the ultrastructure of the hippocampus after OTA exposure.

Thus, on the one hand, the effect of OTA on the function of the hippocampus should be further examined and behavioral studies should be performed. On the other hand, no work has yet been done to study the recovery of the hippocampus after OTA treatment lasting longer than three days. Therefore, it will be important to discover whether the hippocampus can reverse the damage done by OTA in order to understand the potential neurotoxicity of OTA.

4. Materials and Methods

4.1. Experimental Animals

Twenty-one male C57BL/6 mice (weighing 21 ± 4 g, 5–6 months old) obtained from Harlan (Barcelona, Spain) were used. The animals were housed five per cage under controlled temperature (24 ± 1 °C) and humidity ($60 \pm 1\%$) conditions, with a 12 h light/dark cycle and water and food ad libitum. Both hippocampal regions of each mouse were used. All animal procedures were conducted in accordance with Spanish legislation (RD 1201/05) and the guidelines of the CEU Cardenal Herrera University Animal Care Committee, approval number 117021.

4.2. Ochratoxin A Administration in the In Vivo Assay

OTA, as shown in Figure 1 (1 mg/mL, stock solution, 99% purity), was purchased from Sigma-Aldrich (St. Louis, MO, USA), dissolved in 100% ethanol, and further diluted to a concentration of 0.1 M (pH 7.2) in phosphate-buffered saline (PBS). Seventeen mice received an i.p. dose of 3.5 mg OTA/kg body weight (equivalent to approximately 10% of the LD50) in a volume of 2.8 µL/g body weight. This dose was selected because it is in the range of doses used by other studies; in this way, the obtained results could be compared easily [30,43]. The animals were divided into four experimental groups with n = 4–5 animals each: OTA1, OTA2, OTA3, and OTA6, which each received one, two, three, or six doses, respectively. Because the hippocampal half-life of OTA is 42.5 h [27], each dose was separated by three days to minimize its cumulative toxic effect. A control group of mice was injected i.p. with the vehicle from one to six times, corresponding with the treated mice. Three days after the last OTA dose administration, all the mice were humanely euthanized by cervical dislocation.

4.3. Immunohistochemistry

The animals were transcranial perfused with 4% paraformaldehyde. Their brains were removed and postfixed overnight in the same fixative solution, and then washed and cryoprotected by immersion in 30% sucrose dissolved in PBS for 2 days at 4 °C. Coronal sections (20 µm) were serially obtained using a cryostat (Leica, Wetzlar, Germany), mounted on glass slides, and stored at −20 °C. The brain sections between the Bregma zone (1.46 mm) and the interaural area (2.34 mm) and the Bregma zone (2.32 mm) and the interaural area (1.50 mm) were selected using the Paxinos G atlas [44].

The brain sections were washed in PBS and blocked for 2 h with blocking buffer (BB; 10% fetal bovine serum in PBS-triton X-100 0.1%) at room temperature.

Next, the sections were incubated overnight at 4 °C with either rabbit polyclonal anti-Glial fibrillary acidic protein (GFAP: Dako, Denmark), anti-microtubule-associated protein 2 (MAP2: Millipore, California), or rabbit polyclonal anti-neural migration protein doublecortin primary antibody (DCX: Abcam, Cambridge, UK) diluted in BB (GFAP, 1:500; Map2, 1:200; DCX, 1:300). After rinsing with PBS-triton X-100 0.1%, the binding of the primary antisera was visualized with Alexa Fluor™ 488 goat anti-rabbit and Alexa Fluor™ 594-conjugated donkey anti-rabbit (1:200, both Invitrogen, Barcelona, Spain) diluted in BB. Secondary antibodies were applied for 2 h at room temperature in the dark. Afterwards, the sections were mounted with DAPI Vectashield (Vector Labs, Peterborough, UK) and coverslipped. In the case of Map2 and DCX staining, the tissue sections were incubated at 100 °C in 10 mM citrate (pH 8.0) for 20 min to expose the immunoreactive sites before adding the primary antibody. The antibodies and specifications are shown in Table 3.

Table 3. Primary and secondary antibodies used for fluorescence analysis and counting.

Product	Antibody	Dilution	Specificity	Wavelength	Reference	Company
H-1200	4′,6-diamidino-2-fenylindool (DAPI)	-	Binds with chromatin-DNA/RNA	461 nm (blue)	Chazotte 2011	Vector Labs, UK
Z0334	Glial fibrillary acidic protein (GFAP)	1:500	Astroglial lineage	-	Eng et al., 2000	Dako, Denmark
AB18723	Doublecortin (DCX)	1:300	Neuronal precursor cells and immature neurons	-	Gleave et al.	Abcam, UK
AB5622	Microtubule-associated protein 2 (MAP2)	1:200	Mature neurons	-	Lyck et al.	Millipore, CA, USA
A11008	Alexa Fluor™ 488 goat anti-rabbit	1:200	Secondary antibody	495–519 (green)	Borg et al.	Invitrogen, Spain
A1102	Alexa Fluor™ 594-conjugated donkey anti-rabbit	1:200	Secondary antibody	590–617 (red)	Purkartova et al.	Invitrogen, Spain

The combinations used were: GFAP—Alexa Fluor™ 488 (green), DCX—Alexa Fluor™ 594 (red), and MAP2—Alexa Fluor™ 488 (green). All the slides were mounted with DAPI (blue) in order to identify and count the nuclei.

4.4. Cell Quantification and Statistical Analysis

The cells were counted in the granular and polymorphic layers of the dentate gyrus using digital images of the hippocampus obtained with a DS-Fi-1 (Nikon, Spain) digital camera coupled with a Leica DM2000 microscope. Four animals from each group were analyzed using four serial coronal sections per animal at 100 µm intervals. Only cells stained positively for both DAPI and one other marker were included. The counting area was measured using ImageJ 1.46r software. The data are expressed as the total number of immunostained cells/mm^2.

To quantify the number and length of the branches, the fluorescence photomicrographs were converted into skeletonized images and analyzed using the ImageJ software plugins AnalyzeSkeleton and FracLac, according to the method described by Young and Morrison [45]. One-way ANOVA was used to analyze the statistical significance of any between- and within-group differences using SPSS statistics software (Version 17.0, SPSS Inc., Chicago, IL, USA). When variances were found to be homogenously distributed, a least significant difference test (Fisher's LSD) was used. When variances were not homogenously distributed, as was the case with counting the MAP2 cells, a Games–Howell test was used. Any differences were considered significant at $p < 0.05$.

Author Contributions: J.M.S., M.A.G.E. and M.M. contributed to the concept and design of the study. R.P.B.T., E.M. and A.C. performed the experiments and established the database. M.A.G.E. performed the statistical analysis. J.M.S., R.P.B.T. and M.A.G.E. wrote the first draft of the manuscript. J.M.S., M.A.G.E., A.M.N. and M.M. wrote sections of the manuscript. All authors contributed to manuscript revision and read and approved the submitted version. All authors have read and agreed to the published version of the manuscript.

Funding: This work was supported by grants from Cardenal Herrera CEU University and Fundación Universitaria San Pablo CEU (JMS: CEU-UCH 2020-2021 INDI20/49, CEU-UCH 2021-2022, INDI21/55. MAGE: CEU-UCH 2021-2022, reference INDI21/19). This work was also supported by the University of Valencia (Ministerio de Economía y Competitividad and Ministerio de Ciencia, Innovación y Universidades (Spanish Government) through Projects AGL2014-53928-C2-1-R and RTI2018-097593-B-C22).

Institutional Review Board Statement: The animal study protocol was approved by the University CEU Cardenal Herrera animal unit.

Data Availability Statement: The data presented in this study are available on request from the corresponding author.

Acknowledgments: We acknowledge the University CEU Cardenal Herrera and the University of Valencia for their help, support, and use of facilities. We thank Misericordia Jimenez at the University of Valencia for her help and support of this project.

Conflicts of Interest: The authors declare no conflict of interest.

References

1. Walker, R. Risk assessment of ochratoxin: Current views of the european scientific committee on food, the JECFA and the CODEX committee on food additives and contaminants. *Adv. Exp. Med. Biol.* **2002**, *504*, 249–255. [CrossRef] [PubMed]
2. Galtier, P. Pharmacokinetics of ochratoxin A in animals. *IARC Sci. Publ.* **1991**, *115*, 187–200.
3. Schrenk, D.; Bodin, L.; Chipman, J.K.; del Mazo, J.; Grasl-Kraupp, B.; Hogstrand, C.; Hoogenboom, L.; Leblanc, J.C.; Nebbia, C.S.; Nielsen, E.; et al. Risk assessment of ochratoxin A in food. *EFSA J.* **2020**, *18*, e06113. [CrossRef]
4. JECFA. *Safety Evaluation of Certain Mycotoxins in Food*; FAO/WHO JECFA: Geneva, Switzerland, 2001.
5. Pickova, D.; Ostry, V.; Malir, J.; Toman, J.; Malir, F. A Review on Mycotoxins and Microfungi in Spices in the Light of the Last Five Years. *Toxins* **2020**, *12*, 789. [CrossRef] [PubMed]
6. Pleadin, J.; Frece, J.; Markov, K. Mycotoxins in food and feed. *Adv. Food Nutr. Res.* **2019**, *89*, 297–345. [CrossRef]
7. Dey, D.K.; Kang, J.I.; Bajpai, V.K.; Kim, K.; Lee, H.; Sonwal, S.; Simal-Gandara, J.; Xiao, J.; Ali, S.; Huh, Y.S.; et al. Mycotoxins in food and feed: Toxicity, preventive challenges, and advanced detection techniques for associated diseases. *Crit. Rev. Food Sci. Nutr.* **2022**, *2022*, 2059650. [CrossRef]
8. Bui-Klimke, T.R.; Wu, F. Ochratoxin A and human health risk: A review of the evidence. *Crit. Rev. Food Sci. Nutr.* **2015**, *55*, 1860–1869. [CrossRef]
9. Creppy, E.E.; Kane, A.; Dirheimer, G.; Lafarge-Frayssinet, C.; Mousset, S.; Frayssinet, C. Genotoxicity of ochratoxin A in mice: DNA single-strand break evaluation in spleen, liver and kidney. *Toxicol. Lett.* **1985**, *28*, 29–35. [CrossRef]
10. Pfohl-Leszkowicz, A.; Manderville, R.A. Ochratoxin A: An overview on toxicity and carcinogenicity in animals and humans. *Mol. Nutr. Food Res.* **2007**, *51*, 61–99. [CrossRef]
11. Kőszegi, T.; Poór, M. Ochratoxin A: Molecular Interactions, Mechanisms of Toxicity and Prevention at the Molecular Level. *Toxins* **2016**, *8*, 111. [CrossRef]
12. Malir, F.; Ostry, V.; Pfohl-Leszkowicz, A.; Novotna, E. Ochratoxin A: Developmental and reproductive toxicity-an overview. *Birth Defects Res. B Dev. Reprod. Toxicol.* **2013**, *98*, 493–502. [CrossRef] [PubMed]
13. Heussner, A.H.; Bingle, L.E.H. Comparative ochratoxin toxicity: A review of the available data. *Toxins* **2015**, *7*, 4253–4282. [CrossRef] [PubMed]
14. Cramer, B.; Osteresch, B.; Muñoz, K.A.; Hillmann, H.; Sibrowski, W.; Humpf, H.U. Biomonitoring using dried blood spots: Detection of ochratoxin A and its degradation product 2'R-ochratoxin A in blood from coffee drinkers. *Mol. Nutr. Food Res.* **2015**, *59*, 1837–1843. [CrossRef] [PubMed]
15. Coronel, M.B.; Sanchis, V.; Ramos, A.J.; Marin, S. Review. Ochratoxin A: Presence in human plasma and intake estimation. *Food Sci. Technol. Int.* **2010**, *16*, 5–18. [CrossRef]
16. Paradells, S.; Rocamonde, B.; Llinares, C.; Herranz-Pérez, V.; Jimenez, M.; Garcia-Verdugo, J.M.; Zipancic, I.; Soria, J.M.; Garcia-Esparza, M.A. Neurotoxic effects of ochratoxin A on the subventricular zone of adult mouse brain. *J. Appl. Toxicol.* **2015**, *35*, 737–751. [CrossRef]
17. Behrens, M.; Hüwel, S.; Galla, H.J.; Humpf, H.U. Efflux at the Blood-Brain Barrier Reduces the Cerebral Exposure to Ochratoxin A, Ochratoxin α, Citrinin and Dihydrocitrinone. *Toxins* **2021**, *13*, 327. [CrossRef]

18. Belmadani, A.; Tramu, G.; Betbeder, A.M.; Creppy, E.E. Subchronic effects of ochratoxin A on young adult rat brain and partial prevention by aspartame, a sweetener. *Hum. Exp. Toxicol.* **1998**, *17*, 380–386. [CrossRef]
19. Yoon, S.; Cong, W.T.; Bang, Y.; Lee, S.N.; Yoon, C.S.; Kwack, S.J.; Kang, T.S.; Lee, K.Y.; Choi, J.K.; Choi, H.J. Proteome response to ochratoxin A-induced apoptotic cell death in mouse hippocampal HT22 cells. *Neurotoxicology* **2009**, *30*, 666–676. [CrossRef]
20. Erceg, S.; Mateo, E.M.; Zipancic, I.; Jiménez, F.J.R.; Aragó, M.A.P.; Jiménez, M.; Soria, J.M.; Garcia-Esparza, M.Á. Assessment of Toxic Effects of Ochratoxin A in Human Embryonic Stem Cells. *Toxins* **2019**, *11*, 217. [CrossRef]
21. Pei, X.; Zhang, W.; Jiang, H.; Liu, D.; Liu, X.; Li, L.; Li, C.; Xiao, X.; Tang, S.; Li, D. Food-Origin Mycotoxin-Induced Neurotoxicity: Intend to Break the Rules of Neuroglia Cells. *Oxid. Med. Cell. Longev.* **2021**, *2021*, 9967334. [CrossRef]
22. Zhang, X.; Boesch-Saadatmandi, C.; Lou, Y.; Wolffram, S.; Huebbe, P.; Rimbach, G. Ochratoxin A induces apoptosis in neuronal cells. *Genes Nutr.* **2009**, *4*, 41. [CrossRef] [PubMed]
23. Sava, V.; Velasquez, A.; Song, S.; Sanchez-Ramos, J. Adult hippocampal neural stem/progenitor cells in vitro are vulnerable to the mycotoxin ochratoxin-A. *Toxicol. Sci.* **2007**, *98*, 187–197. [CrossRef] [PubMed]
24. Vukovic, J.; Borlikova, G.G.; Ruitenberg, M.J.; Robinson, G.J.; Sullivan, R.K.P.; Walker, T.L.; Bartlett, P.F. Immature doublecortin-positive hippocampal neurons are important for learning but not for remembering. *J. Neurosci.* **2013**, *33*, 6603–6613. [CrossRef] [PubMed]
25. Sava, V.; Reunova, O.; Velasquez, A.; Harbison, R.; Sánchez-Ramos, J. Acute neurotoxic effects of the fungal metabolite ochratoxin-A. *Neurotoxicology* **2006**, *27*, 82–92. [CrossRef]
26. Belmadani, A.; Tramu, G.; Betbeder, A.M.; Steyn, P.S.; Creppy, E.E. Regional selectivity to ochratoxin A, distribution and cytotoxicity in rat brain. *Arch. Toxicol.* **1998**, *72*, 656–662. [CrossRef]
27. Sava, V.; Reunova, O.; Velasquez, A.; Sanchez-Ramos, J. Can low level exposure to ochratoxin-A cause parkinsonism? *J. Neurol. Sci.* **2006**, *249*, 68–75. [CrossRef]
28. Kamp, H.G.; Eisenbrand, G.; Schlatter, J.; Würth, K.; Janzowski, C. Ochratoxin A: Induction of (oxidative) DNA damage, cytotoxicity and apoptosis in mammalian cell lines and primary cells. *Toxicology* **2005**, *206*, 413–425. [CrossRef]
29. Damiano, S.; Longobardi, C.; Andretta, E.; Prisco, F.; Piegari, G.; Squillacioti, C.; Montagnaro, S.; Pagnini, F.; Badino, P.; Florio, S.; et al. Antioxidative effects of curcumin on the hepatotoxicity induced by ochratoxin a in rats. *Antioxidants* **2021**, *10*, 125. [CrossRef]
30. Schwerdt, G.; Kopf, M.; Gekle, M. The impact of the nephrotoxin ochratoxin a on human renal cells studied by a novel co-culture model is influenced by the presence of fibroblasts. *Toxins* **2021**, *13*, 219. [CrossRef]
31. Fukui, Y.; Hayasaka, S.; Itoh, M.; Takeuchi, Y. Development of neurons and synapses in ochratoxin A-induced microcephalic mice: A quantitative assessment of somatosensory cortex. *Neurotoxicol. Teratol.* **1992**, *14*, 191–196. [CrossRef]
32. Wilk-Zasadna, I.; Minta, M. Developmental Toxicity of Ochratoxin A in Rat Embryo Midbrain Micromass Cultures. *Int. J. Mol. Sci.* **2009**, *10*, 37–49. [CrossRef] [PubMed]
33. Gage, F.H. Mammalian neural stem cells. *Science* **2000**, *287*, 1433–1438. [CrossRef] [PubMed]
34. Doetsch, F.; Caille, I.; Lim, D.A.; Garcia-Verdugo, J.M.; Alvarez-Buylla, A. Subventricular zone astrocytes are neural stem cells in the adult mammalian brain. *Cell* **1999**, *97*, 703–716. [CrossRef]
35. Mirzadeh, Z.; Merkle, F.T.; Soriano-Navarro, M.; Garcia-Verdugo, J.M.; Alvarez-Buylla, A. Neural stem cells confer unique pinwheel architecture to the ventricular surface in neurogenic regions of the adult brain. *Cell Stem Cell* **2008**, *3*, 265–278. [CrossRef] [PubMed]
36. Ponti, G.; Obernier, K.; Guinto, C.; Jose, L.; Bonfanti, L.; Alvarez-Buylla, A. Cell cycle and lineage progression of neural progenitors in the ventricular-subventricular zones of adult mice. *Proc. Natl. Acad. Sci. USA* **2013**, *110*, E1054. [CrossRef]
37. Zurich, M.G.; Lengacher, S.; Braissant, O.; Monnet-Tschudi, F.; Pellerin, L.; Honegger, P. Unusual astrocyte reactivity caused by the food mycotoxin ochratoxin A in aggregating rat brain cell cultures. *Neuroscience* **2005**, *134*, 771–782. [CrossRef] [PubMed]
38. Janzer, R.C.; Raff, M.C. Astrocytes induce blood-brain barrier properties in endothelial cells. *Nature* **1987**, *325*, 253–257. [CrossRef]
39. Barcia, C.; Sanderson, N.S.R.; Barrett, R.J.; Wawrowsky, K.; Kroeger, K.M.; Puntel, M.; Liu, C.; Castro, M.G.; Lowenstein, P.R. T Cells' Immunological Synapses Induce Polarization of Brain Astrocytes In Vivo and In Vitro: A Novel Astrocyte Response Mechanism to Cellular Injury. *PLoS ONE* **2008**, *3*, e2977. [CrossRef]
40. Acuña Castroviejo, E.; Martínez Arias de Reyna, L.; Andrés de la Calle, I.; Armengol y Butrón de Mújica, J.A.; Artigas Pérez, F. *Manual de Neurociencias*, 1st ed.; Sintesis: Madrid, Spain, 1998; ISBN 9788477386001.
41. Sullivan, S.M.; Björkman, S.T.; Miller, S.M.; Colditz, P.B.; Pow, D.V. Structural remodeling of gray matter astrocytes in the neonatal pig brain after hypoxia/ischemia. *Glia* **2010**, *58*, 181–194. [CrossRef]
42. Morel, G.R.; Andersen, T.; Pardo, J.; Zuccolilli, G.O.; Cambiaggi, V.L.; Hereñú, C.B.; Goya, R.G. Cognitive impairment and morphological changes in the dorsal hippocampus of very old female rats. *Neuroscience* **2015**, *303*, 189–199. [CrossRef]
43. Le, G.; Yuan, X.; Hou, L.; Ge, L.; Liu, S.; Muhmood, A.; Liu, K.; Lin, Z.; Liu, D.; Gan, F.; et al. Ochratoxin A induces glomerular injury through activating the ERK/NF-κB signaling pathway. *Food Chem. Toxicol.* **2020**, *143*, 111516. [CrossRef] [PubMed]
44. Franklin, K.B.J.; George, P. *Paxinos and Franklin's the Mouse Brain in Stereotaxic Coordinates*, 7th ed.; Academic Press Inc.: Cambridge, MA, USA, 2019; ISBN 9780128161593.
45. Young, K.; Morrison, H. Quantifying Microglia Morphology from Photomicrographs of Immunohistochemistry Prepared Tissue Using ImageJ. *J. Vis. Exp.* **2018**, *2018*, e57648. [CrossRef] [PubMed]

 *toxins*

Article

Lactic Acid Bacteria as Potential Agents for Biocontrol of Aflatoxigenic and Ochratoxigenic Fungi

Eva María Mateo [1,*], Andrea Tarazona [2], Misericordia Jiménez [2] and Fernando Mateo [3]

[1] Departamento de Microbiología y Ecología, Facultad de Medicina y Odontología, Universitat de Valencia, E-46100 Burjasot, Valencia, Spain
[2] Departamento de Microbiología y Ecología, Facultad de Ciencias Biológicas, Universitat de Valencia, E-46100 Burjasot, Valencia, Spain
[2] Departamento de Ingeniería Electrónica, ETSE, Universitat de Valencia, E-46100 Burjasot, Valencia, Spain
* Correspondence: eva.mateo@uv.es

Abstract: Aflatoxins (AF) and ochratoxin A (OTA) are fungal metabolites that have carcinogenic, teratogenic, embryotoxic, genotoxic, neurotoxic, and immunosuppressive effects in humans and animals. The increased consumption of plant-based foods and environmental conditions associated with climate change have intensified the risk of mycotoxin intoxication. This study aimed to investigate the abilities of eleven selected LAB strains to reduce/inhibit the growth of *Aspergillus flavus*, *Aspergillus parasiticus*, *Aspergillus carbonarius*, *Aspergillus niger*, *Aspergillus welwitschiae*, *Aspergillus steynii*, *Aspergillus westerdijkiae*, and *Penicillium verrucosum* and AF and OTA production under different temperature regiments. Data were treated by ANOVA, and machine learning (ML) models able to predict the growth inhibition percentage were built, and their performance was compared. All factors LAB strain, fungal species, and temperature significantly affected fungal growth and mycotoxin production. The fungal growth inhibition range was 0–100%. Overall, the most sensitive fungi to LAB treatments were *P. verrucosum* and *A. steynii*, while the least sensitive were *A. niger* and *A. welwitschiae*. The LAB strains with the highest antifungal activity were *Pediococcus pentosaceus* (strains S11sMM and M9MM5b). The reduction range for AF was 19.0% (aflatoxin B1)-60.8% (aflatoxin B2) and for OTA, 7.3–100%, depending on the bacterial and fungal strains and temperatures. The LAB strains with the highest anti-AF activity were the three strains of *P. pentosaceus* and *Leuconostoc mesenteroides* ssp. *dextranicum* (T2MM3), and those with the highest anti-OTA activity were *Leuconostoc paracasei* ssp. *paracasei* (3T3R1) and *L. mesenteroides* ssp. *dextranicum* (T2MM3). The best ML methods in predicting fungal growth inhibition were multilayer perceptron neural networks, followed by random forest. Due to anti-fungal and anti-mycotoxin capacity, the LABs strains used in this study could be good candidates as biocontrol agents against aflatoxigenic and ochratoxigenic fungi and AFL and OTA accumulation.

Keywords: lactic acid bacteria; biocontrol; *Aspergillus* spp.; *Penicillium verrucosum*; aflatoxins; ochratoxin A; fungal growth; machine learning

Key Contribution: The selected LAB strains showed efficacy to reduce or inhibit the growth of toxigenic *Aspergillus* spp. and *Penicillium verrucosum* and reducing mycotoxin (aflatoxins or ochratoxin A) production in a dual culture medium of MRS agar-Czapek yeast 20% and sucrose agar at 20, 25, and 30 °C. The LAB strains showing the highest antifungal activity were two strains of *P. pentosaceus*, followed by *L. mesenteroides* ssp. *dextranicum* and *Companilactobacillus farciminis*, whereas the least effective were two strains of *L. sakei* ssp. *carnosus*. Various ML models were designed to predict the percentage of fungal growth inhibition, and the best algorithms were multilayer perceptron neural networks and random forest.

Citation: Mateo, E.M.; Tarazona, A.; Jiménez, M.; Mateo, F. Lactic Acid Bacteria as Potential Agents for Biocontrol of Aflatoxigenic and Ochratoxigenic Fungi. *Toxins* **2022**, *14*, 807. https://doi.org/10.3390/toxins14110807

Received: 27 October 2022
Accepted: 17 November 2022
Published: 19 November 2022

Publisher's Note: MDPI stays neutral with regard to jurisdictional claims in published maps and institutional affiliations.

1. Introduction

Mycotoxins are secondary metabolites of filamentous fungi that can cause diseases in humans and animals. Although filamentous fungi can collectively produce hundreds of mycotoxins, in the EU, only a few of them are subject to regulation [1]. In terms of acute and chronic toxicity for humans and animals, the most relevant mycotoxins are aflatoxins (AF). AF are produced by various species of *Aspergillus* subgenus *Circumdati* section *Flavi*, but the two species of greatest concern are *Aspergillus flavus* and *Aspergillus parasiticus* [2]. *A. flavus* produces aflatoxin B1 (AFB1) and B2 (AFB2), and *A. parasiticus* can produce AFB1, AFB2, aflatoxins G1 (AFG1), and G2 (AFG2), although some reports indicate that *A. flavus* strains can also synthesize the G-type AF [3]. AF are present in very important food with many foods, such as cereals and nuts [4–6], breakfast cereals [7], infant foods [8], cocoa [9], legumes [10], or milk [11], among others. For a long time, AFB1 was listed by the International Agency for Research on Cancer (IARC) [12] as carcinogenic to humans. In addition to hepatocellular carcinoma, AF are associated with occasional outbreaks of acute aflatoxicosis, which lead to death shortly after exposure. Approximately 4500 million people living in developing countries are chronically exposed to AF through contaminated diets. In Kenya, acute AF intoxication leads to liver failure and, ultimately, death in around 40% of the cases [13]. Thus, the presence of these fungi and AF in food and feed is an important concern for manufacturers, consumers, researchers, and regulatory agencies.

After AF, ochratoxins are the most relevant mycotoxins produced by *Aspergillus* spp., particularly ochratoxin A (OTA). This mycotoxin is produced mainly by *Aspergillus steynii* and *Aspergillus westerdijkiae* [14], *Aspergillus niger*, *Aspergillus welwitschiae*, and *Aspergillus carbonarius* [15,16], and *Penicillium* species, mainly *Penicillium verrucosum* [17,18]. OTA is frequently found in cereals [19–22], beer and wine [23], grapes [24], cheeses [25], cured meat products [26], bread [27], coffee [28], cocoa [29], dried fruits [30], and spices [31]. OTA acts as a potent nephrotoxin [32] and also exhibits teratogenic, embryotoxic, genotoxic, neurotoxic, and immunosuppressive effects [33,34]. It has been classified as a possible human carcinogen by the IARC [12].

A. flavus and *A. niger* are reported, after *Aspergillus fumigatus*, as the second leading cause of invasive aspergillosis and the most common cause of superficial infection. It is estimated that 1.5 to 2 million people die of a fungal infection each year, many of which are caused by these *Aspergillus* spp. [35–37].

Some European countries are major producers and exporters of cereals, but their economies and trade have been severely affected because of contaminated products by mycotoxins in recent decades. According to the Food and Agriculture Organization, 25% of the world's food crops are badly affected by mycotoxins either during growth or storage [38,39]. Moreover, globally, fungal infections of the most cultivated food crops were assessed to annihilate about 125 million tons of products annually [40]. Fungal attacks on cereals entail an annual loss of about $60 billion in global agricultural production [41]. In this context, the design of strategies capable of controlling the growth of toxigenic fungi and minimizing the presence of mycotoxins in food, particularly in cereals, is an urgent need. Though different physical and chemical methods are used [42–44], the data on economic losses, illnesses, and death in the world warn that the risks associated with contamination by fungi and mycotoxins in food remain unsolved.

Temperature is a key environmental factor that influences both the rate of fungal spoilage and the production of mycotoxins. Based on the predictive model developed for *A. flavus* growth and AFB1 production linked to crop phenology data, the risk of AF contamination was assessed to have a chance of increasing in maize in the future, due to the climate change trend. In the +2 °C climate change scenario, there is a clear increase in AF risk in areas such as central and southern Spain, southern Italy, Greece, northern and southeastern Portugal, Bulgaria, Albania, Cyprus, and European Turkey, as compared to the actual current temperature [45,46].

It has been reported that, in the last few years, mycotoxin levels in cereals have increased in Europe, which is probably due to the adaptive abilities of the toxigenic fungi, especially *A. flavus*, to the higher temperatures associated with climate change [45].

Lactic acid bacteria (LAB) have been broadly used in conventional food fermentations since ancient times and are considered GRAS, and many of them are classified as Qualified Presumption of Safety by the American Food and Drug Agency and the European Food Safety Authority, respectively [47]. Numerous LAB are considered "Green preservatives", due to their potential to retard fungal contamination in food [48]. Currently, consumers have an increasing demand for 'natural' and 'healthy' foods, and LAB are a promising alternative for the biocontrol of toxigenic fungi [48]. The role of LAB is not limited to inhibiting fungal growth, but some LAB strains can interact with mycotoxins, causing their inactivation or removal [48–50].

Few reports have shown the antifungal effect of selected LAB strains against aflatoxigenic or ochratoxigenic fungi; only in a few of them, the effect of LAB on the production of AF or OTA is studied, and in none of them, the effect of temperature on the growth and mycotoxin production in dual culture bacteria-fungus has been investigated; in addition, predictive models have not been applied to these systems [48,49].

Machine learning (ML) is a subfield under artificial intelligence that constitutes a challenge and a great opportunity in numerous scientific, technical, and clinical disciplines. ML investigates algorithms able to learn autonomously, directly from the data. ML refers to the process of fitting predictive models to data or identifying informative clusters within data. The field of ML essentially attempts to approximate or mimic the human ability to recognize patterns using computation [51]. ML methods can be supervised and unsupervised. Supervised ML concerns the fitting of a model to data that have been labeled, where there exists some ground truth property, which is experimentally measured [51]. Supervised learning can be grouped into regression and classification. The dataset where ML works is composed of many data points, each of which consists of an experimental observation. For regression tasks, in the dataset, there are continuous or categorical input variables and output variables associated with the former, which are the focus of the researcher's interest (targets). Most of the dataset (inputs and the related outputs) is used to train the ML, another part is used for model validation, and another part (the test set) is used to evaluate the built model's performance, with data unseen during training. K-fold cross-validation is usually performed, with K being the number of groups in which the training set is divided to train/validate the model. ML algorithms iteratively make predictions of the output on the training data to establish optimum algorithm parameters. The calculated output values differ from the experimental values, and a loss function computes the average difference between them. Upon changing the values of their parameters, the algorithm tries to minimize that difference, but the process must avoid the problem of overfitting the data, which may prevent the generalization from making predictions ahead of the data used. The process ends when the loss function reaches a satisfactory average minimum difference between the predicted and experimental values. The number of applications of ML methods in most research areas is rapidly increasing. ML has been applied to every microbiology research [51,52]. In the field of predictive mycology, the most applied ML algorithms are artificial neural networks (NN), especially multilayer perceptrons (MLP), random forest (RF), support vector machines (SVM), or gradient boosting algorithms, such as extreme gradient boosting trees (XGBoost).

The objectives of the present study were (i) to evaluate the ability of eleven previously selected LAB strains, to inhibit the development of eight relevant species of aflatoxigenic or ochratoxigenic fungi in vitro under different temperature regimes, (ii) to assess the ability of the LAB strains to reduce AF and OTA levels in the media at the assayed temperatures, and (iii) to test the ability of models developed using various ML algorithms to predict the efficacy of LAB strains to inhibit or reduce the growth of the assayed fungi.

2. Results

2.1. Effect of LAB on Fungal Growth in Dual Medium MRS-CYA20S

In the controls, fungal growth became visible after a lag phase ranging from 1.5 to 3 days of incubation, depending on the temperature and fungal species.

In treatments, all LAB strains partially inhibited fungal growth. The measurements of the inhibition halos of fungal growth (transverse and longitudinal) were the same at 5 and 10 days of incubation. In the absence of LAB (controls), all fungal isolates grew and filled the Petri dishes within the incubation period at any of the three temperatures tested (20, 25, and 30 °C). Thus, the growth inhibition percentage (GIP) in the controls was zero.

Figure 1 shows the results of the GIP, relative to the pertinent controls for *A. flavus, A. parasiticus, A. carbonarius, A. niger, A. welwitschiae, A. steynii, A. westerdijkiae,* and *P. verrucosum* in dual medium (MRS agar-Czapek Yeast 20% Sucrose (CY20S) agar) in treatments with LAB strains. The values of these GIP depended on the incubation temperature, so that, for *Aspergillus* spp., the usual order was 20 °C > 25 °C > 30 °C. However, in the cultures of *A. westerdijkiae,* the GIP was higher at 30 °C than at 25 °C. In the cultures of *P. verrucosum,* there was not a large influence of the incubation temperature on inhibition data, although on average, the fungus grew better at 25 °C and worse at 30 °C. In other fungi with some LAB, the GIP at 30 °C was a bit higher than at 25 °C. Regardless of temperature and LAB strains, the GIP ranges were 5–25% for *A. flavus,* 0–35% for *A. parasiticus,* 0–40% for *A. carbonarius,* 0–22% for *A. niger,* 2–20% for *A. welwitschiae,* 15–70% for *A. steynii,* 9–47% for *A. westerdijkiae,* and 10–100% for *P. verrucosum* (Figure 1).

The inhibition data were treated by multifactor ANOVA, which revealed that all the main factors (fungal species, incubation temperature, and LAB strain) significantly influenced the GIP ($p < 0.001$). First- and second-order interactions between these factors were also significant ($p < 0.05$). The influence of the fungal species on the GIP is manifested by the arrangement of the assayed fungi in six homogeneous groups, according to *post-hoc* Duncan's test (Table 1).

Table 1. Arrangement of *Aspergillus* spp. and *P. verrucosum* isolates in homogeneous groups, in terms of their overall susceptibility to the tested LAB strains, according to Duncan's multiple range test ($\alpha = 0.05$).

| Fungal Species | Homogeneous Groups [1] | | | | | |
| | Lower | | ◀—Susceptibility—▶ | | Higher | |
	A	B	C	D	E	F
A. flavus		X				
A. parasiticus		X				
A. carbonarius			X			
A. niger	X					
A. welwitschiae	X					
A. steynii					X	
A. westerdijkiae				X		
P. verrucosum						X

[1] The susceptibility increased from group A to group F. There were no statistically significant differences ($\alpha = 0.05$) between the fungi showing an X within the same column.

In general, the most susceptible fungal species to the LAB strains was *P. verrucosum* and, in second place, *A. steynii.* The least susceptible species were *A. niger* and *A. welwitschiae* (they did not differ significantly from each other). *A. flavus* and *A. parasiticus* were scarcely susceptible, with no significant difference between them.

The influence of the bacterial strain on the growth of the studied fungi is shown in Figure 1. Only *P. verrucosum* growth was totally inhibited by two strains of *P. pentosaceus.* There were statistical differences among the LAB strains ($p < 0.001$). *Post-hoc* Duncan's test arranged these bacteria into eight homogeneous groups (Table 2).

In general, considering all the tested fungi, the LAB strain with the highest efficacy in inhibiting the growth of the tested fungal species was *P. pentosaceus* (S11sMM1), followed by *P. pentosaceus* (M9MM5b), and then followed by *L. mesenteroides* ssp. *dextranicum* and *C. farciminis*. Conversely, the least effective LAB strains were the two strains of *L. sakei* ssp. *carnosus*, followed by *L. brevis* (Table 2). The efficacy of each LAB strain against each of the tested fungi was studied by ANOVA.

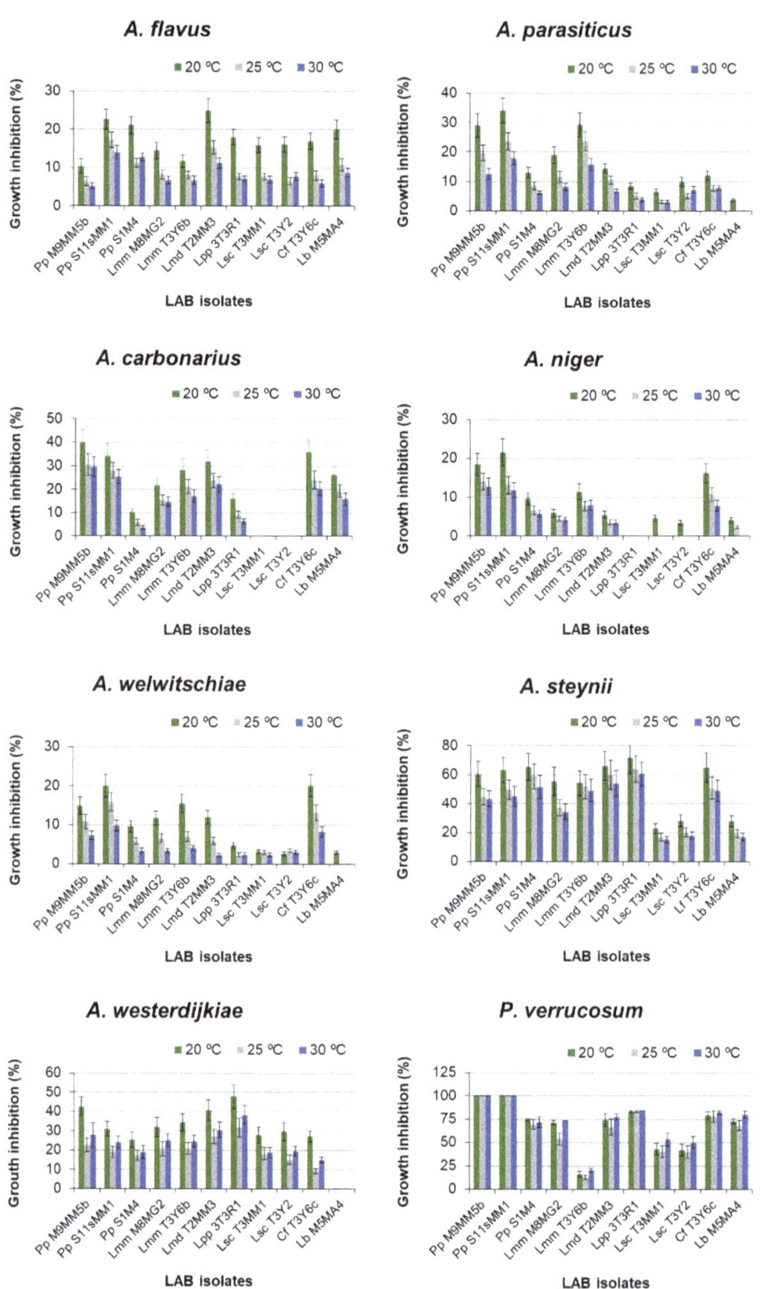

Figure 1. Growth inhibition percentages (GIP) of *Aspergillus* spp. and *P. verrucosum* in dual medium MRS agar-Czapek Yeast 20% Sucrose (MRS-CY20S) agar in the presence of LAB strains and cultured at 20, 25, and 30 °C. Error bars represent standard deviations. Incubation time: 10 days. For abbreviations, see the Abbreviation list of the LAB species. Pp, *Pediococcus pentosaceus*; Lmm, *Leuconostoc mesenteroides* ssp. *mesenteroides*; Lmd, *Leuconostoc mesenteroides* ssp. *dextranicum*; Lpp, *Lacticaseibacillus paracasei* ssp. *paracasei*; Lsc, *Latilactobacillus sakei* ssp. *carnosus*; Cf, *Companilactobacillus farciminis*; Lb, *Levilactobacillus brevis*. The number after the abbreviation is the strain number.

Table 2. Arrangement of LAB strains in homogeneous groups, in terms of their overall efficacy in inhibiting fungal growth of the assayed fungi, according to Duncan's multiple range test ($\alpha = 0.05$). For LAB abbreviations, see the legend of Figure 1 and the Abbreviation list of the LAB species.

| LAB Species (Strain) | Homogeneous Groups [1] | | | | | | | |
| | Lower | | | ◄—Efficacy—► | | Higher | | |
	A	B	C	D	E	F	G	H
Pp (M9MM5b)							X	
Pp (S11sMM1)								X
Pp (S1M4)				X				
Lmm (M8MG2)				X				
Lmm (T3Y6b)			X					
Lmd (T2MM3)						X		
Lpp (3T3R1)					X			
Lsc (T3MM1)	X							
Lsc (T3Y2)	X							
Cf (T3Y6c)					X	X		
Lb (M5MA4)		X						

[1] The order of efficacy increases from group A to group H. There are no statistically significant differences between the strains showing an X within the same column. The presence of more than one X in the same row indicates overlap.

For every tested fungal species, both the factors LAB strain and incubation temperature significantly influenced the GIP. Therefore, the efficacy of each LAB depended on the fungal species and the incubation temperature. The results of *post-hoc* Duncan's analysis are shown in Table 3. The LAB strain that showed the highest efficacy against one fungal species was not always the most effective against the other species.

Table 3. Arrangement of LAB strains in homogeneous groups, in terms of their degree of efficacy to inhibit the growth of toxigenic *Aspergillus* spp. and *P. verrucosum*, according to *post-hoc* Duncan's test ($\alpha = 0.05$) [1].

| LAB Strain | Fungal Species | | | | | | | |
	A. flavus	*A. parasiticus*	*A. carbonarius*	*A. niger*	*A. welwitschiae*	*A. steynii*	*A. westerdijkiae*	*P. verrucosum*
Pp (M9MM5b)	1	7	8	7	5	3	6	6
Pp (S11sMM1)	6	9	7	7	7	3, 4	4, 5	6
Pp (S1M4)	5	4, 5	2	4	3	4, 5	2, 3	4
Lmm (M8MG2)	2, 3	6	4	3	3	2	5	3
Lmm (T3Y6b)	2	8	5	5	4	3, 4	5	1
Lmd (T2MM3)	6	5	6	3	3	5, 6	6	4
Lpp (3T3R1)	3	2, 3	3	1	2	6	7	5
Lsc (T3MM1)	2, 3	2	1	2	2	1	3, 4	2
Lsc (T3Y2)	2, 3	3, 4	1	1, 2	2	1	3, 4	2
Cf (T3Y6c)	2, 3	4, 5	6, 7	6	6	3, 4, 5	2	5
Lb (M5MA4)	4	1	5	2	1	1	1	4

[1] Groups are denoted by a digit that increases with increasing efficacy. The presence of two or more digits separated by a comma indicates the overlapping of groups. Within a column, the LAB strains having the same digit(s) are not statistically different. For LAB abbreviations, see the legend of Figure 1 and the Abbreviation list of the LAB species.

Temperature also significantly influenced the growth inhibition effectiveness of LAB strains on the tested fungi. Overall, the highest effectiveness of LAB strains occurred at 20 °C, with a mean GIP close to 29%, while no significant difference was observed between 25 and 30 °C.

2.2. Machine Learning Approach to Model Fungal Growth Inhibition

Different ML algorithms were applied to the dataset made of input variables (fungal species, LAB strains, and incubation temperatures) and the correspondent output variable $\log_{10}(x + 1)$, where x is the mean GIP, in order to explore the ability of such algorithms to build models able to predict the GIP as accurately as possible. The categorical variables were converted into numerical "dummy" variables before computation. The multiple linear regression (MLR), MLP, RF, and XGBoost algorithms were comparatively tested.

The training was performed by 10-fold cross-validation, using 75% of the samples randomly taken from the whole dataset. The criterion for selecting the best model was based on minimizing the chosen loss function, which was the minimum root mean square error (RMSE) value for a test set that is not used to create the model during the training/cross-validation step. The R^2 value obtained in the predictive task on the test set was also considered because a high value (near 1) indicates that the predictor variables in the regression model can well-predict the variance in the output variable.

Table 4 lists the results of the best models designed by the algorithms applied for predicting the GIP on the same test set made of 72 blind samples (25% of the dataset). An MLP with a single layer of seven hidden nodes (neurons) and *decay* of 0.01 provided the best model, with an RMSE of 0.1999 and an R^2 value of 0.9232. An RF algorithm with *mtry* = 3 (the maximum number of selected predictor variables) and *ntry* = 500 gave the second better model, in order of performance. The XGBoost proved the worst algorithm, in spite of its large number of parameters to be tuned. RMSE and R^2 were calculated on the basis of the log-transformed values of the GIP.

Table 4. Best machine learning (ML) model performance for predicting the percentage of fungal growth inhibition (GIP) in dual cultures (MRS-CYA20S), based on all fungal isolates and LAB strains assayed and incubation temperature. The model performance was attained on the same test set.

ML Algorithm [1]	Assayed Parameters	Best Model Parameters	RMSE [2]	R-Squared
MLR	Regression coefficients	Found by the least-square method	0.2714	0.7629
MLP	*size*: 1–20; *decay*: 0.01, 0.05, 1.00	*size* = 7; *decay* = 0.01	0.1999	0.9232
RF	*mtry*: 2, 3; *ntry*: 500	*mtry* = 3	0.2268	0.8623
XGBoost	*max-depth*: 2–7; *eta*: 0.1–0.5); *subsample*: 0.5, 0.75, 1	*max_depth* = 5, *eta* = 0.2, *subsample* = 1	0.2828	0.7767

[1] MLR: multiple linear regression; MLP: multilayer perceptron (neural network); RF: random forest; XGBoost: extreme gradient boosted trees (the following parameters of this algorithm had constant values: *nrounds* = 150, *gamma* = 0, *colsample_bytree* = 1, *min_child_weight* = 0.5). [2] RMSE: root mean square error.

2.3. Effect of the LAB Strains on Mycotoxin Production

2.3.1. Results of the Validation of the Method for Mycotoxin Determination

Table 5 shows the retention times and the limits of detection and quantification of the chromatographic method followed. The mean values of the recovery at three different levels for each analyte and the relative standard deviation of the recovery at these levels are also given.

Table 5. Retention time, limits of detection (LOD), limits of quantification (LOQ), mean recoveries (%), and mean relative standard deviations of recoveries (RSD) for mycotoxins analyzed by UPLC-MS/MS in dual solid medium (MRS-CYA20S).

Mycotoxin [1]	Retention Time (min)	LOD (ng/g)	LOQ (ng/g)	Mean Recovery (%)	Mean RSD of Recoveries (%)
AFB1	8.30	0.78	2.34	85.2	9.5
AFB2	8.04	0.8	2.4	87.3	11
AFG1	7.80	1.18	3.5	83.0	8.2
AFG2	7.50	0.4	1.2	90.5	8.7
OTA	10.52	0.8	2.4	91.1	7.6

[1] AFB1: aflatoxin B1; AFB2: aflatoxin B2; AFG1: aflatoxin G1; AFG2: aflatoxin G2; OTA: ochratoxin A.

2.3.2. Mycotoxin Production in Controls and Cultures Containing LAB

1. Aflatoxins

The mean concentrations and standard deviations of AFB1 and AFB2 for the *A. flavus* and *A. parasiticus* isolates, as well as those of AFG1 and AFG2 for the *A. parasiticus* isolate (*A. flavus* did not produce AFG1 or AFG2), were determined by UPLC-MS/MS in both controls and treatments, with LAB strains on the last day of incubation. The results appear in Figure 2. The levels of these mycotoxins were lower in cultures in which LAB strains were present than in the controls without LAB strains incubated at the same temperature. *A. parasiticus* produced higher AFB1 and AFB2 concentrations than *A. flavus* at the same temperatures. The maximum average AFB1 production by *A. flavus* was 4220 ng/g in the control culture grown at 25 °C, whereas *A. parasiticus* produced 12,120 ng/g under the same conditions. The highest levels of AFB2 in the *A. flavus* and *A. parasiticus* control cultures were also recorded at 25 °C and were 240 and 409 ng/g, respectively. ANOVA indicated that both species differed significantly in AFB1 and AFB2 production.

In treatments, the factors LAB strain and temperature showed a significant influence on AF level in *A. flavus* and *A. parasiticus* ($p < 0.01$). The interaction between these two factors was also significant. Concerning AFB1 production in the treatment cultures of both species, the LAB strains were arranged into four homogeneous groups by Duncan's test (Figure 2, denoted by letters A to D). Considering AFB2, there were three homogeneous groups of LAB (Figure 2, denoted by letters A to C). The cluster of LAB strains giving rise to lower AFB1 production in *A. flavus* (Figure 2, group A) was constituted by the three strains of *P. pentosaceus* and *L. mesenteroides* ssp. *dextranicum* (T2MM3). Therefore, these strains were the most effective for reducing AFB1 production.

The AFG1 level exceeded the AFG2 level at all the temperatures, even in the controls, where the highest AFG1 concentration was 1740 ng/g at 25 °C, whereas the AFG2 concentration was 95 ng/g, also at 25 °C. In treatments, the AFG1 levels were always below 1560 ng/g, and those of AFG2 were below 75 ng/g. The LAB strains were arranged in three and four groups, respectively, by Duncan's test. In the case of AFG2, there was a high degree of overlapping (Figure 2).

P. pentosaceus (S11sMM1 and M9MM5b) and *L. mesenteroides* ssp. *mesenteroides* (T3Y6B) hindered AFG1 and AFG2 accumulation more than the remaining strains. The temperature significantly influenced the production of AF by the two fungi assayed, and the order of AF production was 25 °C > 30 °C > 20 °C, which is contrary to mycelium development. Table 6 lists the minimum and maximum percentages of AF reduction, compared to the controls in the LAB treatments of *A. flavus* and *A. parasiticus*. They depended on the incubation temperature and the fungal and LAB species. The percentage of mycotoxin reduction was calculated as $100 \times$ [(mycotoxin level in control − mycotoxin level in treatment)/mycotoxin level in control].

For AFB1, the lowest reduction took place in *A. parasiticus* cultures at 30 °C (19.0%), and the maximum reduction occurred in *A. parasiticus* cultures at 20 °C (55.0%). For AFB2, the minimum reduction was observed in *A. flavus* cultures at 25 °C (16.2%), and the maximum reduction occurred in *A. parasiticus* cultures at 30 °C (60.8%). For AFG1 and AFG2, the minimum and maximum reductions occurred at 30 °C and 20 °C, respectively.

2. Ochratoxin A

The production of OTA in control and treatment cultures at the three tested temperatures by *A. carbonarius*, *A. niger*, *A. welwitschiae*, *A. steynii*, *A. westerdijkiae*, and *P. verrucosum* are shown in Figure 3. The OTA levels were always higher in the controls than in the treatments under the same temperatures. The OTA level was the maximum in the control cultures of *A. steynii*, where the highest concentration (near 1400 ng/g) was attained at 25 °C. The following fungus, in order of OTA level, was *A. westerdijkiae* (about 670 ng/g at 25 °C in the control culture). *P. verrucosum* provided the lowest OTA concentrations.

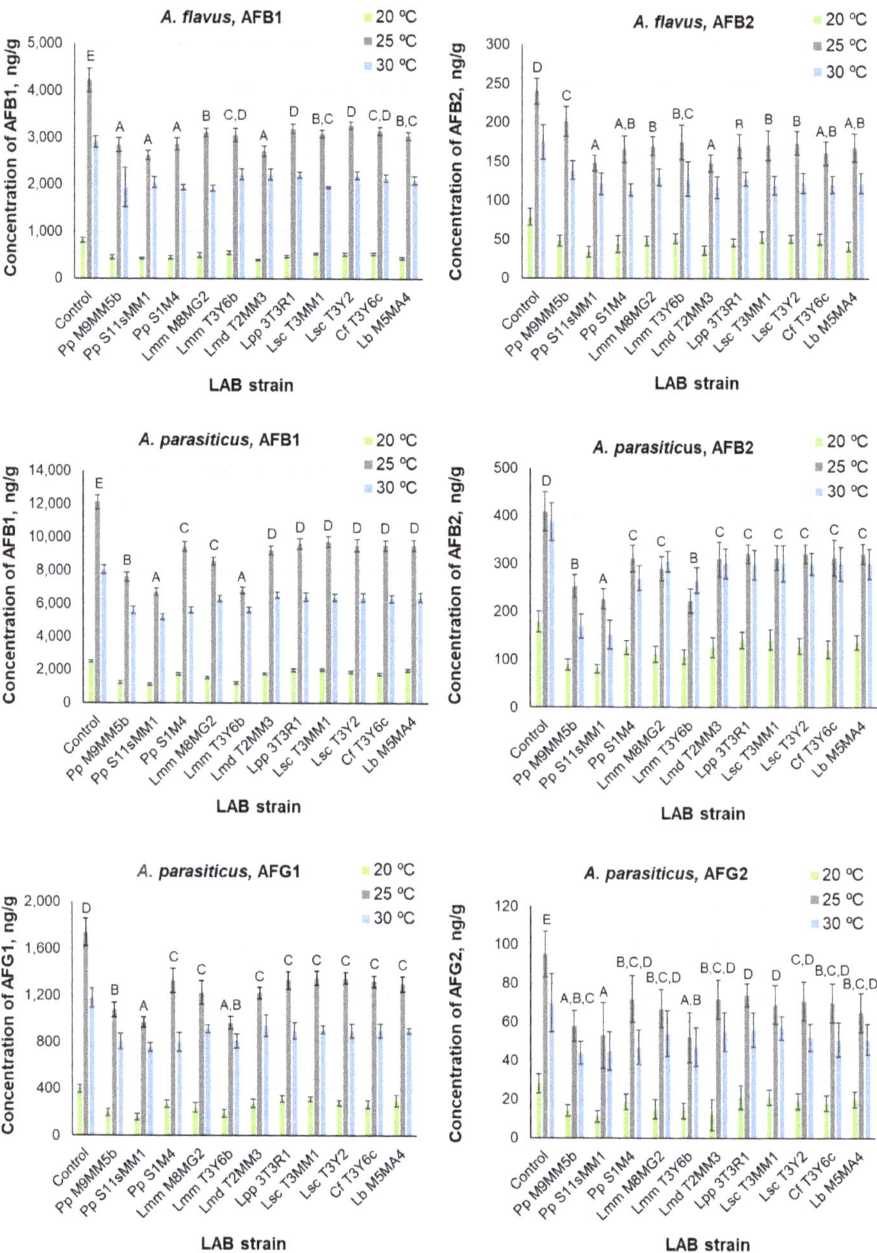

Figure 2. Mean concentrations of aflatoxins produced by *A. flavus* and *A. parasiticus* in dual medium MRS-CY20S agar in the presence of LAB strains and cultured at 20, 25, and 30 °C. Error bars represent standard deviations. Incubation time: 10 days. Capital letters above the bars indicate Duncan's homogeneous groups of LAB strains. The presence of more than one letter separated by commas means that a LAB strain may be included in more than one group. For LAB abbreviations, see the legend of Figure 1 and the Abbreviation list of the LAB species.

Table 6. Ranges of mean reduction of aflatoxin concentration in cultures of *A. flavus* and *A. parasiticus* treated with LAB strains, compared to untreated controls.

| | | Fungi | | | |
| | | *A. flavus* | | *A. parasiticus* | |
Mycotoxin	Temperature (°C)	Minimum Reduction (%)	Maximum Reduction (%)	Minimum Reduction (%)	Maximum Reduction (%)
AFB1	20	32.7	52.3	20.0	55.0
	25	22.8	37.9	19.7	44.7
	30	23.8	34.0	19.0	35.2
AFB2	20	32.9	57.0	20.7	55.3
	25	16.2	38.3	21.5	45.5
	30	20.6	34.9	21.4	60.8
AFG1	20	-	-	21.7	59.7
	25	-	-	22.4	44.4
	30	-	-	19.8	37.1
AFG2	20	-	-	25.0	59.7
	25	-	-	22.1	45.3
	30	-	-	18.6	35.5

ANOVA revealed that the three factors (fungal species, LAB strain, and temperature) significantly influenced the OTA level in the cultures ($p < 0.01$). The mutual interactions between these factors were also significant. If we consider the whole dataset, the overall order of OTA production varied with temperature as follows: 25 °C > 30 °C > 20 °C. However, the order of OTA production with temperature was different, depending on the fungal species assayed. For the black aspergilli (*A. carbonarius*, *A. niger*, and *A. welwitschiae*), it was: 25 °C > 20 °C > 30 °C, but for *A. steynii*, *A. westerdijkiae*, and *P. verrucosum*, the order was: 25 °C > 30 °C > 20 °C. There were significant differences among the fungal species. Table 7 shows the arrangement of fungi in homogeneous groups obtained by Duncan's test.

Table 7. Arrangement of ochratoxigenic *Aspergillus* spp. and *P. verrucosum* isolates in homogeneous groups, regarding OTA production in cultures of the tested LAB strains, according to Duncan's multiple range test ($\alpha = 0.05$). OTA levels increased from group A to F.

| Fungal Species | Homogeneous Groups | | | | | |
	A	B	C	D	E	F
A. carbonarius				X		
A. niger			X			
A. welwitschiae		X				
A. steynii						X
A. westerdijkiae					X	
P. verrucosum	X					

Considering the mean OTA levels in all treatment cultures of ochratoxigenic fungi at the three temperatures, the LAB strains were clustered in four homogeneous groups by Duncan's test. They were in the order of increasing OTA levels, as follows: group A: *L. paracasei* ssp. *paracasei* (3T3R1) and *L. mesenteroides* ssp. *dextranicum* (T2MM3); group B: *P. pentosaceus* (S11sMM1 and M9MM5b), *C. farciminis* (T3Y6c), *P. pentosaceus* (S1M4), and *L. mesenteroides* ssp. *mesenteroides* (T3Y6b); group C: *L. mesenteroides* ssp. *mesenteroides* (M8MG2), and group D: *L. brevis* (M5MA4) and *L. sakei* ssp. *carnosus* (T3MM1 and T3Y2).

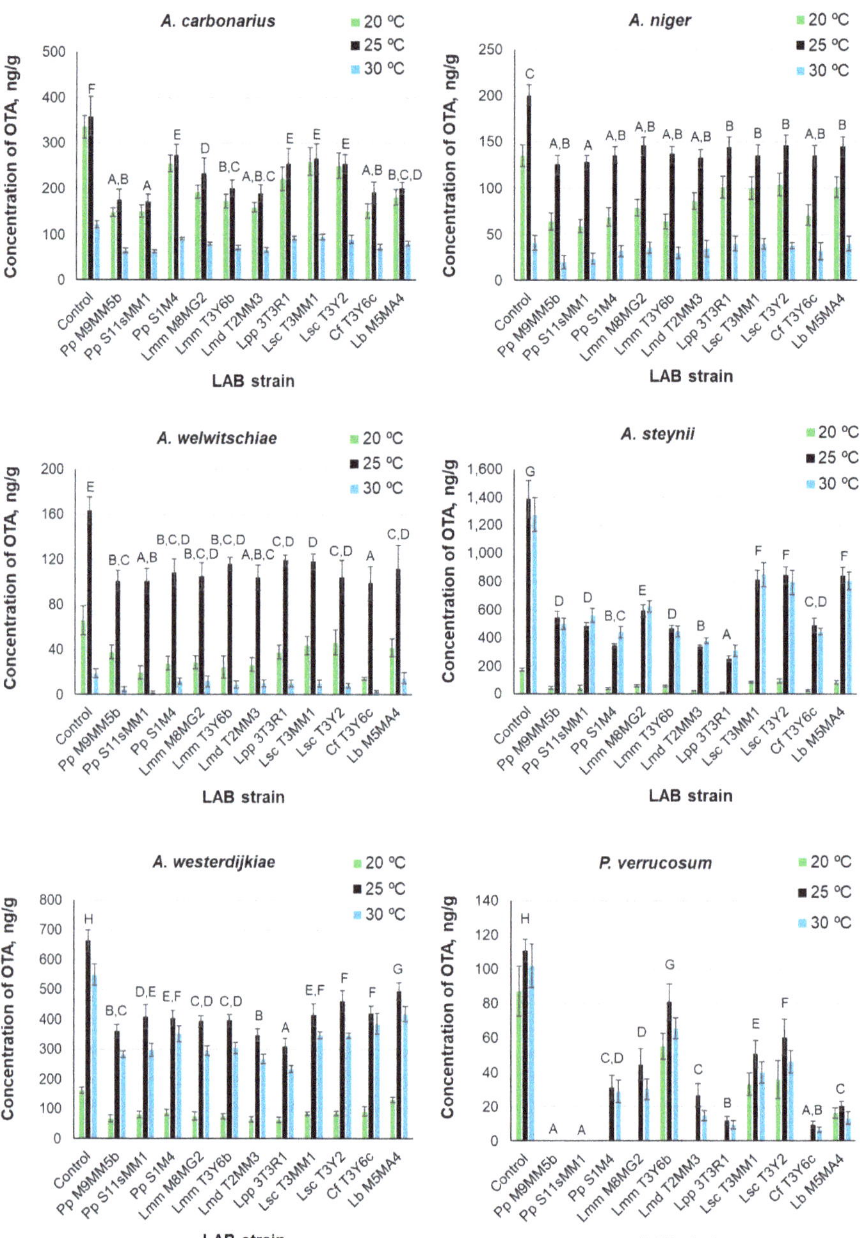

Figure 3. Concentrations of ochratoxin A (OTA) produced by various ochratoxigenic *Aspergillus* spp. and *P. verrucosum* in dual medium MRS agar-Czapek Yeast 20% Sucrose agar in the presence of LAB strains and cultured at 20, 25, and 30 °C. Error bars represent standard deviations. Incubation time: 10 days. Capital letters above the bars indicate Duncan's homogeneous groups of LAB strains. The presence of more than one letter separated by commas means that a LAB strain may be included in more than one group. For LAB abbreviations, see the legend of Figure 1 and the Abbreviation list of the LAB species.

Each fungus was placed in a single group (A to F) without overlapping. The arrangement of LAB strains into homogeneous groups, denoted by capital letters, can be observed in Figure 3. All strains in group A hampered OTA production more than those placed in the remaining groups. *P. pentosaceus* (S11sMM1) belonged to group A in cultures of *A. carbonarius*, *A. niger*, *A. welwitschiae*, and *P. verrucosum*; however, this strain was included in group D in cultures of *A. steynii* and *A. westerdijkiae*. *P. pentosaceus* (M9MM5b) behaved similarly. On the contrary, *L. paracasei* ssp. *paracasei* (3t3R1) belonged to group E in cultures of *A. carbonarius*, to group C in cultures of *A. welwitschiae* and *P. verrucosum*, and to group B in cultures of *A. niger*, but it was included in group A in cultures of *A. steynii* and *A. westerdijkiae*. Table 8 lists the minimum and maximum percentages of OTA reduction in the LAB treatments of the ochratoxigenic fungi, compared to the controls. These values changed with temperature and were dependent on the LAB strains. The minimum reduction was attained in *A. niger* cultures at 30 °C and occurred in treatments with the two strains of *L. sakei* ssp. *carnosus*, *L. brevis* (M5MA4), and *L. paracasei* ssp. *paracasei* (3T3R1). This agrees with the low ability of these LAB strains to hinder *A. niger* growth (Figure 1 and Table 3). The maximum reduction (100%) was reached in *P. verrucosum* cultures, with *P. pentosaceus* (M9MM5b and S11sMM1) at all the assayed temperatures, which also agrees with their high efficacy to hinder the growth of this fungus (Table 3).

Table 8. Ranges of mean reduction of ochratoxin A (OTA) concentration in cultures of ochratoxigenic fungi treated with LAB strains, compared to untreated controls.

Fungi	Temperature (°C)	OTA	
		Minimum Reduction (%)	Maximum Reduction (%)
A. carbonarius	20	22.6	55.6
	25	23.7	52.1
	30	22.1	48.4
A. niger	20	23.0	56.3
	25	27.0	37.0
	30	7.3	51.2
A. welwitschiae	20	28.8	77.9
	25	27.9	39.3
	30	21.1	87.4
A. steynii	20	44.5	94.8
	25	39.2	81.8
	30	33.5	75.6
A. westerdijkiae	20	18.7	61.1
	25	25.6	53.2
	30	23.8	57.3
P. verrucosum	20	36.7	100.0
	25	27.0	100.0
	30	35.6	100.0

3. Discussion

Research approaches for bio-control agents have mainly focused on LAB, due to their antifungal effects and their GRAS status. However, in toxigenic fungi research, it is necessary to also study the effect of bacteria on mycotoxin production, since LAB modulate the growth environment of fungi, which may affect growth and mycotoxin production differently [53]. Likewise, environmental factors, mainly temperature, can play a fundamental role in these possible interactions that are established in mixed bacteria-fungi cultures. In the present study, eleven strains of different LAB species previously selected for their antifungal activity were tested, and their effects on the biocontrol of relevant aflatoxigenic or ochratoxigenic fungal species and the production of AF and OTA under different temperature regimes were studied. The assays were performed in dual-layer cultures composed of MRS agar (to promote bacterial growth) and CYA20S (to promote fungal growth and mycotoxin production).

It is known that the characteristics of the medium/food and microorganisms (both fungi and LAB strains) contribute to the detoxification properties of LAB [48,49]. However, a comparative study of our results with those found in the bibliography is difficult because of the different species/microbial strains and environmental conditions. Moreover, some studies analyzed the effects of LAB strains on the growth of aflatoxigenic or ochratoxigenic fungi or on the production of one or more AF or OTA, but not on both processes at the same time, and the interactive effect of environmental temperature was not studied either [48,49]. Therefore, we will try to discuss and compare the results obtained in the present study with those previously reported for other strains/species of LAB and aflatoxigenic or ochratoxigenic fungi, regardless of the experimental conditions employed

Some authors have reported the antifungal activity of the selected LAB strains against aflatoxigenic or ochratoxigenic fungi; for example, *Lactobacillus fermentum* (YML014) isolated from Nigerian fermented foods was active against *A. flavus* and *A. niger* (reduction of fungal mycelia by ~50% in liquid medium) [54]. The mycelial dry weight of *A. flavus* was reduced to 73 and 85% using *L. casei* CRL 431 (isolated from human feces) and *L. rhamnosus* CRL 1224 (isolated from yogurt), respectively [55]. *Lactobacillus brevis* (LPBB03) isolated from coffee fruits inhibited the growth of *A. westerdijkiae* (over 50%) [56]. Other studies have shown that fermentation products of LAB can reduce the growth of aflatoxigenic or ochratoxigenic fungi. In this way, the fermentation products of *Leuconostoc citreum*, *Lactobacillus rossiae*, and *Weissella cibaria*, isolated from Italian durum wheat semolina, with high content in lactic acid and acetic acid, inhibited the growth of *A. niger*, *Penicillium roqueforti*, and *Endomyces fibuliger*, near 100% [57]. Bread spoilage caused by *A. niger* was remarkably decreased using a strain of *Lactobacillus reuteri* isolated from whole wheat sourdough. The bioactive ingredients produced by the bacterium were identified as n-decanoic acid, 3-hydroxydecanoic acid, and 3-hydroxydodecanoic acid [58]. The use of phenyl lactic acid (PLA) produced by *Pediococcus acidilactici* (CRL 1753), together with calcium propionate (CP), in bread-making as a bio-preserver significantly increased the shelf life of bread. At 18 days of storage, no molds were observed, while 70% of pieces of bread treated with CP alone were spoiled by *A. niger* and other fungi [59]. In our study, the results show that all assayed LAB strains reduced the growth of *A. flavus* (Af2225), *A. parasiticus* (Ap02), *A. carbonarius* (Ac12g), *A. niger* (An07g), *A. welwitschiae* (Aw11g), *A. steynii* (As1w), *A. westerdijkiae* (Aw019), and *P. verrucosum* (Pv10w) in percentages ranging from 0 (*A. steynii*) to 100% (*P. verrucosum*). Overall, the LAB strain encompassing the most antifungal activity was *P. pentosaceus* (S11sMM1), followed by *P. pentosaceus* (M9MM5b) and, in the third place, *L. mesenteroides* ssp. *dextranicum* (T2MM3) and *C. farciminis* (T3Y6c). The least effective LAB strains were *L. sakei* ssp. *carnosus* (T3MM1 and T3Y2). Additionally, the effectivity of each LAB strain may be quite different, depending on the fungal isolate. The fungal isolates most resistant to LAB treatments were *A. niger* and *A. welwitschiae*, followed by *A. flavus* and *A. parasiticus*. Conversely, *P. verrucosum* was the most sensitive fungus, while the remaining isolates showed intermediate susceptibility.

Concerning the use of ML to develop models able to predict, as accurately as possible, the output values (GIP) as a function of the input variables, it has been found that an MLP (NN) gave the best results because of the lowest RMSE (0.1999) and the highest R^2 value (0.9232) for the test set. After the detransformation of the logarithmic approach, the RMSE was 0.585%. It was followed by RF, then by MLR, and the XGBoost, which proved to be the worst algorithm, in spite of its major complexity. There are few reports concerning ML applications in the field of predictive mycology. Models designed by RF were superior to those built by MLP, MLR, and XGBoost to predict the growth rate (GR) of *Fusarium culmorum* and *F. proliferatum* cultured on partly milled maize with ethylene-vinyl alcohol copolymer (EVOH) films containing pure active components of essential oils [60]. However, XGBoost was better than other assayed algorithms (RF, MLR, MLP, SVM) for predicting the GR of *F. sporotrichioides* in oat grain under the influence of EVOH films containing pure active components of essential oils [61]. XGBoost also performed better than other algorithms (RF, MLR, MLP) for predicting the GR of *F. culmorum* and *F. proliferatum* in cultures carried out in a maize extract medium treated with antifungal formulations at two temperatures and two water activities [62]. These differences may be due to differences in the structure of the datasets. Thus, it is difficult to ensure a priori which ML algorithm will have the average minimum error in predicting the GR or the GIP of the fungi. It is necessary to compare them using the same dataset. There are not many reports concerning the application of ML to predict the growth of fungi in food or food-related substrates. Wawrzyniak [63] found that the MLP can be applied to develop a model able to predict the fungal population levels in bulk-stored rapeseeds at various temperatures (12–30 °C) and water activity (0.75–0.90) with high levels of generalization capability and prediction accuracy. This agrees with the study of Panagou and Kodogiannis [64] for predicting the maximum specific GR of *Monascus ruber*. However, the assessment of ML models for predicting the growth of *Aspergillus* spp. or *P. verrucosum* in the presence of LAB strains has been carried out for the first time in the present study.

These and other research, among which the present study is an important contribution, demonstrate that both the selected LAB strains and their metabolism products can be extraordinarily useful in the biocontrol of aflatoxigenic and ochratoxigenic species.

In our study, all the tested LAB strains produced a reduction of AF levels, compared to the controls. For each fungus, the reduction rate was dependent on the temperature and LAB strain. For AFB1, the ranges of reduction of the concentration in the medium were 19.0–55.0% for *A. parasiticus* and 22.8–52.3% for *A. flavus*. For AFB2, the ranges of reduction of the concentration in the medium were 20.7–60.8% for *A. parasiticus* and 16.2–57.0% for *A. flavus*. The most efficient LAB strains in reducing AF levels in the medium were *P. pentosaceus* (S11sMM1) for AFB1 in cultures of *A. parasiticus* and for AFB2 in cultures of both species and *L. mesenteroides* ssp. *dextranicum* (T2MM3) for AFB1 in cultures of *A. flavus*. This agrees with the effectiveness of these strains, especially the former, for reducing the growth of *A. flavus* and *A. parasiticus*. The reduction of OTA levels ranged from 7.3% for *A. niger* to 100% for *P. verrucosum*. The most efficient LAB strains for reducing OTA in the medium were *L. paracasei* ssp. *paracasei* (3T3R1) and *L. mesenteroides* ssp. *dextranicum* (T2MM3), whereas the least efficient were the two strains of *L. sakei* ssp. *carnosus* and *L. brevis* (M5MA4).

Some previous reports focused on the study of the effect of the selected LAB strains on the inhibition or elimination of AF or OTA in the media. Thus, *L. casei* (L30) bound up to 49.2% of the available AFB1 in aqueous solution [65]. *L. casei* Shirota (mainly live bacteria) bound AFB1 to their cell components with an efficacy of up to 90% [66]. *Lactococcus lactis* ssp. *cremoris*, *Lactobacillus rhamnosus*, and *L. lactis* ssp. *lactis* exhibited high binding capacities to remove AFM1 in milk at levels of 81.4, 56.8, and 50.8%, respectively [67], and selected isolates of *L. fermentum* (LC3/a, LC4/c, LC/5a, and LM13/b), from curd samples, were effective in removing AFB1 from culture media above 75% [68]. Some assays have shown the elimination of AF during fermentation. *L. kefiri* (KFLM3), isolated from kefir grains, was able to adsorb 80–100% AFB1, zearalenone, and OTA when cultivated in milk [69]. *L. plantarum* (R2014 and EQ12), *L. buchneri* (R1102), and *P. acidilactici* (R2142 and EQ01) linearly decreased the initial concentration of AFB1 (30 µg/kg) in silage corn forage after 3 days to <0.35 µg/kg [70].

Regarding the effect of LAB on OTA production by ochratoxigenic fungi, or their elimination from the medium, the present work provides relevant information that complements the previous studies and broadens the spectrum of the possible LAB species/strains that are useful in the food industry. In a liquid medium, *Lactobacillus acidophilus* (VM 20) produced a decrease of OTA level $\geq$ 95% by adsorption of the toxin [71]. Different *Oenococcus oeni* strains removed OTA from the medium at levels > 60%, with significant differences depending on the strain, incubation period, initial OTA level, and pH. Toxin removal was independent of bacterial viability [72,73]. *L. plantarum*, *L. brevis*, and *L. sanfranciscensis* reduced the OTA level from 16.9% to 35% in the MRS medium after 24 h of contact. OTA binding was even higher in the case of thermally inactivated bacterial biomass [74]. A high percentage of OTA reduction by *L. plantarum* (LabN10), *L. graminis* (LabN11) (>97%), and *P. pentosaceus* (>81.5%) was recorded. Factors such as temperature, pH, and bacterial biomass showed a significant effect on the growth of *A. carbonarius* (ANC89) and OTA levels in the medium [75]. These studies show the high efficacy of LAB in the control of AF and/or OTA in culture media and in food, as well as their high ability as detoxification agents.

Furthermore, no stimulation of AF or OTA production was detected in any treatment with the LAB strains used in the present work, with respect to the controls. On the contrary, in all treatments, a positive relationship between the GIP and the reduction of AF or OTA levels in the cultures was observed (Figure 1 and Tables 6 and 8). Moreover, the percentage of mycotoxin reduction was always higher than the GIP under the same conditions. It suggests the diffusion in the solid medium of possible antifungal and, especially, "antimycotoxin" metabolites, such as lactic acid, benzoic acid, propionic acid, formic acid, butyric acid, hexanoic, caproic acids, phenyl lactic acid, H_2O_2, monohydroxy octadecenoic acid, CO_2, cyclic dipeptides, phenolics, bacteriocins, fungicins, reuterine, ethanol, diacetyl, hydroxyl fatty acids, etc. [76,77]. Our results agree with the previous reports, where the efficacy of selected strains of other LAB species on the control of the growth of aflatoxigenic or ochratoxigenic fungi and the production of AF or OTA were analyzed at the same time. Taheur et al. [78], using the strain of *L. kefiri* (FR7), found levels of growth reduction of *A. flavus* and *A. carbonarius* of 51.67% and 45.56%, respectively. However, *L. kefiri* (FR7) more deeply impacted mycotoxin suppression, with reduction percentages reaching 97.22%, 95.27%, and 75.26% for AFB1, AFB2, and OTA, respectively. These results suggest that, in addition to the space competitivity fungi-LAB, the liberation of bacterial metabolites can contribute to the inhibition of mycotoxin production. Although in the present work, natural media are not used, the previous results point out that reduction in mycotoxin production also happens in natural matrices. Thus, inoculation of *L. kefiri* (FR7) in almonds artificially contaminated with *A. flavus* decreased 85.27% of AFB1 and 83.94% of AFB2 content after 7 incubation days. Application of *L. kefiri* (FR7) in peanuts artificially contaminated with *A. carbonarius* reduced OTA content to 25%. Similar results were obtained by Oliveira de Almeida-Moller et al. [79], who found that LAB strains, such as *L. brevis* (2QB422), *Levilactobacillus* spp. (2QB383), *L. brevis* (2QB446), and *Levilactobacillus* spp. (3QB398), affected the growth of *A. parasiticus*, and they were also able to reduce the production of at least three of the four tested AF by 50%. Surprisingly, despite no clear inhibition effect of the LAB strains on fungal growth, strong inhibition potential (>50%) on AF was shown by strains *L. plantarum* (3QB350), *L. plantarum* (1QB314), and *Levilactobacillus* spp. (3QB167). Ghanbari et al. [80] reported the effect of *L. plantarum* and *L. delbrueckii* ssp. *lactis* on the simultaneous reduction of the growth of *A. parasiticus* (ATCC15517), AF production (mainly AFG2), and the level of aflR gene expression. Gomaa et al. [81] found that *L. brevis* reduced the growth of *A. flavus* and AFB1 production by about 96%. The reduction effect was also confirmed at the transcriptional level, as well, where 80% less expression of the omt-A gene was observed, as compared to the control [82]. Strains of *P. pentosaceus* and *L. plantarum* isolated from grapes showed good antifungal activity against the *A. niger* aggregate and *A. carbonarius*. *P. pentosaceus* (RG7B) showed promising potential probiotic characteristics and had a high ability for OTA removal after 48 h of incubation in MRS (84%).

4. Conclusions

The selected LAB strains used in the present study can be excellent tools for the simultaneous biocontrol of the main aflatoxigenic and ochratoxigenic fungal species that affect cereals and grapes, as well as for the reduction/inhibition of the production of AF and OTA in the medium. The environmental temperature is a parameter with a very significant influence on these processes. In treatments with the assayed LAB strains, no stimulation of AF or OTA production, compared with the controls, was observed. Due to the anti-fungal and anti-mycotoxin capacity, the LAB strains used in this study could be good bio-preservative candidates for many food commodities. Further studies, using, in the first place, the natural matrices from which the fungal strains have been isolated (cereal grain and grapes), will be carried out in the future.

5. Materials and Methods

5.1. Reagents and Standards

Standards of mycotoxins AFB1, AFB2, AFG1, AFG2, and OTA were purchased from Sigma-Aldrich (Alcobendas, Spain). Glycerol was from Panreac Química (Castellar del Vallés, Barcelona, Spain). Potato dextrose agar (PDA) was from Scharlab (Barcelona, Spain). Agar, yeast extract, and De Man–Rogosa–Sharpe agar (MRS) were from Oxoid (Basingstoke, UK). Tween 80 was from Merck (Darmstadt, Germany). All reagents supplied were of analytical grade. Acetonitrile (ACN) and formic acid (all LC grade) were from J.T. Baker (Deventer, the Netherlands). Pure water was obtained from a Milli-Q Plus apparatus (Millipore, Billerica, MA, USA).

5.2. Microbial Strains and Culture Conditions

Eleven LAB strains previously selected among several hundreds of strains, in view of their antifungal activity, were assayed against relevant toxigenic and/or phytopathogenic *Aspergillus* spp. and *P. verrucosum* isolated from cereals and grapes grown in Spain. These LAB strains were: *Pediococcus pentosaceus* (M9MM5b, S11sMM1 and S1M4), *Leuconostoc mesenteroides* ssp. *mesenteroides* (M8MG2 and T3Y6b), *Leuconostoc mesenteroides* ssp. *dextranicum* (T2MM3), *Lacticaseibacillus paracasei* ssp. *paracasei* (3T3R1), *Latilactobacillus sakei* ssp. *carnosus* (T3MM1 and T3Y2), *Companilactobacillus farciminis* (T3Y6c), and *Levilactobacillus brevis* (M5MA4). All LAB strains were previously isolated from food samples in Argentina and Peru and characterized as described by Elizaquível et al. [83] and Jiménez et al. [84]. LAB strains were held at the IATA-UVEG/RA Collection (https://www.microbiospain.org/portfolio-item/iata-uveg-ra/ (accessed on 2 February 2022) and stored in MRS liquid medium containing 20% (*v/v*) glycerol at −80 °C. Before being included in this study, they were tested for authenticity by MALDI-TOF MS profiles determination using a Microflex LT MALDI-TOF MS device, software, and database for identification (Bruker Daltonics, Germany) at the Spanish-Type Culture Collection (CECT, University of Valencia, Spain).

Nine isolates of toxigenic fungi were previously selected for their phytopathogenic and/or toxigenic capacity. They were: *A. flavus* (Af2225) and *A. parasiticus* (Ap02) (from maize), *A. carbonarius* (Ac12g), *A. niger* (An07g) and *A. welwitschiae* (Aw11g) (from grapes), *A. steynii* (As1w), and *A. westerdijkiae* (Aw019) (from wheat) and *P. verrucosum* (Pv10w). Spore suspensions of pure cultures were stored at −20 °C in glycerol/saline solution (25/75, *v/v*). Fungal isolates are held at the Mycology and Mycotoxins Group Culture Collection of Valencia University (Spain).

5.3. Antifungal Assays

5.3.1. Inoculum of Bacterial and Fungal Preparation

Before carrying out the study, LAB isolates were grown in MRS broth at 28 °C for 3–5 days, then they were transferred to MRS agar and incubated under the same conditions. From these fresh cultures, bacterial suspensions on a saline solution (1.5×10^8 CFU/mL, 0.5 McFarland turbidity standard) were prepared. Two µL of this suspension were used as inoculum in the antifungal assays.

Fungal isolates were grown on PDA at 28 °C for 7 days. From these fresh fungi cultures, a suspension of spores containing 1×10^6 spores/mL was prepared in sterile pure water modified with Tween 80 (0.005%). One mL of this suspension was used as inoculum in the antifungal assays. All LAB and fungal suspensions were prepared immediately before each antifungal experiment.

5.3.2. Preparation, Inoculation, and Incubation of Dual Cultures

LAB strains were assayed for antifungal activity against toxigenic fungi using a dual culture overlay assay, according to Magnusson et al. [85], with some modifications. MRS agar was prepared, autoclaved (115 °C for 30 min), and poured into Petri dishes (15.0 mL/dish). A fresh suspension of each LAB strain (as described in Section 5.3.1) was inoculated on plates containing MRS agar in two 2-cm lines (parallel and centered at $\frac{1}{4}$ and $\frac{3}{4}$ of the same diameter) (2 µL per line) and allowed to grow at 28–30 °C for 48–72 h. The plates were then overlaid with 10 mL of Czapek yeast 20% sucrose agar (CYA20S) (yeast extract 5.0 g, sucrose 200.0 g, sodium nitrate 3.0 g, dipotassium hydrogen phosphate 1.0 g, potassium chloride 0.5 g, magnesium sulfate heptahydrate 0.5 g, ferrous sulfate heptahydrate 0.01 g, agar 15.0 g, and deionized water 1000 mL). Before being poured into the plate, this medium was autoclaved at 115 °C for 30 min, allowed to cool until 45 °C, and inoculated with 1 mL of the previously prepared suspension of fungal spores (Section 5.3.1). The dual cultures (LAB plus fungus) and control cultures (without LAB) were incubated at 20, 25, and 30 °C for 5 days. After this first incubation period, the inhibition zone of fungal growth around the previously grown bacteria was measured. The inhibition halo from each LAB inoculation line was approximately elliptical, so two measurements of the halo axes (longitudinal and transverse) at right angles were taken. Then, the plates were incubated again at the same temperatures for another 5 days. At the end of the incubation period (10 days), the size of the inhibition halo was measured again. In treatments, the x and y axes of the two elliptical zones of fungal growth inhibition on each Petri dish were measured with a rule and averaged. The surface of the inhibition area was calculated using the formula of the area inside the ellipse. The GIP was calculated from the ratio between the area of the inhibition zone and the inner area of the dish. The upper limits for the x and y axes of each elliptical inhibition zone on the 90 mm diameter plate were estimated as 44 mm and 76.2 mm, respectively. When the axes reached these values, the inhibition zone was considered to be equal to the entire inner surface of the dish, and the GIP was supposed to be 100%. At the end of the incubation period, mycotoxin levels were determined in all controls and LAB treatments. Experiments were run in triplicate and repeated twice. The results were averaged.

5.4. Mycotoxin Determination

Before the determination of the mycotoxin concentrations in MRS agar–CYA20S cultures by ultra-high-performance liquid chromatography with detection by triple quadrupole mass spectrometry (UPLC–MS/MS), calibration and validation assays of the method were performed. The analytical method described by Romera et al. [86] was followed, with minor modifications concerning the substrate and the UPLC–MS/MS equipment.

5.4.1. Calibration Solutions

Standards of AF and OTA were dissolved in ACN/water (50/50, *v/v*) to obtain stock concentrated solutions, which were maintained at −20 °C when not in use. They were appropriately diluted with ACN/water (50/50, *v/v*) to prepare the diluted standard solutions of different concentrations. Portions of non-inoculated solid control medium (MRS Agar–CYA20S) (3:2, *v/v*) were homogenized in a stomacher. Four g of the homogenate was mixed with 16 mL of ACN/water/formic acid (80:19:1, *v/v/v*) in a Falcon tube and shaken in an orbital shaker for 1 h. The mixture was centrifuged at $4260\times g$ for 5 min, and aliquots of the supernatant (2.0 mL) were transferred to vials and evaporated to dryness at 40 °C under a slight stream of N_2. To perform matrix-matched calibration, appropriate volumes of mixtures of mycotoxin standard solutions were added to the residues and then diluted (if necessary) with ACN/water/formic acid (80:19:1, *v/v/v*) up to 2.0 mL to attain the desired concentrations. Before being injected into the UPLC–MS/MS system, these standards were filtered using a syringe filter (0.22 µm, PTFE). For each mycotoxin, a calibration line was obtained by linear regression (weighting by 1/x) of the peak area from the quantifier ion vs. mycotoxin concentration. The concentration ranges of working calibration solutions (ng/mL) were as follows: AFB1 (0.57–20), AFB2 (0.57–20), AFG1 (0.57–9.25), AFG2 (0.3–9.25), and OTA (1.9–62.5).

5.4.2. Mycotoxin Recovery

Method validation was carried out by analysis of blank MRS–CYA20S agar spiked with standards of the four AF and OTA ($n = 5$) at different concentrations, which was achieved by the addition of aliquots of mycotoxin standard solutions to Erlenmeyer flasks containing 10 g of autoclaved MRS–CYA20S agar (3/2 *v/v*) allowed to cool to around 45 °C. The level ranges (ng of mycotoxin/g medium) for recovery studies were 4–70 for AFB1, 4–70 for AFB2, 4–32 for AFG1, 3–32 for AFG2, and 10–100 for OTA. Once homogenized, the spiked medium was poured into Petri dishes and allowed to cool at room temperature. After solvent evaporation, the solid medium was cut into small pieces and homogenized using a stomacher. The spiked homogenate (2.0 ± 0.1 g) was extracted in a capped Falcon tube with 8 mL ACN/water/formic acid (80:19:1, *v/v/v*) in an orbital shaker for 1 h. The extracts were treated as described in 5.4.1. Concentrations were determined by interpolation of the signals in the calibration lines. Then, the mean recovery rates and the mean relative standard deviations were calculated.

5.4.3. Determination of Mycotoxins in Dual Cultures MRS-CYA20S

To determine the effect of LAB on mycotoxin production, the dual cultures (substrate plus biomass of microorganisms) of 10 days in MRS agar-CYA20S agar used for antifungal assays were cut into small pieces, removed, weighed, homogenized in a stomacher, and analyzed, as described in Section 5.4.2, for spiked media to determinate the levels of AFB1 and AFB2 (*A. flavus* cultures), AFB1, AFB2, AFG1, and AFG2 (*A. parasiticus* cultures), and OTA (*A. carbonarius, A. niger, A. welwitschiae, A. steynii, A. westerdijkiae*, and *P. verrucosum* cultures). Mycotoxin analysis was performed as described in Section 5.4.2.

The mycotoxins (AF and OTA) accumulated in MRS–CYA20S agar by the strains of *Aspergillus* spp. and *P. verrucosum* were determined at the end of the incubation period. The whole culture in Petri dishes (substrate plus biomass of microorganisms) was cut into small pieces, removed, weighed, homogenized in a stomacher, and analyzed, as described above (Section 5.4.2), for spiked media. Extracts filtered using 0.22-μm filters were injected into the UPLC–MS/MS system. When mycotoxin levels in cultures were too high for the linear calibration range, extracts were appropriately diluted with the same solvent and injected again. Concentrations were determined by interpolation in the calibration lines obtained with standards (Section 5.4.1).

5.4.4. UPLC-MS/MS Conditions

The instrument used for the separation and detection of the mycotoxins was an Exion LC AD coupled to an MS/MS Triple Quad 6500+ System, provided with an electrospray ionization source (ESI) (AB Sciex, Foster City, CA, USA) and operated in positive ion mode (ESI+). Chromatograms were obtained by multiple reaction monitoring, and the two main product ions, one quantifier (q1) and one qualifier (q3), were monitored. The injection volume was 5 μL. Separation was performed at 30 °C in an Acquity UPLC BEH C18 column (50 × 2.1 mm, 1.7 μm particle size) (waters). The mobile phase was a time-programmed gradient of solvent A (water containing 0.15 mM ammonium formate and 0.1% formic acid) and solvent B (methanol) at a constant flow rate of 0.35 mL/min. The program was 95% solvent A for 2.0 min; 0% solvent A at min 13.0 hold to min 15.0, followed by a return to the initial conditions (95% solvent A) at min 15.1 and a stabilization period up to min 18.0. The ESI and other MS/MS conditions are in Table 9. Other ESI-source parameters were as follows: source temperature: 350 °C; curtain gas: 206.843 kPa; nebulizer gas (GS1): 379.212 kPa; heating gas (GS2): 379.212 kPa; ion spray voltage: 4500 V; target cycle time: 1 s; collision gas pressure (pure nitrogen): high.

Table 9. MS/MS conditions for the detection of mycotoxins.

Mycotoxin	ESI Polarity	Molecular Mass (Da)	Precursor Ion	*m/z* (Da)	Product Ion (*m/z*) (Da)	DP(V)	EP(V)	CE(V)	CXP(V)
AFB1	+	312.063	[M + H]$^+$	313.1	285.2 [1]	106	10	33	16
					128.1 [2]	106	10	91	10
AFB2	+	314.079	[M + H]$^+$	315.1	287.2 [1]	96	10	37	18
					259.2 [2]	96	10	43	18
AFG1	+	328.058	[M + H]$^+$	329.1	243.1 [1]	86	10	39	14
					200.0 [2]	86	10	59	12
AFG2	+	330.074	[M + H]$^+$	331.1	313.2 [1]	111	10	35	18
					245.2 [2]	111	10	43	14
OTA	+	403.082	[M + H]$^+$	404.0	239.0 [1]	91	10	37	16
					102.0 [2]	91	10	105	14

DP: declustering potential; EP: Entrance potential; CE: Collision energy; CXP: Collision cell exit potential.[1] Quantifier ion (q1); [2] Qualifier ion (q3).

5.5. Statistics

Three replications were conducted for each treatment, and the treatments were repeated two times. The results were expressed as the mean value with standard error. Data were analyzed by multifactor analysis of variance (ANOVA) using Statgraphics Centurion XV.II statistical package (StatPoint, Inc., Warrenton, VA, USA). *Post-hoc* Duncan's multiple range test (α = 0.05) was used to find homogenous groups when significant differences between means were revealed by ANOVA. For calculation purposes, undetectable mycotoxin levels were considered to be zero.

5.6. Method for the Design of Predictive ML Models for Growth Inhibition Percentage

Various ML models were comparatively tested using the same dataset for modeling the GIP in cultures of the different fungal species, carried out at three temperatures and treated with the different LAB strains. Control cultures were also considered. They were multilayer perceptrons (MLP), which is a kind of NN (we used a single layer perceptron), random forest (RF), extreme gradient boosting trees (XGBoost using the *xgbTree* method), and multiple linear regression (MLR). The software used was R and the 'classification and regression training' (caret) package [87]. The parameters to be tuned in MLP were *size* (the number of hidden nodes in the hidden layer) and *decay* (the weight decay rate set at 0.01 to 1.0); in RF, they were *ntry* (number of trees to grow, usually 500) and *mtry* (number of variables randomly sampled as candidates at each split, here 2 or 3); in XGBoost, there were more parameters, and to avoid very large computation times, some of them were kept constant, while other were tuned (Table 4). The input variables were three (temperature, LAB strain, and fungal species). The output variable was $\log_{10}$(GIP + 1). The output values were averaged, as only one output value was associated with a set of input values. Variables were preprocessed before mathematical treatment. The categorical input variables (LAB strain and fungal species) were transformed into numerical variables by recoding them as "dummy" variables [60,61].

Before training the ML models, the dataset was randomly split into training (75%) and test (25%) sets. The training set was used to develop the ML models using 10-fold cross-validation, whereas the test set was used to assess model performance using independent data hidden during the training/validation. MLR used all independent variables to predict the output value, and the coefficients were chosen by the least square method. This algorithm assumes that there is a linear relationship between the dependent and independent variables. The metric used for obtaining the best parameters and for evaluating the performance of the ML models (the loss function) was the RMSE over the test set. The model parameters were tuned to optimize their performance (minimum RMSE) for each task. The RMSE is measured in the same units as the observed/predicted values and is severely affected by large error values. The coefficient of determination (R-squared), which evaluates how much of the output variance can be explained by the variation in the input variables, was also obtained. The goal was to maximize it and minimize the RMSE. The best predictive ML models obtained after training/cross-validation (minimum RMSE) or by the least square method, in the case of MLR, were further tested and compared against the same test set. The RMSE and R-squared values for the test sets evaluated the model performance working as a comparison tool to choose the best model.

Author Contributions: Conceptualization, E.M.M. and M.J.; methodology, A.T.; software, E.M.M. and F.M.; validation, E.M.M., F.M. and M.J.; formal analysis, E.M.M. and A.T.; investigation, E.M.M., A.T. and F.M.; resources, M.J.; data curation, E.M.M. and F.M.; writing—original draft preparation. E.M.M.; writing—review and editing, E.M.M. and F.M.; visualization, E.M.M. and F.M.; supervision, E.M.M. and M.J.; project administration, M.J.; funding acquisition, M.J. All authors have read and agreed to the published version of the manuscript.

Funding: This research was co-funded by the European Regional Development Fund (ERDF) and 'Ministerio de Economía y Competitividad' and 'Ministerio de Ciencia, Innovación y Universidades' (Spanish Government) through Project RTI2018-097593-B-C22.

Institutional Review Board Statement: Not applicable.

Informed Consent Statement: Not applicable.

Data Availability Statement: The data supporting reported results other than those included in Tables and Figures in the main text are available, on request, from the corresponding Author.

Acknowledgments: R. Aznar (Spanish CECT and the University of Valencia) is highly acknowledged for providing and characterizing the LAB strains used in this study.

Conflicts of Interest: The authors declare no conflict of interest.

Abbreviation List of the LAB Species

Full Name	Abbreviation
Pediococcus pentosaceus	Pp
Leuconostoc mesenteroides ssp. *mesenteroides*	Lmm
Leuconostoc mesenteroides ssp. *dextranicum*	Lmd
Lacticaseibacillus paracasei ssp. *paracasei*	Lpp
Latilactobacillus sakei ssp. *carnosus*	Lsc
Companilactobacillus farciminis	Cf
Levilactobacillus brevis	Lb

References

1. European Commission. Commission Regulation (EC) No. 1881/2006 of 19 December 2006 setting maximum levels for certain contaminants in foodstuffs. *Off. J. Eur. Union* **2006**, *L 364*, 5–24.
2. Marasas, W.F.O.; Gelderblom, W.C.A.; Shephard, G.S.; Vismer, H.F. Mycotoxins: A global problem. In *Mycotoxins: Detection Methods, Management, Public Health and Agricultural Trade*; Leslie, J.F., Bandyopadhyay, R., Visconti, A., Eds.; CABI: Wallingford, UK, 2008; pp. 29–40.
3. Frisvad, J.C.; Hubka, V.; Ezekiel, C.N.; Hong, S.-B.; Novakova, A.; Chen, A.J.; Arzanlou, M.; Larsen, T.O.; Sklenar, F.; Mahakarn-chanakul, W.; et al. Taxonomy of *Aspergillus* section *Flavi* and their production of aflatoxins, ochratoxins and other mycotoxins. *Stud. Mycol.* **2019**, *93*, 1–63. [CrossRef] [PubMed]
4. Taniwaki, M.H.; Pitt, J.I.; Magan, N. *Aspergillus* species and mycotoxins: Occurrence and importance in major food commodities. *Curr. Opin. Food Sci.* **2018**, *23*, 38–43. [CrossRef]
5. Naeem, I.; Ismail, A.; Rehman, A.U.; Ismail, Z.; Saima, S.; Naz, A.; Faraz, A.; de Oliveira, C.A.F.; Benkerroum, N.; Aslam, M.Z.; et al. Prevalence of aflatoxins in selected dry fruits, impact of storage conditions on contamination levels and associated health risks on Pakistani consumers. *Int. J. Environ. Res. Public Health* **2022**, *19*, 3404. [CrossRef] [PubMed]
6. Tarazona, A.; Gómez, J.V.; Mateo, F.; Jiménez, M.; Romera, D.; Mateo, E. M: Study on mycotoxin contamination of maize kernels in Spain. *Food Control* **2020**, *118*, 107370. [CrossRef]
7. Ibáñez-Vea, M.; Martínez, R.; González-Peñas, E.; Lizarraga, E.; López de Cerain, A. Co-occurrence of aflatoxins, ochratoxin A and zearalenone in breakfast cereals from Spanish market. *Food Control* **2011**, *22*, 1949–1955. [CrossRef]

8.   Bashiry, M.; Javanmardi, F.; Sadeghi, E.; Shokri, S.; Hossieni, H.; Oliveira, C.A.F.; Khaneghah, A.M. The prevalence of aflatoxins in commercial baby food products: A global systematic review, meta-analysis, and risk assessment study. *Trends Food Sci. Technol.* **2021**, *114*, 100–115. [CrossRef]

9.   Copetti, M.V.; Iamanaka, B.T.; Pereira, J.L.; Fungaro, M.H.; Taniwaki, M.H. Aflatoxigenic fungi and aflatoxin in cocoa. *Int. J. Food Microbiol.* **2011**, *148*, 141–144. [CrossRef]

10. Lutfullah, G.; Hussain, A. Studies on contamination level of aflatoxins in some cereals and beans of Pakistan. *Food Control* **2012**, *23*, 32–36. [CrossRef]

11. Mollayusefian, I.; Ranaei, V.; Pilevar, Z.; Cabral-Pinto, M.M.S.; Rostami, A.; Nematolahi, A.; Khedher, K.M.; Thai, V.M.; Fakhri, Y.; Khaneghah, A.M. The concentration of aflatoxin M1 in raw and pasteurized milk: A worldwide systematic review and meta-analysis. *Trends Food Sci. Technol.* **2021**, *115*, 22–30. [CrossRef]

12. IARC (International Agency for Research on Cancer). Evaluation of Carcinogenic Risks of Chemicals to Humans. Some Naturally Occurring Substances: Food Items and Constituents, Heterocyclic Aromatic Amines and Mycotoxins. In *Aflatoxins: IARC Monographs*; IARC: Lyon, France, 1993; Volume 56, pp. 245–395.

13. Center for Disease Control and Prevention (CDC). Outbreak of Aflatoxin Poisoning—Eastern and Central Provinces, Kenya 3 September 2004. Available online: https://www.cdc.gov/nceh/hsb/chemicals/aflatoxin.htm (accessed on 20 July 2022).

14. Gil-Serna, J.; Patiño, B.; Cortes, L.; González-Jaén, M.T.; Vázquez, C. *Aspergillus steynii* and *Aspergillus westerdijkiae* as potential risk of OTA contamination in food products in warm climates. *Food Microbiol.* **2015**, *46*, 168–175. [CrossRef] [PubMed]

15. Hong, S.-B.; Lee, M.; Kim, D.-H.; Varga, J.; Frisvad, J.C.; Perrone, G.; Gomi, K.; Yamada, O.; Machida, M.; Houbraken, J.; et al. *Aspergillus luchuensis*, an industrially important black *Aspergillus* in East Asia. *PLoS ONE* **2013**, *8*, e63769. [CrossRef]

16. Perrone, G.; Stea, G.; Epifani, F.; Varga, J.; Frisvad, J.C.; Samson, R.A. *Aspergillus niger* contains the cryptic phylogenetic species *A. awamori*. *Fungal Biol.* **2011**, *115*, 1138–1150. [CrossRef] [PubMed]

17. Amézqueta, S.; Schorr-Galindo, S.; Murillo-Arbizu, M.; González-Peñas, E.; López de Cerain, A.; Guiraud, J.P. OTA-producing fungi in foodstuffs: A review. *Food Control* **2012**, *26*, 259–268. [CrossRef]

18. Geisen, R.; Schmidt-Heydt, M.; Touhami, N.; Himmelsbach, A. New aspects of ochratoxin A and citrinin biosynthesis in *Penicillium*. *Curr. Opin. Food Sci.* **2018**, *23*, 23–31. [CrossRef]

19. Kara, G.N.; Ozbey, F.; Kabak, B. Co-occurrence of aflatoxins and ochratoxin A in cereal flours commercialised in Turkey. *Food Control* **2015**, *54*, 275–281. [CrossRef]

20. Lai, X.; Liu, R.; Ruan, C.; Zhang, H.; Liu, C. Occurrence of aflatoxins and ochratoxin A in rice samples from six provinces in China. *Food Control* **2015**, *50*, 401–404. [CrossRef]

21. Mateo, E.M.; Gil-Serna, J.; Patiño, B.; Jiménez, M. Aflatoxins and ochratoxin A in stored barley grain in Spain and impact of PCR-based strategies to assess the occurrence of aflatoxigenic and ochratoxigenic *Aspergillus* spp. *Int. J. Food Microbiol.* **2011**, *149*, 118–126. [CrossRef]

22. Tarazona, A.; Gómez, J.V.; Mateo, F.; Jiménez, M.; Mateo, E.M. Potential Health Risk Associated with Mycotoxins in Oat Grains Consumed in Spain. *Toxins* **2021**, *13*, 421. [CrossRef]

23. Mateo, R.; Medina, A.; Mateo, F.; Mateo, E.M.; Jiménez, M. An overview of ochratoxin A in beer and wine. *Int. J. Food Microbiol.* **2007**, *119*, 79–83. [CrossRef]

24. Mehri, F.; Esfahani, M.; Heshmati, A.; Jenabi, E.; Khazaei, S. The prevalence of ochratoxin A in dried grapes and grape-derived products: A systematic review and meta-analysis. *Toxin Rev.* **2022**, *41*, 347–356. [CrossRef]

25. Anelli, P.; Haidukowski, M.; Epifani, F.; Cimmarusti, M.T.; Moretti, A.; Logrieco, A.; Susca, A. Fungal mycobiota and mycotoxin risk for traditional artisan Italian cave cheese. *Food Microbiol.* **2019**, *78*, 62–72. [CrossRef] [PubMed]

26. Stefanello, A.; Gasperini, A.M.; Copetti, M.V. Ecophysiology of OTA-producing fungi and its relevance in cured meat products. Review. *Curr. Opin. Food Sci.* **2022**, *45*, 100838. [CrossRef]

27. Duarte, S.C.; Pena, A.; Lino, C.M. A review on ochratoxin A occurrence and effects of processing of cereal and cereal derived food products. *Food Microbiol.* **2010**, *27*, 187–198. [CrossRef]

28. Benites, A.J.; Fernandes, M.; Boleto, A.R.; Azevedo, S.; Silva, S.; Leitão, A.L. Occurrence of ochratoxin A in roasted coffee samples commercialized in Portugal. *Food Control* **2017**, *73*, 1223–1228. [CrossRef]

29. Copetti, M.V.; Iamanaka, B.T.; Nester, M.A.; Efraim, P.; Taniwaki, M.H. Occurrence of ochratoxin A in cocoa by-products and determination of its reduction during chocolate manufacture. *Food Chem.* **2013**, *136*, 100–104. [CrossRef]

30. Bircan, C. Incidence of ochratoxin A in dried fruits and co-occurrence with aflatoxins in dried figs. *Food Chem. Toxicol.* **2009**, *47*, 1996–2001. [CrossRef]

31. Prelle, A.; Spadaro, D.; Garibaldi, A.; Gullino, M.L. Co-occurrence of aflatoxins and ochratoxin A in spices commercialized in Italy. *Food Control* **2014**, *39*, 192–197. [CrossRef]

32. Khoi, C.S.; Chen, J.H.; Lin, T.Y.; Chiang, C.K.; Hung, K.Y. Ochratoxin A-Induced Nephrotoxicity: Up-to-Date Evidence. *Int. J. Mol. Sci.* **2021**, *22*, 11237. [CrossRef]

33. Heussner, A.H.; Bingle, L.E.H. Comparative ochratoxin toxicity: A review of the available data. *Toxins* **2015**, *7*, 4253–4282. [CrossRef]

34. Gan, F.; Zhou, Y.; Hou, L.; Qian, G.; Chen, X.; Huang, K. Ochratoxin A induces nephrotoxicity and immunotoxicity through different MAPK signaling pathways in PK15 cells and porcine primary splenocytes. *Chemosphere* **2017**, *182*, 630–637. [CrossRef]

35. Hedayati, M.T.; Pasqualotto, A.C.; Warn, P.A.; Bowyer, P.; Denning, D.W. *Aspergillus flavus*: Human pathogen, allergen and mycotoxin producer. *Microbiology* **2007**, *153*, 1677–1692. [CrossRef]

36. Wong, S.S.W.; Venugopalan, L.P.; Beaussart, A.; Karnam, A.; Mohammed, M.R.S.; Jayapal, J.M.; Bretagne, S.; Bayry, J.; Prajna, L.; Kuppamuthu, D.; et al. Species-specific immunological reactivities depend on the cell-wall organization of the two *Aspergillus*, *Aspergillus fumigatus* and *A. flavus*. *Front. Cell. Infect. Microbiol.* **2021**, *11*, 643312. [CrossRef]

37. Denning, D.M.; Bromley, M.J. Infectious Disease. How to bolster the antifungal pipeline. *Science* **2015**, *347*, 1414–1416. [CrossRef] [PubMed]

38. United States Department of Agriculture (USDA). *Grain, Fungal Diseases and Mycotoxin Reference*; United States Grain Inspection, Packers and Stockyards Administration: Washington, DC, USA, 2016. Available online: https://www.ams.usda.gov/sites/default/files/media/FungalDiseaseandMycotoxinReference2017.pdf (accessed on 4 February 2022).

39. Winter, G.; Pereg, L. A review on the relation between soil and mycotoxins: Effect of aflatoxin on field, food and finance. *Eur. J. Soil Sci.* **2019**, *70*, 882–897. [CrossRef]

40. Savary, S.; Ficke, A.; Aubertot, J.N.; Hollier, C. Crop losses due to diseases and their implications for global food production losses and food security. *Food Secur.* **2012**, *4*, 519–537. [CrossRef]

41. Varsha, K.K.; Devendra, L.; Shilpa, G.; Priya, S.; Pandey, A.; Nampoothiri, K.M. 2,4-Di-*tert*-butyl phenol as the antifungal, antioxidant bioactive purified from a newly isolated *Lactococcus* sp. *Int. J. Food Microbiol.* **2015**, *211*, 44–50. [CrossRef]

42. Spadaro, D.; Garibaldi, A. Containment of mycotoxins in the food chain by using decontamination and detoxification techniques. In *Practical Tools for Plant and Food Biosecurity*; Gullino, M., Stack, J., Fletcher, J., Mumford, J., Eds.; Springer International Publishing: Cham, Switzerland, 2017; Volume 8, pp. 163–177. [CrossRef]

43. James, A.; Zikankuba, V.L. Mycotoxins contamination in maize alarms food safety in sub-Sahara Africa. *Food Control* **2018**, *90*, 372–381. [CrossRef]

44. Karlovsky, P.; Suman, M.; Berthiller, F.; De Meester, J.; Eisenbrand, G.; Perrin, I.; Dussort, P. Impact of food processing and detoxification treatments on mycotoxin contamination. *Mycotoxin Res.* **2016**, *32*, 179–205. [CrossRef]

45. Battilani, P.; Toscano, P.; Van der Fels-Klerx, H.J.; Moretti, A.; Camardo Leggieri, M.; Brera, C.; Rortais, A.; Goumperis, T.; Robinson, T. Aflatoxin B1 contamination in maize in Europe increases due to climate change. *Sci. Rep.* **2016**, *6*, 24328. [CrossRef]

46. Moretti, A.; Pascale, M.; Logrieco, A.F. Mycotoxin risks under a climate change scenario in Europe. *Trends Food Sci. Technol.* **2019**, *84*, 38–40. [CrossRef]

47. Leuschner, R.G.; Robinson, T.P.; Hugas, M.; Cocconcelli, P.S.; Richard-Forget, F.; Klein, G.; von Wright, A. Qualified presumption of safety (QPS): A generic risk assessment approach for biological agents notified to the European food safety authority (EFSA). *Trends Food Sci. Technol.* **2010**, *21*, 425–435. [CrossRef]

48. Bangar, S.P.; Sharma, N.; Kumar, M.; Ozogul, F.; Purewal, S.S.; Trif, M. Recent developments in applications of lactic acid bacteria against mycotoxin production and fungal contamination. *Food Biosci.* **2021**, *44*, 101444. [CrossRef]

49. Sadiq, F.A.; Yan, B.; Tian, F.; Zhao, J.; Zhang, H.; Chen, W. Lactic acid bacteria as antifungal and anti-mycotoxigenic agents: A comprehensive review. *Compr. Rev. Food Sci. Food Saf.* **2019**, *18*, 1403–1436. [CrossRef]

50. Muhialdin, B.J.; Saari, N.; Hussin, A.S.M. Review on the biological detoxification of mycotoxins using lactic acid bacteria to enhance the sustainability of foods supply. *Molecules* **2020**, *25*, 2655. [CrossRef]

51. Greener, J.G.; Kandathil, S.M.; Moffat, L.; Jones, D.T. A guide to machine learning for biologists. *Nat. Rev. Mol. Cell Biol.* **2022**, *23*, 40–55. [CrossRef]

52. Goodswen, S.J.; Barratt, J.L.N.; Kennedy, P.J.; Kaufer, A.; Calarco, L.; Ellis, J.T. Machine learning and applications in microbiology. *FEMS Microbiol. Rev.* **2021**, *45*, fuab015. [CrossRef]

53. Asurmendi, P.; Barberis, C.; Pascual, L.; Dalcero, A.; Barberis, L. Influence of *Listeria monocytogenes* and environmental abiotic factors on growth parameters and aflatoxin B1 production by *Aspergillus flavus*. *J. Stored Prod. Res.* **2015**, *60*, 60–66. [CrossRef]

54. Adedokun, E.O.; Rather, I.A.; Bajpai, V.K.; Park, Y.H. Biocontrol efficacy of *Lactobacillus fermentum* YML014 against food spoilage moulds using the tomato puree model. *Front. Life Sci.* **2016**, *9*, 64–68. [CrossRef]

55. Bueno, D.J.; Silva, J.O.; Oliver, G.; Gonzalez, S.N. *Lactobacillus casei* CRL 431 and *Lactobacillus rhamnosus* CRL 1224 as biological controls for *Aspergillus flavus* strains. *J. Food Prot.* **2006**, *69*, 2544–2548. [CrossRef]

56. Pereira, G.V.D.; Beux, M.; Pagnoncelli, M.G.B.; Soccol, V.T.; Rodrigues, C.; Soccol, C.R. Isolation, selection and evaluation of antagonistic yeasts and lactic acid bacteria against ochratoxigenic fungus *Aspergillus westerdijkiae* on coffee beans. *Lett. Appl. Microbiol.* **2016**, *62*, 96–101. [CrossRef] [PubMed]

57. Valerio, F.; Favilla, M.; De Bellis, P.; Sisto, A.; de Candia, S.; Lavermicocca, P. Antifungal activity of strains of lactic acid bacteria isolated from a semolina ecosystem against *Penicillium roqueforti*, *Aspergillus niger* and *Endomyces fibuliger* contaminating bakery products. *Syst. Appl. Microbiol.* **2009**, *32*, 438–448. [CrossRef] [PubMed]

58. Sadeghi, A.; Ebrahimi, M.; Mortazavi, S.A.; Abedfar, A. Application of the selected antifungal LAB isolate as a protective starter culture in pan whole-wheat sourdough bread. *Food Control* **2019**, *95*, 298–307. [CrossRef]

59. Bustos, A.Y.; de Valdez, G.F.; Gerez, C.L. Optimization of phenyllactic acid production by *Pediococcus acidilactici* CRL 1753. Application of the formulated bio-preserver culture in bread. *Biol. Control* **2018**, *123*, 137–143. [CrossRef]

60. Tarazona, A.; Mateo, E.M.; Gómez, J.V.; Gavara, R.; Jiménez, M.; Mateo, F. Machine learning approach for predicting *Fusarium culmorum* and *F. proliferatum* growth and mycotoxin production in treatments with ethylene-vinyl alcohol copolymer films containing pure components of essential oils. *Int. J. Food Microbiol.* **2021**, *338*, 109012. [CrossRef]

61. Mateo, E.M.; Gómez, J.V.; Tarazona, A.; García-Esparza, M.Á.; Mateo, F. Comparative Analysis of Machine Learning Methods to Predict Growth of *F. sporotrichioides* and Production of T-2 and HT-2 Toxins in Treatments with Ethylene-Vinyl Alcohol Films Containing Pure Components of Essential Oils. *Toxins* **2021**, *13*, 545. [CrossRef] [PubMed]

62. Tarazona, A.; Mateo, E.M.; Gómez, J.V.; Romera, D.; Mateo, F. Potential use of machine learning methods in assessment of *Fusarium culmorum* and *Fusarium proliferatum* growth and mycotoxin production in treatments with antifungal agents. *Fungal Biol.* **2021**, *125*, 123–133. [CrossRef] [PubMed]

63. Wawrzyniak, J. Application of Artificial Neural Networks to Assess the Mycological State of Bulk Stored Rapeseeds. *Agriculture* **2020**, *10*, 567. [CrossRef]

64. Panagou, E.; Kodogiannis, V. Application of neural networks as a non-linear modelling technique in food mycology. *Expert Syst. Appl.* **2009**, *36*, 121–131. [CrossRef]

65. Hernández-Mendoza, A.; García, H.S.; Steele, J.L. Screening of *Lactobacillus casei* strains for their ability to bind aflatoxin B1. *Food Chem. Toxicol.* **2009**, *47*, 1064–1068. [CrossRef]

66. Liew, W.-P.-P.; Nurul-Adilah, Z.; Than, L.T.L.; Mohd-Redzwan, S. The binding efficiency and interaction of *Lactobacillus casei* Shirota toward aflatoxin B1. *Front. Microbiol.* **2018**, *9*, 1503. [CrossRef] [PubMed]

67. Muaz, K.; Riaz, M.; Rosim, R.E.; Akhtar, S.; Corassin, C.H.; Gonçalves, B.L.; Fernandes-Oliveira, C.A. In vitro ability of nonviable cells of lactic acid bacteria strains in combination with sorbitan monostearate to bind to aflatoxin M1 in skimmed milk. *LWT Food Sci. Technol.* **2021**, *147*, 111666. [CrossRef]

68. Kumara, S.S.; Bashisht, A.; Venkateswaran, G.; Hariprasad, P.; Gayathri, D. Characterization of novel *Lactobacillus fermentum* from curd samples of indigenous cows from Malnad Region, Karnataka, for their aflatoxin B1 binding and probiotic properties. *Probiot. Antimicrob. Proteins* **2019**, *11*, 1100–1109. [CrossRef] [PubMed]

69. Taheur, F.B.; Fedhila, K.; Chaieb, K.; Kouidhi, B.; Bakhrouf, A.; Abrunhosa, L. Adsorption of aflatoxin B1, zearalenone and ochratoxin A by microorganisms isolated from Kefir grains. *Int. J. Food Microbiol.* **2017**, *251*, 1–7. [CrossRef] [PubMed]

70. Ma, Z.X.; Amaro, F.X.; Romero, J.J.; Pereira, O.G.; Jeong, K.C.; Adesogan, A.T. The capacity of silage inoculant bacteria to bind aflatoxin B1 in vitro and in artificially contaminated corn silage. *J. Dairy Sci.* **2017**, *100*, 7198–7210. [CrossRef]

71. Fuchs, S.; Sontag, G.; Stidl, R.; Ehrlich, V.; Kundi, M.; Knasmüller, S. Detoxification of patulin and ochratoxin A, two abundant mycotoxins, by lactic acid bacteria. *Food Chem. Toxicol.* **2008**, *46*, 1398–1407. [CrossRef]

72. Mateo, E.M.; Medina, A.; Mateo, F.; Valle-Algarra, F.M.; Pardo, I.; Jiménez, M. Ochratoxin A removal in synthetic media by living and heat-inactivated cells of *Oenococcus oeni* isolated from wines. *Food Control* **2010**, *21*, 23–28. [CrossRef]

73. Mateo, E.M.; Medina, A.; Mateo, R.; Jiménez, M. Effect of ethanol on the ability of *Oenococcus oeni* to remove ochratoxin A in synthetic wine-like media. *Food Control* **2010**, *21*, 935–941. [CrossRef]

74. Piotrowska, M. The adsorption of ochratoxin A by *Lactobacillus* Species. *Toxins* **2014**, *6*, 2826–2839. [CrossRef]

75. Belkacem-Hanfi, N.; Fhoula, I.; Semmar, N.; Guesmi, A.; Perraud-Gaime, I.; Ouzari, H.L.; Roussos, S. Lactic acid bacteria against post-harvest moulds and ochratoxin A isolated from stored wheat. *Biol. Control* **2014**, *76*, 52–59. [CrossRef]

76. Cortés-Zavaleta, O.; López-Malo, A.; Hernández-Mendoza, A.; García, H.S. Antifungal activity of lactobacilli and its relationship with 3-phenyllactic acid production. *Int. J. Food Microbiol.* **2014**, *173*, 30–35. [CrossRef] [PubMed]

77. Ruggirello, M.; Nucera, D.; Cannoni, M.; Peraino, A.; Rosso, F.; Fontana, M.; Cocolin, L.; Dolci, P. Antifungal activity of yeasts and lactic acid bacteria isolated from cocoa bean fermentations. *Food Res. Int.* **2019**, *115*, 519–525. [CrossRef] [PubMed]

78. Taheur, F.B.; Mansour, C.; Kouidhi, B.; Chaieb, K. Use of lactic acid bacteria for the inhibition of Aspergillus flavus and *Aspergillus carbonarius* growth and mycotoxin production. *Toxicon* **2019**, *166*, 15–23. [CrossRef] [PubMed]

79. Oliveira de Almeida-Moller, C.; Freire, L.; Rosim, R.E.; Pereira-Margalho, L.; Fasura-Balthazar, C.; Tuanny-Franco, L.; de Souza-Sant'Ana, A.; Corassin, C.H.; Rattray, F.P.; Fernandes de Oliveira, C.A. Effect of lactic acid bacteria strains on the growth and aflatoxin production potential of *Aspergillus parasiticus*, and their ability to bind aflatoxin B1, ochratoxin A, and zearalenone in vitro. *Front. Microbiol.* **2021**, *12*, 899.

80. Ghanbari, R.; Aghaee, E.M.; Rezaie, S.; Khaniki, G.J.; Alimohammadi, M.; Soleimani, M.; Noorbakhsh, F. The inhibitory effect of lactic acid bacteria on aflatoxin production and expression of aflR gene in *Aspergillus parasiticus*. *J. Food Saf.* **2018**, *38*, e12413. [CrossRef]

81. Gomaa, E.Z.; Abdelall, M.F.; El-Mahdy, O.M. Detoxification of aflatoxin B1 by antifungal compounds from *Lactobacillus brevis* and *Lactobacillus paracasei*, isolated from Dairy Products. *Probiot. Antimicrob. Proteins* **2018**, *10*, 201–209. [CrossRef] [PubMed]

82. Taroub, B.; Salma, L.; Manel, Z.; Ouzari, H.I.; Hamdi, Z.; Moktar, H. Isolation of lactic acid bacteria from grapefruit: Antifungal activities, probiotic properties, and in vitro detoxification of ochratoxin A. *Ann. Microbiol.* **2019**, *69*, 17–27. [CrossRef]

83. Elizaquível, P.; Pérez-Cataluña, A.; Yépez, A.; Aristimuño, C.; Jiménez, E.; Cocconcelli, P.S.; Vignolo, G.; Aznar, R. Pyrosequencing vs. culture-dependent approaches to analyze lactic acid bacteria associated to chicha, a traditional maize-based fermented beverage from Northwestern Argentina. *Int. J. Food Microbiol.* **2015**, *198*, 9–18. [CrossRef]

84. Jiménez, E.; Yépez, A.; Pérez-Cataluña, A.; Ramos Vásquez, E.; Zúñiga Dávila, D.; Vignolo, G.; Aznar, R. Exploring diversity and biotechnological potential of lactic acid bacteria from tocosh—traditional Peruvian fermented potatoes—by high throughput sequencing (HTS) and culturing. *LWT-Food Sci. Technol.* **2018**, *87*, 567–574. [CrossRef]

85. Magnusson, L.; Ström, K.; Roos, S.; Sjögren, J.; Schnürer, J. Broad and complex antifungal activity among environmental isolates of lactic acid bacteria. *FEMS Microbiol. Lett.* **2003**, *219*, 129–135. [CrossRef]

36. Romera, D.; Mateo, E.M.; Mateo-Castro, R.; Gómez, J.V.; Gimeno-Adelantado, J.V.; Jiménez, M. Determination of multiple mycotoxins in feedstuffs by combined use of UPLC–MS/MS and UPLC–QTOF–MS. *Food Chem.* **2018**, *267*, 140–148. [CrossRef] [PubMed]
37. Kuhn, M.; Wing, J.; Weston, S.; Williams, A.; Keefer, C.; Engelhardt, A.; Cooper, T.; Mayer, Z.; Kenkel, B.; Benesty, M.; et al. Classification and Regression Training. Version 6.0-93. 2022. Available online: https://cran.r-projet.org.//web/pakages/caret/caret.pdf (accessed on 1 September 2022).

Article

Evaluation of the Effectiveness of Charcoal, *Lactobacillus rhamnosus*, and *Saccharomyces cerevisiae* as Aflatoxin Adsorbents in Chocolate

Gamal M. Hamad [1], Amr Amer [2], Baher El-Nogoumy [3], Mohamed Ibrahim [4], Sabria Hassan [1], Shahida Anusha Siddiqui [5,6], Ahmed M. EL-Gazzar [7], Eman Khalifa [8], Sabrien A. Omar [9], Sarah Abd-Elmohsen Abou-Alella [2], Salam A. Ibrahim [10], Tuba Esatbeyoglu [11,*] and Taha Mehany [1,*]

[1] Department of Food Technology, Arid Lands Cultivation Research Institute, City of Scientific Research and Technological Applications, New Borg El-Arab 21934, Egypt

[2] Department of Food Hygiene and Control, Faculty of Veterinary Medicine, Alexandria University, Alexandria 21544, Egypt

[3] Department of Botany and Microbiology, Faculty of Science, Kafrelsheikh University, Kafr El Sheikh 33516, Egypt

[4] Department of Food Toxicology and Contaminants, National Research Centre, Dokki, Cairo 12622, Egypt

[5] Technical University of Munich Campus Straubing for Biotechnology and Sustainability, Essigberg 3, 94315 Straubing, Germany

[6] German Institute of Food Technologies (DIL e.V.), Prof.-von-Klitzing-Straße 7, 49610 Quakenbrück, Germany

[7] Department of Veterinary Forensic Medicine and Toxicology, Faculty of Veterinary Medicine, Alexandria University, Alexandria 21544, Egypt

[8] Department of Microbiology, Medicine/Alexandria University Branch, Matrouh University, Marsa Matruh 51511, Egypt

[9] Department of Microbiology, Faculty of Agriculture, Mansoura University, Mansoura 35516, Egypt

[10] Food and Nutritional Sciences, North Carolina Agricultural and Technical State University, E. Market Street 1601, Greensboro, NC 24711, USA

[11] Department of Food Development and Food Quality, Institute of Food Science and Human Nutrition, Gottfried Wilhelm Leibniz University Hannover, Am Kleinen Felde 30, 30167 Hannover, Germany

* Correspondence: esatbeyoglu@lw.uni-hannover.de (T.E.); tmehany@srtacity.sci.eg (T.M.); Tel.: +49-5117625589 (T.E.); +20-1028065903 (T.M.)

Citation: Hamad, G.M.; Amer, A.; El-Nogoumy, B.; Ibrahim, M.; Hassan, S.; Siddiqui, S.A.; EL-Gazzar, A.M.; Khalifa, E.; Omar, S.A.; Abd-Elmohsen Abou-Alella, S.; et al. Evaluation of the Effectiveness of Charcoal, *Lactobacillus rhamnosus*, and *Saccharomyces cerevisiae* as Aflatoxin Adsorbents in Chocolate. *Toxins* **2023**, *15*, 21. https://doi.org/10.3390/toxins15010021

Received: 30 November 2022
Revised: 18 December 2022
Accepted: 23 December 2022
Published: 28 December 2022

Abstract: The high incidence of aflatoxins (AFs) in chocolates suggests the necessity to create a practical and cost-effective processing strategy for eliminating mycotoxins. The present study aimed to assess the adsorption abilities of activated charcoal (A. charcoal), yeast (*Saccharomyces cerevisiae*), and the probiotic *Lactobacillus rhamnosus* as AFs adsorbents in three forms—sole, di- and tri-mix—in phosphate-buffered saline (PBS) through an in vitro approach, simulated to mimic the conditions present in the gastrointestinal tract (GIT) based on pH, time and AFs concentration. In addition, the novel fortification of chocolate with A. charcoal, probiotic, and yeast (tri-mix adsorbents) was evaluated for its effects on the sensory properties. Using HPLC, 60 samples of dark, milk, bitter, couverture, powder, and wafer chocolates were examined for the presence of AFs. Results showed that all the examined samples contained AFs, with maximum concentrations of 2.32, 1.81, and 1.66 µg/kg for powder, milk, and dark chocolates, respectively. The combined treatment demonstrated the highest adsorption efficiency (96.8%) among all tested compounds. Scanning electron microscope (SEM) analysis revealed the tested adsorbents to be effective AF-binding agents. Moreover, the novel combination of tri-mix fortified chocolate had a minor cytotoxicity impact on the adsorptive abilities, with the highest binding at pH 6.8 for 4 h, in addition to inducing an insignificant effect on the sensory attributes of dark chocolate. Tri-mix is thus recommended in the manufacturing of dark chocolate in order to enhance the safety of the newly developed product.

Keywords: aflatoxin; biocontrol; gastrointestinal tract; natural adsorbent; probiotic; *Saccharomyces cerevisiae*; *Lactobacillus rhamnosus*; chocolate contamination; chocolate safety

Key Contribution: This work highlights the effects of *S. cerevisiae*, *L. rhamnosus*, and A. charcoal on AF adsorption in chocolate, both individually and in combined forms. The combined treatment exhibited the highest adsorption efficiency towards AFs, and the detoxification effect was improved at neutral pH. In sum, this research markedly reduced AFs (i.e., AFB_1, AFB_2, AFG_1, and AFG_2) in a chocolate model with good sensory attributes.

1. Introduction

The history of chocolate dates back more than 4000 years, with the primary ingredient being cocoa powder produced from cocoa (*Theobroma cacao*) beans. Fungi can easily contaminate cocoa beans during the product-handling stages that include harvesting, drying, fermentation, roasting, preparation, transport, and storage. However, these steps are crucial for the formation of a distinctive chocolate flavour. Although this process is adequate for the elimination of harmful bacteria and molds, it has been demonstrated that mycotoxins are able to maintain their stability during most heat processing. Moreover, unhygienic practices can occur during the manufacturing of cocoa powder and chocolates [1]. Mycotoxin contamination from mycotoxigenic fungi such as *Penicillium*, *Aspergillus*, *Fusarium* and *Alternaria*, is thus a critical hazard to food quality and safety, with their occurrence in foodstuffs estimated to be 60–80% and resulting in over US $ 932 million in economic losses each year in the agro-food sector [2]. Aflatoxins (AFs) are the most frequently identified mycotoxins in cocoa, and the optimal conditions for AFs production by *Aspergillus* species in cocoa are at 33 °C and 0.99 a_w [3]. The ingestion of foods contaminated with AFs can result in severe side effects, including genetic disorders, tumors, mutagenicity, carcinogenic effects, and cytotoxicity [4].

According to Li et al. [5], a technical adsorbent is described as an agent that triggers the elimination of mycotoxins via adsorption or binding. Thus, the addition of adsorbents to contaminated food is a progressive and secure method for reducing the detrimental health impacts of AFs. However, the majority of adsorbents have significant practical restrictions. Activated charcoal (A. charcoal) is used as an antidote against severe poisoning. In this regard, and according to the German Federal Statistical Office (GFSO) in 2016, 178,425 cases of intoxication poisoning were treated in German hospitals. A. charcoal was recommended in 4.37% of cases. A. charcoal employment plays a major role in both primary and secondary detoxification. Moreover, in 2013, A. charcoal was used in treatment in 0.89% of cases of poisoning in children, for cases registered in the US. Therefore, A. charcoal is indicated to treat moderately severe to life-threatening intoxication [6]. A. charcoal has an excellent adsorbent property due to its porosity and large surface area. As a result, it can remove harmful pollutants such as harmful gases, heavy metals, mycotoxins, pesticides, and other chemicals from aqueous solutions [7].

The maximum ratios for mycotoxins in foods are very low owing to their severe toxicities. For instance, the maximum levels for AFs set by the Codex in various grains, seeds, nuts, milk and dried figs are in the range between 0.5–15 µg/kg (a 1 µg is 1 billionth of a kilogram) [8]. Moreover, Brazilian Sanitary Surveillance Agency has set limits of 10 µg/kg for cocoa beans and 5 µg/kg for cocoa products and chocolate sold in Brazil, for both ochratoxin A and total aflatoxins [9].

Chemical agents such as curcumin can be utilized to eliminate the mycotoxins in vivo and could antagonize the deleterious effects of AFB_1 on the kidney in mice. This impact is realized by inhibiting Bax/Bcl-2–Cyt-c signaling cascade-mediated apoptosis and modulating the Keap1–Nrf2 signal mechanism in order to improve renal antioxidant activity. These findings confirmed the application of curcumin as a natural food additive to degrade AFB_1 [10]. Moreover, enzymes could be a promising degrading candidate against aflatoxins toxicity. In this regard, Zhou et al. [11] developed a new laccase which purified from the white-rot fungus (*Cerrena unicolor*) in order to catalyze AFB_1 degradation. AFB_1 elimination by laccase was performed at 45 °C/24 h and pH 7.0 in vitro. The half-life of AFB_1 degradation catalyzed by laccase was 5.16 h; 2,2′-azino-bis-(3-ethylbenzothiazoline-6-sulfonic acid), Syringaldehyde, and Acetosyringone, at 1 mM concentration, seemed to be similar mediators for greatly improving AFB_1 detoxification by laccase. These results are promising for a potential application of laccase as a novel aflatoxin oxidase in degrading AFB_1 in both feeds and foods.

Recently, there has been a growing interest in the use of probiotics and other natural alternatives that do not rely on chemical additives to counteract the toxins and pollutants in foods. For example, probiotics are useful in food bioremediation due to their excellent antioxidant potential and antimicrobial properties [12,13], in addition to being an eco-friendly, highly effective, valuable technology in food processing. Probiotics are defined as microorganisms that have a positive effect on human health when consumed in sufficient quantities. *Lactobacillus rhamnosus* and *Saccharomyces cerevisiae* are examples of generally recognized as safe (GRAS) microorganisms; consequently, they can be used as feed additives with minimal risks [14]. Moreover, for decontamination purposes, lactic acid bacteria (LAB) may be effective at decreasing mycotoxins in alcoholic drinks during processing [15]. Similarly, *S. cerevisiae* has been considered to be a food additive for decades as it plays a fundamental role in providing vitamin B, minerals, and proteins [16]. *S. cerevisiae* is currently of considerable importance with regard to several biotechnological applications. For example, the biotechnology benefit of *S. cerevisiae* is inherent in its unique biological properties, such as its fermentation ability, conveyed by the production of CO_2 and alcohol, and its flexibility to opposing circumstances of low pH and osmolarity. The most prominent applications involving the utility of *S. cerevisiae* are in the food, beverage, and biofuel production industries [17].

LAB can thus eliminate or remove mycotoxins in food by either physical attachment or bio-transforming mechanisms [18]. The adsorption efficiency of A. charcoal and lactic acid bacteria in the removal of AFs from liquid solutions was recently determined in several studies [19,20]. In another study, it was found that certain probiotics can adsorb AFs in doogh (cultured dairy product) during fermentation and storage. The adsorption rate depends on the type of probiotics. Thus, the use of *Lactobacillus* strains such as *L. acidophilus*, *L. rhamnosus*, and *L. casei* for the ripening of cocoa has been evidenced to be an approach with excellent prospects [21].

Currently, there is no commercially available chocolate containing tri-mix (charcoal, probiotic, and yeast) and nothing in the current literature addresses such food formulation. In addition, to the best of our knowledge, no investigation has described the use of adsorbent combination models for sequestering AFs in food matrices. Consequently, the aim of this study was to evaluate and compare the effectiveness of A. charcoal, *L. rhamnosus*, and *S. cerevisiae* as aflatoxin adsorbents in three forms—sole, di- and tri-mix—in phosphate buffered saline (PBS) using an in vitro approach simulated to mimic the conditions found in the gastrointestinal tract (GIT) based on pH, time and AFs concentration. In addition, the adsorptive activity of tri-mix was studied using an in vitro dark chocolate model, and its effect on the sensory properties of novel tri-mix-fortified chocolate was evaluated.

2. Results

2.1. Screening of Aflatoxins in Various Chocolate Kinds

A total of 60 local chocolate samples of six chocolate types (10 samples each) were surveyed for the presence of total and individual AFs. The occurrence and concentrations of the tested AFs in chocolate samples are summarized in Table 1. The results demonstrated that all chocolate samples were contaminated by AFs. The highest concentration of total AFs was observed in chocolate powder when compared with other chocolate types in the following order: Chocolate powder: 2.32 > milk: 1.81 > dark: 1.66 > bitter: 0.704 > wafer: 0.674 > coverture: 0.290 µg/kg. The contamination rates of AFs in powder, milk and dark chocolates are significantly higher ($p < 0.05$) than those in other chocolates. The concentrations of AFB were higher than AFG in all examined chocolates. In order, the AFB_1 contents in milk, dark, wafer, and bitter chocolates were 0.966, 0.963, 0.414, and 0.369 µg/kg, respectively, while the least AFB_1 concentration presented in coverture chocolate, with 0.112 µg/kg. The contamination rates of AFB_1 in powder, dark, and milk chocolates are significantly different ($p < 0.05$) from other chocolates. Regarding the AFB_2 level, its concentration reached the peak in chocolate powder with a mean of 0.593 µg/kg, while its concentrations were slightly lower in the dark, bitter, milk, and wafer, at 0.573, 0.304, 0.239, and 0.207 µg/kg, respectively. The lowest concentration was reported in coverture chocolate (0.071 µg/kg). In this study, the magnitude of AFG demonstrated different levels among examined chocolates with lower values than AFB. In this context, milk chocolate showed the highest mean concentrations of AFG_1 and AFG_2, which were 0.425 and 0.178 µg/kg, respectively. The levels of AFG_1 and AFG_2 in chocolate powder were: 0.399 and 0.164 µg/kg > dark chocolate; 0.069 and 0.052 µg/kg > coverture chocolate; and 0.067 and 0.041 µg/kg, respectively. In contrast, the lowest AFG levels were reported in wafer and bitter chocolate to be 0.043 and 0.013 µg/kg for AFG_1 and 0.010 and 0.18 µg/kg for AFG_2, respectively.

Table 1. Occurrence and concentration (µg/kg) of aflatoxin residues in chocolate products.

Chocolate Products	Number of Samples	AFB_1		AFB_2		AFG_1		AFG_2		Total AFs	
		Min–Max	Mean ± SD	Min–Max	Mean ± SD	Min–Max	Mean ± SD	Min–Max	Mean ± SD	Min–Max	Mean ± SD
Couverture chocolate	10	0.015–0.542	0.112 ± 0.19 [c]	0.006–0.262	0.071 ± 0.095 [c]	0.014–0.096	0.067 ± 0.007 [b]	0.013–0.086	0.041 ± 0.027 [b,c]	0.106–0.776	0.290 ± 0.231 [d]
Dark chocolate	10	0.06–1.87	0.963 ± 0.689 [a]	0.076–0.972	0.573 ± 0.313 [a]	0.036–0.098	0.069 ± 0.020 [b]	0.037–0.062	0.052 ± 0.007 [b]	0.247–2.669	1.658 ± 0.735 [b]
Milk chocolate	10	0.270–1.60	0.966 ± 0.376 [a]	0.130–0.390	0.239 ± 0.063 [b]	0.280–0.650	0.425 ± 0.112 [a]	0.130–0.280	0.178 ± 0.045 [a]	1.230–2.450	1.808 ± 0.333 [b]
Chocolate powder	10	0.08–1.85	1.116 ± 0.832 [a]	0.280–0.820	0.593 ± 0.183 [a]	0.230–0.560	0.399 ± 0.107 [a]	0.120–0.240	0.164 ± 0.037 [a]	1.460–3.060	2.322 ± 0.489 [a]
Bitter chocolate	10	0.26–0.52	0.369 ± 0.079 [b]	0.140–0.420	0.304 ± 0.099 [b]	0.001–0.029	0.013 ± 0.009 [b]	0.004–0.0360	0.018 ± 0.011 [b,c]	0.526–0.875	0.704 ± 0.101 [c]
Chocolate wafer	10	0.23–0.93	0.414 ± 0.254 [b]	0.130–0.280	0.207 ± 0.058 [b,c]	0.032–0.054	0.043 ± 0.008 [b]	0.004–0.017	0.010 ± 0.005 [c]	0.421–1.278	0.674 ± 0.274 [c]

Min–Max: minimum–maximum; SD: standard deviation; [a, b, c, d] Mean values carrying different superscripts small letter on the same column are significantly different ($p < 0.05$).

2.2. Aflatoxin Adsorption Efficiency in Phosphate Buffer Solution

The results of adsorption efficiency of (A. charcoal + *L. rhamnosus* + *S. cerevisiae*) in sole, di-mixed, and tri-mixed PBS solution against AFs (B_1, B_2, G_1, and G_2) under pH conditions (3 and 6.8) and at 2-time intervals (2 and 4 h) are illustrated in Table 2. The results were compared with the negative control (PBS), which showed the AF adsorption and residual levels of zero and positive control (PBS + AFs), with AF adsorption (zero%) and residual levels (0.99 and 1.00 µg/mL). All the tested compounds were able to adsorb AFs in PBS with varying degrees. In this study, the combined formula (tri-mix) showed higher adsorption efficiencies ($p < 0.05$) and lowered residual AFs levels than individual tested constituents. This finding reflects the synergistic actions between adsorbents against AFs in the combined tri-mix. Accordingly, A. charcoal + *L. rhamnosus* + *S. cerevisiae* tri-mix revealed the greatest AF adsorption of 96.8% and 97.7% at pH 3 and 6.8 for 2 h, respectively, while the values were 98.1% and 99.7% at pH 3 and 6.8 for 4 h, respectively. On the contrary, the concentrations of residual AFs were (0.032–0.019 µg/mL) at pH 3 and (0.023–0.003 µg/mL) at pH 6.8 after 2 and 4 h, respectively. Concerning the di-mix tested compounds, we noticed that A. charcoal + *S. cerevisiae* di-mixed PBS demonstrated the highest AF adsorption (89% and 91% for 2 and 4 h, respectively, at all pH conditions) as well as the lowest residual AFs levels (0.10 and 0.09 µg/mL for 2 and 4 h, respectively, at all pH conditions), compared with *L. rhamnosus* + *S. cerevisiae* and A. charcoal + *L. rhamnosus* di-mixed PBS. The AF adsorption values for *L. rhamnosus* + *S. cerevisiae* vis A. charcoal + *L. rhamnosus* di-mixed buffer were 83% and 85% vis 80% and 78% at pH 3, while at pH 6.8, the values were 85% and 84% versus 80% after 2 and 4 h, respectively. In contrast, the residual AFs levels were 0.17 and 0.15 versus 0.20 and 0.22 µg/mL at pH 3, as well as 0.15 and 0.16 versus 0.20 µg/mL at pH 6.8 after 2 and 4 h, respectively, for *L. rhamnosus* + *S. cerevisiae* versus A. charcoal + *L. rhamnosus* di-mixed PBS.

Table 2. Aflatoxin adsorption efficiency in PBS buffer.

Matrix	Time	pH	Aflatoxins Conc. (µg/mL)					
			B_1	B_2	G_1	G_2	Total AFs	Adsorption%
PBS (−Ve control)	2 h	3.0	n.d.	n.d.	n.d.	n.d.	n.d.	0
		6.8	n.d.	n.d.	n.d.	n.d.	n.d.	0
	4 h	3.0	n.d.	n.d.	n.d.	n.d.	n.d.	0
		6.8	n.d.	n.d.	n.d.	n.d.	n.d.	0
PBS + AFs (+Ve control)	2 h	3.0	0.24 ± 0.01 [j]	0.26 ± 0.01 [i]	0.24 ± 0.03 [i]	0.25 ± 0.02 [j]	0.99 ± 0.01 [o]	0
		6.8	0.23 ± 0.02 [j]	0.25 ± 0.01 [h]	0.26 ± 0.01 [j]	0.25 ± 0.01 [j]	0.99 ± 0.02 [o]	0
	4 h	3.0	0.25 ± 0.01 [k]	0.24 ± 0.02 [h]	0.25 ± 0.03 [i]	0.26 ± 0.01 [a]	1.00 ± 0.01 [o]	0
		6.8	0.24 ± 0.02 [j]	0.26 ± 0.01 [i]	0.24 ± 0.01 [i]	0.25 ± 0.01 [j]	0.99 ± 0.02 [o]	0
Activated charcoal	2 h	3.0	0.12 ± 0.01 [i]	0.13 ± 0.02 [g]	0.12 ± 0.01 [g]	0.11 ± 0.02 [h]	0.48 ± 0.01 [m]	52.0 ± 0.31 [l]
		6.8	0.11 ± 0.01 [h]	0.12 ± 0.01 [f]	0.13 ± 0.02 [h]	0.12 ± 0.01 [h]	0.48 ± 0.01 [m]	52.0 ± 0.70 [l]
	4 h	3.0	0.11 ± 0.01 [h]	0.11 ± 0.01 [f]	0.12 ± 0.01 [g]	0.11 ± 0.01 [h]	0.45 ± 0.01 [l]	55.0 ± 0.42 [k]
		6.8	0.12 ± 0.01 [i]	0.11 ± 0.02 [f]	0.13 ± 0.01 [g]	0.12 ± 0.02 [h]	0.48 ± 0.02 [m]	52.0 ± 0.30 [l]
L. rhamnosus	2 h	3.0	0.13 ± 0.03 [i]	0.12 ± 0.01 [f]	0.12 ± 0.02 [g]	0.11 ± 0.02 [h]	0.48 ± 0.01 [m]	52.0 ± 0.11 [l]
		6.8	0.12 ± 0.01 [i]	0.12 ± 0.01 [f]	0.13 ± 0.01 [h]	0.12 ± 0.01 [h]	0.49 ± 0.01 [n]	51.0 ± 0.52 [m]
	4 h	3.0	0.13 ± 0.01 [i]	0.11 ± 0.01 [f]	0.12 ± 0.01 [a]	0.11 ± 0.01 [h]	0.47 ± 0.01 [m]	53.0 ± 0.01 [l]
		6.8	0.13 ± 0.01 [i]	0.12 ± 0.01 [f]	0.11 ± 0.02 [g]	0.13 ± 0.01 [i]	0.49 ± 0.02 [n]	51.0 ± 0.05 [m]
S. cerevisiae	2 h	3.0	0.08 ± 0.01 [g]	0.07 ± 0.02 [e]	0.08 ± 0.01 [f]	0.06 ± 0.01 [f]	0.29 ± 0.01 [k]	71.0 ± 0.04 [j]
		6.8	0.08 ± 0.02 [a]	0.08 ± 0.01 [e]	0.06 ± 0.01 [e]	0.07 ± 0.02 [f,g]	0.29 ± 0.02 [k]	71.0 ± 0.14 [j]
	4 h	3.0	0.07 ± 0.01 [f]	0.06 ± 0.01 [d]	0.08 ± 0.02 [f]	0.06 ± 0.01 [f]	0.27 ± 0.01 [j]	73.0 ± 0.11 [i]
		6.8	0.08 ± 0.01 [g]	0.07 ± 0.02 [d]	0.06 ± 0.01 [e]	0.08 ± 0.02 [g]	0.29 ± 0.01 [a]	71.0 ± 0.07 [j]
A. charcoal + *L. rhamnosus*	2 h	3.0	0.06 ± 0.01 [e]	0.05 ± 0.01 [c]	0.05 ± 0.01 [d]	0.04 ± 0.01 [e]	0.20 ± 0.01 [h]	80.0 ± 0.41 [g]
		6.8	0.05 ± 0.01 [d]	0.04 ± 0.01 [c]	0.06 ± 0.01 [e]	0.05 ± 0.01 [e]	0.20 ± 0.02 [h]	80.0 ± 0.31 [g]
	4 h	3.0	0.05 ± 0.01 [d]	0.06 ± 0.01 [e]	0.05 ± 0.01 [d]	0.06 ± 0.01 [f]	0.22 ± 0.01 [i]	78.0 ± 0.09 [h]
		6.8	0.04 ± 0.01 [d]	0.05 ± 0.01 [d]	0.06 ± 0.01 [e]	0.05 ± 0.01 [e]	0.20 ± 0.01 [h]	80.0 ± 0.14 [g]
A. charcoal + *S. cerevisiae*	2 h	3.0	0.03 ± 0.01 [c]	0.02 ± 0.01 [c]	0.03 ± 0.01 [c]	0.02 ± 0.01 [d]	0.10 ± 0.01 [e]	89.0 ± 0.21 [d]
		6.8	0.03 ± 0.01 [c]	0.03 ± 0.01 [c]	0.02 ± 0.01 [c]	0.02 ± 0.01 [d]	0.10 ± 0.01 [e]	89.0 ± 0.07 [d]
	4 h	3.0	0.02 ± 0.02 [b,c]	0.02 ± 0.01 [c]	0.03 ± 0.03 [c]	0.02 ± 0.02 [d]	0.09 ± 0.02 [e]	91.0 ± 0.22 [c]
		6.8	0.02 ± 0.0 [b,c]	0.03 ± 0.01 [c]	0.02 ± 0.01 [c]	0.02 ± 0.01 [d]	0.09 ± 0.01 [e]	91.0 ± 0.17 [c]
L. rhamnosus + *S. cerevisiae*	2 h	3.0	0.05 ± 0.01 [d]	0.03 ± 0.01 [c]	0.04 ± 0.01 [d]	0.05 ± 0.01 [e]	0.17 ± 0.01 [g]	83.0 ± 0.11 [e,f]
		6.8	0.04 ± 0.01 [d]	0.04 ± 0.01 [d]	0.03 ± 0.01 [c]	0.04 ± 0.01 [e]	0.15 ± 0.02 [f]	85.0 ± 0.06 [e]
	4 h	3.0	0.03 ± 0.01 [c]	0.05 ± 0.02 [d]	0.04 ± 0.01 [d]	0.03 ± 0.01 [d]	0.15 ± 0.01 [f]	85.0 ± 0.32 [e]
		6.8	0.03 ± 0.02 [c]	0.04 ± 0.01 [d]	0.05 ± 0.01 [d]	0.04 ± 0.02 [e]	0.16 ± 0.01 [f]	84.0 ± 0.41 [e]
A. charcoal + *L. rhamnosus* + *S. cerevisiae*	2 h	3.0	0.01 ± 0.01 [b]	0.01 ± 0.01 [b]	0.007 ± ±0.02 [b]	0.005 ± 0.01 [c]	0.032 ± 0.02 [d]	96.8 ± 0.08 [b]
		6.8	0.01 ± 0.01 [b]	0.006 ± 0.02 [a]	0.004 ± 0.01 [a]	0.003 ± 0.01 [a]	0.023 ± 0.01 [c]	97.7 ± 0.45 [a,b]
	4 h	3.0	0.005 ± 0.01 [a]	0.003 ± 0.01 [a]	0.005 ± 0.01 [a]	0.006 ± 0.01 [b]	0.019 ± 0.01 [b]	98.1 ± 0.41 [a]
		6.8	0.003 ± 0.01 [a]	n.d.	n.d.	n.d.	0.003 ± 0.01 [a]	99.7 ± 0.13 [a]

[a, b, c, d, e, f, g, h, i, j, k, l, m, n, o] Mean values carrying different superscripts small letter on the same column are significantly different ($p < 0.05$). n.d.: not detected. PBS: Phosphate buffered saline; −Ve control: negative control (PBS); +Ve control: positive control + AFs; n.d.: Not detected.

2.3. Aflatoxin Adsorption Efficiency in Fortified Dark Chocolate

An experiment on the adsorption of AFs by the tri-mix (A. charcoal + *L. rhamnosus* + *S. cerevisiae*) was carried out in model chocolate to assess the AF adsorption efficiency of the tri-mix. The experiment was conducted at two time intervals (2 and 4 h) under pH conditions (3.0 and 6.8). As depicted in Table 3, the overall results indicate a significant reduction in AFs in supplemented chocolate compared to negative (chocolate) and positive control (chocolate + AFs). In the positive control, there was a consistency in the range of total AFs (0.98–0.99 μg/kg), with zero adsorption throughout the time intervals and pH conditions. In contrast, the adsorption efficiency at simulated gastric pH (3.0) was better, with an adsorption of 95.4% and 96.1% at 2 and 4 h, respectively, while the levels of residual total AFs were 0.046 and 0.039 μg/kg at 2 and 4 h, respectively. The highest adsorption and the lowest residual total AFs were achieved at pH 6.8 after 4 h, at 96.80% and 0.032 μg/kg, respectively. In contrast, the lowest adsorption and the highest residual total AFs were 90.2% and 0.098 μg/kg, respectively, after 2 h at the same pH. Regarding the adsorption effect on the type of AFs, we found that AFG_1 was the most adsorbed AF in supplemented chocolate when compared with other AFs, with residual AFG_1 values of 0.004 and 0.002 μg/kg at pH 3.0 and 0.003, respectively, and 0.001 μg/kg at pH 6.8, when compared with the range of positive control (0.24–0.25 μg/kg). Furthermore, all AFs types were strongly adsorbable at pH 6.8 compared with pH 3.0, for 2 and 4 h.

Table 3. Aflatoxin adsorption efficiency in supplemented chocolate with tri-mix.

Sample	Time	pH	Aflatoxins Conc. (μg/mL)					
			B_1	B_2	G_1	G_2	Total AFs	Adsorption%
Chocolate (−Ve control)	2 h	3.0	n.d.	n.d.	n.d.	n.d.	n.d.	0
		6.8	n.d.	n.d.	n.d.	n.d.	n.d.	0
	4 h	3.0	n.d.	n.d.	n.d.	n.d.	n.d.	0
		6.8	n.d.	n.d.	n.d.	n.d.	n.d.	0
Chocolate + AFs (+Ve control)	2 h	3.0	0.24 ± 0.02 [c]	0.25 ± 0.01 [c]	0.24 ± 0.01 [c]	0.25 ± 0.03 [c]	0.98 ± 0.01 [c]	0
		6.8	0.24 ± 0.02 [c]	0.26 ± 0.01 [d]	0.25 ± 0.02 [c]	0.24 ± 0.06 [c]	0.99 ± 0.02 [c]	0
	4 h	3.0	0.25 ± 0.01 [c]	0.24 ± 0.02 [c]	0.25 ± 0.01 [c]	0.24 ± 0.03 [c]	0.98 ± 0.01 [c]	0
		6.8	0.25 ± 0.03 [c]	0.25 ± 0.01 [c]	0.24 ± 0.03 [c]	0.25 ± 0.01 [c]	0.99 ± 0.01 [c]	0
Chocolate + AF + tri-mix (A. charcoal + *L. rhamnosus* + *S. cerevisiae*)	2 h	3.0	0.021 ± 0.02 [b]	0.016 ± 0.01 [b]	0.005 ± 0.01 [a]	0.004 ± 0.02 [b]	0.046 ± 0.02 [a]	95.40 ± 0.11 [c]
		6.8	0.020 ± 0.01 [b]	0.015 ± 0.01 [b]	0.06 ± 0.02 [b]	0.003 ± 0.02 [a]	0.098 ± 0.01 [b]	90.20 ± 0.24 [c]
	4 h	3.0	0.017 ± 0.02 [a,b]	0.015 ± 0.01 [b]	0.005 ± 0.01 [a]	0.002 ± 0.03 [a]	0.039 ± 0.01 [a]	96.10 ± 0.47 [a]
		6.8	0.014 ± 0.0 [a]	0.013 ± 0.01 [a]	0.004 ± 0.01 [a]	0.001 ± 0.01 [a]	0.032 ± 0.01 [a]	96.80 ± 0.15 [a]

[a, b, c, d] Mean values carrying different superscripts small letter on the same column are significantly different ($p < 0.05$). n.d.: not detected. −Ve control: negative control or/chocolate; +Ve control: positive control or/chocolate + AFs; n.d.: Not detected.

2.4. Scanning Electron Microscope Assessment

The microscopic examination has been applied to determine the surface characteristics of A. charcoal, probiotics, and yeast cells before and after AF adsorption. Drastic changes in the surface morphology of A. charcoal after incubation with AFs mix (B_1, B_2, G_1, and G_2) for 4 h were observed (Figure 1) when compared with non-incubated A. charcoal (negative control).

In the SEM investigation (Figure 2), with 2000, 5000, and 10,000× magnifications, the untreated *S. cerevisiae* cells, having a diploid form and ellipsoid shape, are observed with capsules of a typical size of 1.71 ± 0.03 mm. The treated cells appeared spheroid, with deformations and cavitation in the yeast cell wall.

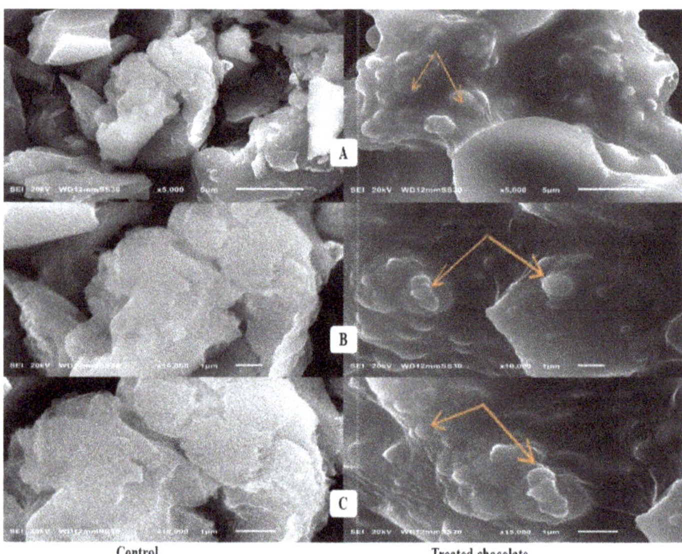

Figure 1. Scanning electron microscopy analysis of activated charcoal + AFs mix. (**A–C**) with 5000, 10,000, and 15,000× magnifications, respectively: control group (untreated charcoal in the left side with honeycomb vacuumed structure) and treatment group (charcoal incubated with 1 mg/mL of AFs for 4 h in the right side). Orange arrows denote the AF surplus upper layer and pimples on the surface of activated charcoal.

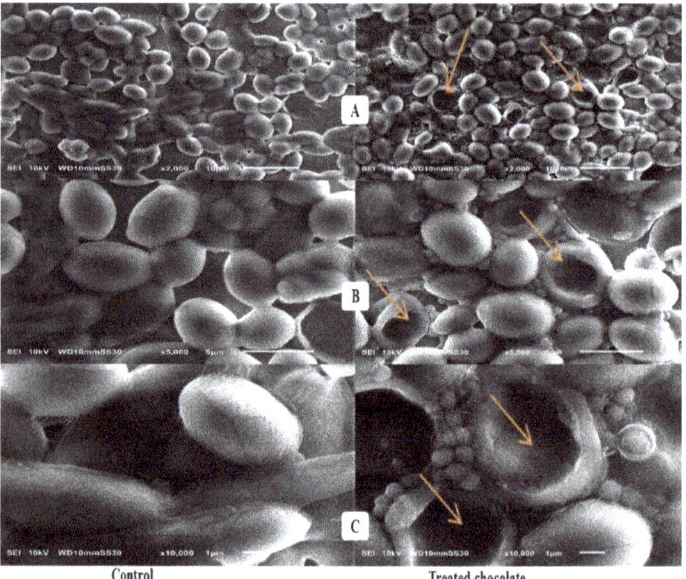

Figure 2. Scanning electron microscopy analysis of *S. cerevisiae* cells + AFs mix. (**A–C**) with 2000, 5000, and 10,000× magnifications, respectively: control group (untreated cells in the left side having diploid form and ellipsoid-shaped) and treatment group (*S. cerevisiae* cells incubated with 1 mg/mL of AFs for 4 h in the right side). Orange arrows indicate structural changes induced by Afs, with an additional upper layer of AFs and their perforations. The yeast cells appeared spheroid with deformations and cavitation in the cell wall.

In the typical micrograph of SEM (Figure 3), with 10,000, 15,000, and 20,000× magnifications, the untreated *L. rhamnosus* cells (control) appeared as small undamaged rods arranged in chains. The SEM imaging reveals a prevalent conformational change probably produced by AFs attached to the cell wall surface.

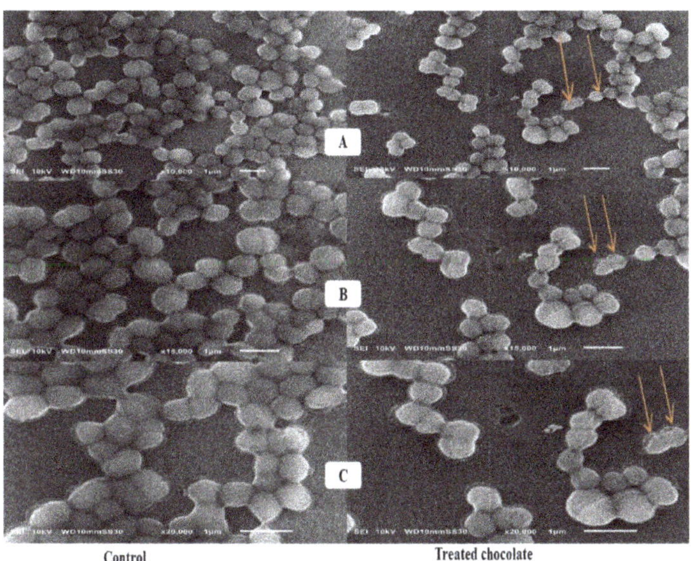

Figure 3. Scanning electron microscopy analysis of *L. rhamnosus* cells + AFs mix. (**A–C**) with 10,000, 15,000, and 20,000× magnifications, respectively: control group (untreated rod-shaped cells in the left side arranged in chains) and treatment group (*L. rhamnosus* cells incubated with 1 mg/mL of AFs for 4 h in the right side). Orange arrows refer to morphological alterations made by AFs; the bacterial cells appeared as bud-like structures.

2.5. Cytotoxicity Assessment of Activated Charcoal

Table 4 illustrates the cytotoxicity assessment of charcoal to determine the IC_{50} (μg/mL) on peripheral blood mononuclear cells (PBMCs). Carbon particles have been proven to be toxic to PBMCs at various concentrations. Consequently, cells exposed to A. charcoal were presented with 94–47% of live cells at concentrations less than 48.8 μg/mL, while at a higher concentration, above 97.5 μg/mL, the cells demonstrated 18–29% of live cells.

Table 4. Estimation of cytotoxicity of charcoal and IC_{50} (μg/mL).

Sample	Concentration (μg/mL)	Viability%	Inhibition%
	390	18	82
	195	26	74
	97.5	29	71
Charcoal	48.75	47	53
	24.37	58	42
	12.18	72	28
	6.09	94	6

IC_{50} for charcoal: 45.9 μg/mL.

2.6. Sensorial Properties of Tri-Mix-Fortified Dark Chocolate

In the sensory evaluation of dark chocolate, taste and texture were the most important characteristics, particularly in the tri-mix-fortified chocolate compared with the control. The results in Table 5 reveal that all sensory scores are not significantly different between chocolates, except for taste and texture. The mean values of taste and texture for tri mix-chocolate were equalized to be 8.0, which was lower than those for the control (8.6). Commercially, A. charcoal has porous and gritty nature, which may affect the texture of chocolate. The mean values of odor and overall acceptance for tri-mix chocolate versus control were 8.1 versus 8.5, and 8.3 versus 8.6, respectively.

Table 5. Sensory properties of tri-mix-fortified dark chocolate.

Treatment/Group	Sensorial Properties Mean $\pm$ SD					
	Color	Odor	Taste	Texture	Appearance	Overall Acceptance
Control	8.4 ± 0.96 [a]	8.5 ± 0.84 [a]	8.6 ± 0.84 [a]	8.6 ± 0.84 [a]	7.9 ± 1.37 [a]	8.6 ± 0.84 [a]
A. charcoal + *L. rhamnosus* + *S. cerevisiae*	8.4 ± 0.94 [a]	8.1 ± 0.99 [a]	8.0 ± 1.05 [b]	8.0 ± 1.05 [b]	7.8 ± 1.31 [a]	8.3 ± 0.78 [a]

[a, b] Mean values carrying different superscript small letter on the same column are significantly different ($p < 0.05$) tri-mix adsorbents: activated charcoal + *L. rhamnosus* + *S. cerevisiae*.

3. Discussion

3.1. Screening of Aflatoxins in Chocolates

AFB_1 was the most frequently detected aflatoxin at the highest concentration, compared with AFB_2, AFG_1, and AFG_2, in all chocolates. Accordingly, the highest mean concentration of AFB_1 was detected in chocolate powder, at 1.12 µg/kg, exceeding the permissible limit according to Codex Standard (1 µg/kg) [22]. On the other hand, our findings regarding AFs in all tested chocolate are within the allowable limits of the Brazilian Sanitary Surveillance Agency, which set limits of 10 µg/kg for cocoa beans and 5 µg/kg for cocoa products and chocolate sold in Brazil, for both ochratoxin A and total aflatoxins [8].

The contamination levels of AFG in powder and milk chocolates are significantly different ($p < 0.05$) from other chocolates. This finding is inconsistent with those reported by Kabak [23], who found that AFG_1 and G_2 contents failed to be detected in all chocolate samples, except milk chocolate, where the concentration of AFG_1 was 0.35 µg/kg.

These results are relatively similar to those obtained by Copetti et al. [24] who detected AFs in approximately 72% of milk chocolates and 100% of bitter and dark chocolate at concentration ranges of 0.11–1.65 and 0.04–0.91 µg/kg, respectively. In a study by Turcotte et al. [25] the AFB_1 was reported in 80% of dark chocolate and 70% of milk chocolate samples at values of 0.63 and 0.18 µg/kg, respectively. In Pakistan, Naz et al. [26] reported that the contents of AFs in dark, milk, and bitter chocolates were 2.27, 1.31, and 0.97 µg/kg, respectively. In contrast, our findings are significantly higher than those reported by Kabak, [23] who counted AFs in 19.6% of milk chocolate, 13.3% of bitter chocolate, and 8.70% of chocolate wafer samples at mean values of 0.839, 0.455, and 0.281 µg/kg, respectively. The authors confirmed that AFB_1 was the highest detected AF in all examined chocolate samples.

To date, there is no maximum threshold for AFs in cocoa and its products involving chocolate established by the European Regulation. The differences in AFs levels amongst examined chocolates may indicate the content and quality of cocoa solids. This finding may explain the high content of total Afs, 2.322 and 1.658 µg/kg, in cocoa powder and dark chocolate samples containing 100% and 35% cocoa solids, respectively. Although milk chocolate sample content was 29–35% cocoa solids [22], the high content of total AFs (1.81 µg/kg) could be attributed to milk solids, based on the presence of AFs in milk products reported recently [27]. The higher contamination levels of chocolates containing AFs may be attributable to the deficiency of quality control and quality assurance during the manufacturing, and unhygienic packaging. Currently, chocolate consumption among

Egyptians, particularly children, is high. Consequently, quantifying the presence of mycotoxins in these products is a significant issue, and additional research is required to determine their mycotoxin concentrations.

3.2. Aflatoxin Adsorption Efficiency in Phosphate Buffer Solution

By comparing the individual tested compounds, it was found that *S. cerevisiae* adsorbed the highest levels of AFs with adsorption vis residual levels of 71% vis 0.29 µg/mL and 73% vis 0.27 µg/mL at pH 3.0, and 71% vis 0.29 µg/mL at pH 6.8, after 2 and 4 h, respectively. This result is higher than those explained by Istiqomah et al. [28], who found that the AFs binding activity of autoclaved *S. cerevisiae* B18 cells was 69.5%. Additionally, Rahaie et al. [29] illustrated that the heat treatment of yeast cells increased their binding abilities to AFs by 56 ± 2%. A. charcoal adsorbed 52–55% (pH 3.0) and 52% (pH 6.8) of AFs versus 52–53% (pH 3.0) and 51% (pH 6.8) for *L. rhamnosus*, after 2 and 4 h, respectively. This finding indicates that *L. rhamnosus*-supplemented PBS demonstrated the highest levels of residual Afs, which were 0.48 and 0.47 µg/mL at pH 3.0, and 0.49 µg/mL at pH 6.8, for 2 and 4 h, respectively.

This investigation suggests that the adsorption efficiency depends on several reasons, e.g., physical, chemical, and biological properties of the adsorbent and the adsorbate (the food or beverage); concentration of the adsorbate in fluids; features of the liquid phase (e.g., pH, temperature) and the residence time. The tri-mix demonstrated the maximum AF adsorption, followed by the di-mix, due to the synergism among the tested adsorbents' physical, chemical, ion-exchanging, and biological mechanisms. Accordingly, A. charcoal has a large surface area and pore volume, providing an excellent adsorption capability. It can sequester molecules, including mycotoxins, via both chemical and physical interactions; hence, it can be used as an adsorbent agent. A. charcoal was recognized to have binding efficiency of more than 90% with AFs.

Lactobacillus strains are Gram-positive bacteria, with a peptidoglycan layer in their cell wall and teichoic acid and polysaccharides (β-D-glucan) on the surface. The divalent cations can lead to alterations in the teichoic acid structure. Furthermore, the adsorption ability of lactobacilli is generally strain-dependent, due to the differences in the structure of biochemical components on the cell wall for most probiotic types [30,31].

In this study, we found that all the PBS-tested compounds containing *S. cerevisiae* demonstrated the best AF adsorption. The adsorption capability mainly depended on yeast composition and mycotoxin. Earlier studies investigated that yeast cells had the ability to attach numerous molecules involving mycotoxins via the polysaccharides (glucans), mannoproteins, and lipids found on their cell wall surface [32]. They can bind to the hydroxyl, ketone, and lactone groups of AFs by hydrogen bonds and van der Waals forces [33]. Furthermore, the natural characteristics of mycotoxins play an integral role in the adsorption activities, such as polarization, solubility, size, shape, cation exchange capacity (CEC), acidity, and relative humidity [34]. This finding agrees with Joannis-Cassan et al. [35], who indicated that the yeast cell wall components played a crucial role in AFs binding by *S. cerevisiae*. Furthermore, *S. cerevisiae* was the most effective microorganism for binding AFs in PBS. Concerning the synergetic action achieved in this study, A. charcoal displays reversible physical adsorption for microorganisms in liquids without removal by simple desorption and permeability, permitting its usage as a bio preservative [3]. A study by Ikegamai et al. [36] revealed that the fermentation activities of *S. cerevisiae* reached 90% in the medium containing A. charcoal, compared to 70% in controls. Besides, attractive interactions between microorganisms and A. charcoal decreased the porosity and negative charges, enhancing the capacity of A. charcoal to adsorb AFs, which might be the chief cause of maximum adsorptive ability for the tri-mixed PBS solution.

As hypothesized, the AF adsorption varied with the type and initial concentration of aflatoxin. Accordingly, AFB_1 was the greatest adsorbed aflatoxin on A. charcoal + *L. rhamnosus* + *S. cerevisiae*-treated PBS, with residual values of 0.003 ± 0.01 µg/mL at pH 6.8 for 4 h, compared with the initial concentration of 0.24 ± 0.01 µg/mL. Furthermore, AFB_2

achieved the best adsorption on A. charcoal + *L. rhamnosus*+ *S. cerevisiae*-treated buffer at pH 3 for 4 h compared to other Afs, with a residual level of 0 µg/mL of an initial concentration of 0.25 ± 0.01 µg/mL. Similar findings were obtained by El-Nezami et al. [37] who found that *L. rhamnosus* strains successfully bound AFB_1 and B_2, rather than G_1 and G_2. In contrast, Liew et al. [20] showed that the cell wall of lactobacilli showed the highest adsorption, at 97% for an AFB_1 concentration of 6 µg/mL. Similarly, Hernandez-Mendoza et al. [38] investigated the feasibility that teichoic acid contributes to AFB-binding by lactobacilli. The ability of bacterial cells to bind with AFs could be enhanced by chemical conditions, as bacterial cell treatment with acid facilitated the AFs' physical binding to the bacterium's molecular components, especially at the cell wall level.

With respect to AFG_2, it was the most efficiently absorbed aflatoxin by *S. cerevisiae*, and A. charcoal + *S. cerevisiae* treated PBS compared to other AFs with residual values of 0.07 ± 0.01 and 0.02 µg/mL, respectively. Moreover, AFG_1 and G_2 showed a relatively lower residual level (0.005 ± 0.002 µg/mL) on A. charcoal + *L. rhamnosus* + *S. cerevisiae* tri-mixed PBS when compared with AFB_1 and B_2 on the same buffer. All AFs showed the same residual values on A. charcoal + *L. rhamnosus*-treated PBS to be 0.04 ± 0.01 µg/mL.

The adsorption efficiency in this study is higher than those recorded by Shetty et al. [39], who examined the adsorption of AFB_1 by 18 species of Saccharomyces at AFB_1 concentrations of 1, 5, and 20 µg/mL. The authors reported that the yeast cells adsorbed 69.1% of the AFB_1 at 1 µg/mL, 41.0% at 5 µg/mL of AFB_1, and 34.0% at 20 µg/mL of AFB_1. In this study, an increase in adsorption was achieved at a reduced initial concentration of each AF, around 0.25 ± 0.01 µg/mL, so the adsorption levels were relatively high. These findings are compatible with Joannis-Cassan et al. [35] who reported that the range of adsorption was 2.5% to 49.3%, based on the AFB_1 concentration and the adsorbent type. Similarly, Gallo and Masoero [40] reported that the adsorption varied from 32% to 54%, with an initial AFB_1 concentration of 0.82 µg/mL.

By modifying surface charges and adsorbent–adsorbate reactions, pH substantially affects a liquid medium's adsorption efficiency. Hence, the adsorption ability of all adsorbents for the AFs was evaluated at 2 pH levels of 3.0 and 6.8. In this regard, we found that the only treatment of PBS either by A. charcoal or L. rhamnosus, or S. cerevisiae achieved the greatest adsorption, ranging from 52% to 73% at acidic pH (3.0). In a study by Joannis-Cassan et al. [35] they demonstrated that acidic pH (3.0) permitted adequate adsorption of the mycotoxins in the buffer. Exchangeable cations are involved in the binding mechanisms of AFs, which can be enhanced after acidulating the adsorbent surface.

In contrast, the di-mixed treatment of PBS by A. charcoal + *L. rhamnosus*, A. charcoal + *S. cerevisiae*, or *L. rhamnosus* + *S. cerevisiae* demonstrated the best adsorption, varying from 80% to 91% at pH 6.8. The tri-mixed PBS by A. charcoal + *L. rhamnosus* + *S. cerevisiae* had the overall maximum adsorption of 97.7% and 99.70% at pH 6.8. These results indicate that the adsorption of AFB_1 and AFB_2 by the combined adsorbents was pH-independent, whereas adsorption of AFG_1 and AFG_2 increased to a certain extent at neutral pH (6.8), as evidenced by the net surface charges on adsorbents and the charge on AFs molecules. Similar findings were reported by Rasheed et al. [41] who found no evident variation in the adsorption for AFB_1 and AFB_2 in buffer at pH 7, while for AFG_1 and AFG_2, adsorption increased enormously. Moreover, this result agrees with Tejada et al. [42] who noticed that the highest adsorption of A. charcoal reached a value of 75.4% in a solution at pH 6.

In the present research, we used two time intervals, and the adsorption efficiency of AFs by all treated buffers was more significant at 4 h than at 2 h. This result is compatible with Rahaie et al. [29], who confirmed that the adsorption capability of AFs in a buffer may occur at 2 h to 3 h of treatment. The variation in the incubation time could be the factor that influences the adsorption efficiency of AFs. In the current study, the incubation times of ranging from 2–4 h were used to reflect the transit time in human GIT. In addition, the absorption of AFs mainly occurs in the small intestine [19]. Generally, incubation time plays a part in the adsorption activity.

3.3. Aflatoxin Adsorption Efficiency in Supplemented Dark Chocolate

Although the results of AF adsorption by the tri-mix in the supplemented chocolate were high, they were relatively lower than in the PBS solution. This decreased percentage suggests chocolate has a lower cation exchange capacity (CEC) than the PBS medium. Additionally, this concept could be acceptable due to polyphenols and alkaloids in the model chocolate having great affinity for less polar AFs and acting as an ideal solvent for them [43], therefore affecting the adsorption efficiency between AFs and adsorbents. Additionally, the hydrophilic proteins involved in the composition of chocolate might compete with AF molecules for adsorption sites on adsorbents, thus decreasing its adsorption capacity. This finding agrees with Barrientos-Velázquez et al. [43]; they found that proteins blocked the hydrophobic sites needed for Afs on the surface of the adsorbent. The results of AF adsorption in model chocolate revealed marginal dissimilarity, which could be due to the presence of a distribution coefficient for AFs amongst aqueous and solid phases compared with PBS.

3.4. Scanning Electron Microscope Assessment

The non-incubated sample exhibited a honeycomb vacuumed morphology similar to the turbostratic shape of the carbon molecules. This finding reflects the large surface area of the carbon particles. The same result was obtained by Kalagatur et al. [44]. The incubated sample showed a highly exfoliated and rougher surface after AF adsorption. The honeycomb shape was replaced by a pimple-like structure, due to an additional upper layer of AFs appearing on the particle surface of A. charcoal.

Vilela et al. [45] informed that such holes can be formed by the action of adsorption, as yeasts are normally characterized by their egg shape and the smoothness of the surface. With a magnification of $15,000\times$, it is noticed that the capsule's core has a slight irregularity with holes along its whole length, which means that the internal morphological structure and the surface of holes were available to the molecules of AFs. Similar results of specific surface area were reported by De Rossi et al. [46].

The appearance of bud-like structures on the surface of bacterial cells was only noticed in the AFs-treated cells. Besides, the morphological changes involved the asymmetrical and coarse surface of the bacterial cell wall. It can be assumed that the interaction between AFs and the teichoic acids surface and beta-d-glucan structure of the *L. rhamnosus* cell wall might cause the structural variations noticed in SEM imaging. The same result of morphological alterations on the bacterial surface against AFB_1 was observed by Liew et al. [20].

3.5. Cytotoxicity Assessment of Activated Charcoal

The results of cytotoxicity in the current research contradicts the findings of Kalagatur et al. [44], who reported the non-toxic effect of A. carbon against neuro-2a cells, displaying 95.46% of live cells. The toxic effect may result from the size and shape, porosity, surface functionality, surface conductivity, or, frequently, the toxic guest species allied with these materials during processing [47]. The IC_{50} value of A. charcoal was 45.9 µg/mL, and this result was considered in the application process, as added A. charcoal did not exceed 40 µg/g in dark chocolate.

3.6. Sensorial Properties of Tri-Mix-Fortified Dark Chocolate

In the present study, dark chocolate exhibited grades of (much-like > 8.0) for all sensory attributes before and after adding tri-mix. This finding disagrees with Ronald and Normalina [48], who found that the supplementation of dark chocolate cake with A. charcoal enhanced its odor and overall interpretation. As hypothesized, A. charcoal acts as a taste- and odor-eliminating agent, so it might adsorb some of the volatile compounds responsible for the aroma and taste perception of dark chocolate. The color and appearance ratings showed the same values of 8.4 and 7.9 in both chocolates, which might be due to the black color of the A. charcoal hidden with the color of the dark chocolate, enhancing its color. This result agrees with Ronald and Normalina [48] who recorded a much-accepted appearance of dark chocolate cake after adding charcoal. This study determined that the tri-mix-fortified chocolate had a negligible impact on the adsorptive abilities of the tri-mix. Therefore, we suggest the application of tri-mix in manufacturing dark chocolate. The amount of A. charcoal used in this evaluation was ideal as it caused a minimal effect on the sensory scoring of dark chocolate. We should consider that A. charcoal is a relatively unspecific adsorbent; hence, essential dark chocolate nutrients could also be adsorbed, particularly if its concentrations in the food are much greater than those of the AFs.

4. Conclusions

In the present study, it was shown that powder, dark, and milk chocolates contained the highest concentrations of aflatoxins, recommending the implementation of good agricultural practices (GAPs) in cocoa farms, and good manufacturing practices (GMPs) in all steps of chocolate processing. In vitro assessment of adsorption abilities of A. charcoal, *S. cerevisiae*, and *L. rhamnosus* revealed that combined adsorbents exhibited higher adsorption efficacy toward AFs than that of individual adsorbents. The highest binding (96.80%) was reported for tri-mix (A. charcoal+ *S. cerevisiae* +*L. rhamnosus*) at neutral pH 6.8 for 4 h. The SEM micrographs showed that AFs caused structural changes on the surfaces of all adsorbents with extraordinary trapping. The dark chocolate newly fortified with tri-mix (A. charcoal+ *S. cerevisiae* +*L. rhamnosus*) displayed a maximum removal of AFs and a high adsorptive power with minimal effects on its sensory evaluation. As a result, tri-mix has significant potential to be extensively utilized as a bio-functional food for enhancing the safety of this newly developed chocolate model. Moreover, further investigations to address various chocolate types produced from different countries and based on spectroscopic techniques combined with big data analyses are highly recommended.

5. Materials and Methods

5.1. Chemicals and Reagents

All reagents and cultural media used in the current study were provided by Fluka chemicals (Fisher Scientific, Cairo, Egypt), and Merck Co (Kenilworth, NJ, USA). AFs and A. charcoal used were bought from SD Fine-Chem company, Maharashtra, India (Product No 43032, Methylene blue adsorption: 270 mg/g, particle size: 300 mesh, pH: 6–7.5, maximum limits of impurities: ash 2.5%, moisture 5%, water-soluble 1.5%, and acid-soluble 2.5%).

5.2. Standards Preparation

The standard solution of AFs was prepared in methanol at concentrations of 1.00, 0.30, 1.00, and 0.30 µg/mL for AFB_1, AFB_2, AFG_1, and AFG_2, respectively. The intermediate multi-standard mixture was dissolved in methanol at concentrations of 0.10 µg/mL for AFB_1 and AFG_1, and 0.03 µg/mL for AFG_2, and AFB_2 was prepared in methanol and kept at $-18\,°C$. Then, these solutions were utilized in the preparation of standards. The detection limits for AFB_1, AFB_2, AFG_1, and AFG_2 were 2.70, 2.50, 1.33, and 1.40 ng/mL, respectively, while the quantification limits were 3.12, 2.92, 1.61, and 1.81 ng/mL, respectively.

5.3. Bacterial Strains and Cultural Conditions

Saccharomyces cerevisiae EMCC 97 and *Lactobacillus rhamnosus* EMCC 1105 were obtained from Microbiological Resources Centre (MIRCEN), Ain Shams University, Cairo, Egypt. *S. cerevisiae* was cultured in yeast peptone dextrose medium broth (YPD) (purchased from Merck KGaA, Darmstadt, Germany), while *L. rhamnosus* was cultivated in de Man Rogosa and Sharpe broth (MRS) (obtained from HIMEDIA, Maharashtra, India) and kept overnight at 37 °C and 100 rpm. The cells were collected by centrifugation at 6000 rpm/15 min, washed with PBS, and distilled water as described by El-Nezami et al. [49]. Cells were finally suspended in 0.1 M PBS at pH (6.8), and the concentration of cells was adjusted at 600 nm to be (2×10^8 CFU/mL) for *S. cerevisiae* and (1×10^9 CFU/mL) for *L. rhamnosus*.

5.4. Screening of Aflatoxins in Various Chocolate Types

Sixty samples of six chocolate types were purchased from various areas in Alexandria, Egypt. Ten samples of each chocolate type, dark, milk, bitter, couverture, powder, and wafer chocolates, were screened. The samples were grounded and saved at $-20\,^{\circ}$C until extraction and examination.

5.5. Extraction and Immunoaffinity Chromatography (IAC) Clean-Up

The extraction of AFs from chocolates was done as previously reported by Copetti et al. [24] with minor alterations. Twenty grams of chocolate were extracted by 120 mL methanol: water (8:2, v/v) and 2 g NaCl and were blended at a speed of 1500 rpm for 2 min, and then were filtered using filter paper. Six milliliters of the filtrate were mixed with 44 mL of PBS and were re-filtered through a 1.5 µm microfibre filter (VWR International, Leuven, Belgium). Sixty milliliters of the filtrate were escaped through AflaTest® IAC fixed on a vacuum manifold (Agilent Technologies, Santa Clara, CA, USA) at a flow of 2 drops/second. At the same time, the column was immersed with 15 mL of Milli Q water (Siemens Ultra Clear, Germany). AFs were separated from the IAC using 0.5 mL of methanol in two stages. The collective eluates were mixed with 1 mL of Milli Q water and kept at 3–7 $^{\circ}$C before HPLC investigation.

5.6. HPLC Analysis

Aflatoxin analysis of chocolate samples was applied using the Waters HPLC system (Model 6000, Milford, MA, USA), as previously reported by Hamad et al. [50]. It consists of a solvent conveyance system controller (model 720) fitted with a fluorescence detector (FLD, model 274) at 360 Ex, and 450 Em. The separation was done using a Waters regularity column ($150 \times 4.6\,\text{mm}^2$ i.d., 5 µm), the reverse phase was utilized at a flow rate of 1 mL/min with an isocratic model consisting of 1% acetic acid: methanol: acetonitrile (55:35:10, $v/v/v$). The calibration curve was made using varied concentrations of AFB_1, AFB_2, AFG_1, and AFG_2.

5.7. Preparation of Adsorbents in Sole, Di, and Tri-Mix Forms

Using A. charcoal, probiotic yeast (*S. cerevisiae*), and lactic acid bacteria (*L. rhamnosus*), AF adsorption tests were conducted. The *S. cerevisiae* cell wall was prepared according to Nathanail et al. [18]. The yeast culture was activated in potato dextrose for mass culture, and then optical density was adjusted at 600 nm (T80 UV/VIS spectrophotometer, Leicestershire, UK) to reach a concentration of 2×10^8 CFU/mL. Yeast pellets were collected by centrifugation at 3000 rpm/20 min (Micro centrifuge, SELECTA, Madrid, Spain), washed three times using sterilized water, then suspended in 0.1 M PBS and autoclaved at 120 $^{\circ}$C/20 min. Finally, centrifugation was carried out at 5000 rpm/20 min, then the supernatant was wasted, and cells were washed with sterilized distilled water and freeze-dried.

The culture of *L. rhamnosus* was activated by inoculating 1 mL into 100 mL of MRS broth (Oxoid, Hampshire, UK) and was kept at 37 $^{\circ}$C for 20 h. Then, 1 mL of this culture was then transferred to 99 mL of MRS broth to obtain a 1% dilution and incubated again was performed at 37 $^{\circ}$C/20 h. The concentration was adjusted using a spectrophotometer at 600 nm to be 1×10^9 CFU/mL. The bacterial cells were deactivated, and their cell walls were separated by incubating them with 4 mL of 2M HCl at 37 $^{\circ}$C/1 h. The cells were then washed with 2 mL of PBS and centrifuged at 5000 rpm/10 min at 10 $^{\circ}$C. The supernatant was then discarded, and bacterial cells were freeze-dried at $-40\,^{\circ}$C. In PBS, concentrations of 40 µg/mL of A. charcoal, lyophilized 2×10^8 CFU/mL of *S. cerevisiae*, and 1×10^9 CFU/mL of *L. rhamnosus* were applied individually, in pairs (di-mix) (A. charcoal + *L. rhamnosus*, A. charcoal + *S. cerevisiae*, and *L. rhamnosus* + *S. cerevisiae*), and all together (tri-mix) in PBS. The doses were prepared as follows: (1) A. charcoal = 40 µg/mL or g (2) *L. rhamnosus* = 110 µg/mL or g and (3) *S. cerevisiae* = 130 µg/mL or g.

5.8. Assessment of Aflatoxin Adsorption in PBS Solution

The adsorption effects of previously prepared adsorbents in the sole, di-, and tri-mix forms were examined in PBS containing aflatoxin in 2 mL tubes. The assessment was conducted at two medium pH (3.0 and 6.8) and at 2 time intervals. The samples were sited in a shaker incubator with a speed of 40 cycles/min at 37 °C/30 min. For HPLC analysis, 1 mL of the supernatant was obtained after centrifugation at 25 °C and 3000 rpm/5 min. The results were compared with those of the positive control (1 μg of AF/mL in the absence of adsorbents) and negative control (adsorbents suspended in pure PBS). Adsorption percentage was calculated according to Ghofrani Tabari et al. [51] by matching the primary aflatoxin concentrations with their concentrations in the existence of adsorbents using the following formula (Equation (1)).

$$\text{Adsorption percentage} = \frac{1 - \text{AF concentrations in the existence of adsorbent}}{\text{AF concentrations in the standard sample}} \times 100 \quad (1)$$

where AF is the aflatoxin.

5.9. Determination of Aflatoxins Residues Using HPLC

A standard solution (26.0 μg/mL) of individual aflatoxin (AFB_1, AFB_2, AFG_1, and AFG_2) was prepared by liquefying 1 mL mixture of HPLC-grade benzene: acetonitrile (97:3 v/v). The standard solution was diluted until reaching a concentration of 1 μg/mL (250 ng/mL for each aflatoxin) using 0.5 mol/L PBS (pH 3.0 and 6.8). Analysis of AFs in PBS was performed according to Hamad et al. [52]. For sample derivatization, 100 μL of trifluoroacetic acid (TFA) + 200 μL of *n*-hexane were added to each sample and were vortexed for 30 s, and samples were kept for 15 min at ambient temperature. Then, 900 μL water: acetonitrile (9:1, v/v) was added and blended by a vortex. The *n*-hexane layer was discarded, and AFs in samples were determined.

5.10. Assessment of Aflatoxin Adsorption by Tri-Mixed Adsorbents in Dark Chocolate Model

Dark chocolate was prepared as follows: dark coverture chocolate was melted in a water bath, then after tempering at 40 °C, divided into three equal portions; (1) negative control (plain chocolate), (2) positive control (chocolate + AFs 1 μg/kg), (3) tri-mix of A. charcoal, lyophilized *L. rhamnosus* (1×10^9 CFU/g) and *S. cerevisiae* (2×10^8 CFU/g) with a ratio of (40:110:130 μg/kg of chocolate) and AFs mix (AFB1, AFB2, AFG1, AFG2) (1 μg/kg of chocolate). Chocolate masses were blended (1500 rpm/5 min), molded, cooled, detached from the form, wrapped in aluminum foil and paper blanks, then kept at 20 °C [53]. The adsorption effect was assessed and compared with control, after 2 and 4 h of incubation at 40 °C, at pH 3.0 and 6.8. The adsorption% and residual AFs were estimated as previously done in PBS.

5.11. Scanning Electron Microscope Analysis

Activated charcoal, probiotics, and yeast cell surface characteristics before and after AF adsorption were visualized by Scanning Electron Microscope (SEM) (SEM—Joel JSM 6360, LA, Tokyo, Japan).

5.12. Cytotoxicity Assessment of Activated Charcoal

Cell viability was examined using peripheral blood mononuclear cells (PBMCs) maintained in the RPMI medium. Blank wells (150 μL PBS), control wells (150 μL PBMCs), and tested wells (150 μL PBMCs) were allocated on a 96-well microtiter plate. Activated charcoal at different concentrations was added to test wells and incubated for 24 h. After adding neutral red (150 μL/wells), they were incubated at 37 °C/2 h, cells were washed, and plates were shaken with de-staining solution (150 μL/well) of (1% acetic acid: 49% deionized water: 50% ethanol). Absorbance was monitored at 540 nm in a spectrophotometer [54]. The percentage of inhibition = 100 − (O.D Control − O.D Treatment / O.D Control)

(Equation (2)); IC_{50} values were calculated online "www.aatbio.com/tools/ic50-calculator (accessed on 14 June 2022)".

$$\text{Inhibition percentage} = 100 - \frac{\text{O.D Control} - \text{O. D Treatment}}{\text{O. D Control}} \qquad (2)$$

where O.D = optical density.

5.13. Application of Tri-Mixed Adsorbents in Dark Chocolate Model

Dark chocolate was prepared according to Mirković et al. [53] as follows: (1) plain chocolate: control; (2) treatment: tri-mix of A. charcoal, lyophilized *L. rhamnosus* (1×10^9 CFU/g), and *S. cerevisiae* (2×10^8 CFU/g) with a ratio of (40:110:130 µg/kg of chocolate).

5.14. Sensory Evaluation

Sensory assessment of control and supplemented chocolate was carried out following the procedure of Senaka Ranadheera et al. [55] with slight modifications. The samples were examined at room temperature using a 9-point Hedonic scale for color, odor, taste, texture, appearance, and overall acceptability. Organoleptic assessment was conducted by 20 panelists (22–58 years old).

5.15. Statistical Analysis

Data were expressed as means of duplicates $\pm$ standard deviation (SD). Data were analyzed by multiple comparisons of one-way analysis of variance (ANOVA) using the Duncan test in IBM SPSS statistics 23 software program (IBM Corp, Armonk, NY, USA), where probability ($p < 0.05$) was considered statistically significant.

Author Contributions: Conceptualization, G.M.H.; methodology, G.M.H.; software, T.M., G.M.H., A.A., B.E.-N., M.I., S.H., S.A.S., A.M.E.-G., E.K., S.A.O. and S.A.-E.A.-A.; validation, T.M., S.A.I., T.E. and S.A.S.; formal analysis, T.M., G.M.H., A.A., B.E.-N., M.I., S.H., S.A.S., A.M.E.-G., E.K., S.A.O. and S.A.-E.A.-A.; investigation, T.M.; resources, G.M.H.; data curation, T.M.; writing—original draft preparation, G.M.H. and T.M.; writing—review and editing, T.M., S.A.I., T.E. and G.M.H.; visualization, T.M., G.M.H., T.E., S.A.I., S.A.S., A.A., B.E.-N., M.I., S.H., A.M.E.-G., E.K., S.A.O. and S.A.-E.A.-A.; supervision, G.M.H.; project administration, G.M.H. and T.M.; funding acquisition, T.E., S.A.I. and T.M. All authors have read and agreed to the published version of the manuscript.

Funding: The open access publication of this article was supported by the Open Access Fund of Leibniz Universität Hannover. This study received no external fundings.

Institutional Review Board Statement: Not applicable.

Informed Consent Statement: Not applicable.

Data Availability Statement: Not applicable.

Acknowledgments: This research was performed in Food Technology Department, Arid Lands Cultivation Research Institute, City of Scientific Research and Technological Applications, Egypt.

Conflicts of Interest: The authors declare no conflict of interest.

References

1. Swanson, K.M. Cocoa, Chocolate and Confectionery. In *Microorganisms in Foods 8*; Foods, I.C., Ed.; Springer: New York, NY, USA, 2011; pp. 241–246.
2. Hamad, G.M.; Mehany, T.; Simal-Gandara, J.; Abou-Alella, S.; Esua, O.J.; Abdel-Wahhab, M.A.; Hafez, E.E. A review of recent innovative strategies for controlling mycotoxins in foods. *Food Control* **2023**, *144*, 109350. [CrossRef]
3. Sánchez-Santillán, P.; Cobos-Peralta, M.A.; Hernández-Sánchez, D.; Álvarado-Iglesias, A.; Espinosa-Victoria, D.; Herrera-Haro, J.G. Uso de carbón activado para conservar bacterias celulolíticas liofilizadas. *Agrociencia* **2016**, *50*, 575–582.
4. Paterson, R.R.M.; Lima, N. Toxicology of mycotoxins. In *Molecular, Clinical and Environmental Toxicology*; Experientia Supplementum (EXS, Volume 100); Luch, A., Ed.; Birkhäuser: Basel, Switzerland, 2010; pp. 31–63.

5. Li, Y.; Tian, G.; Dong, G.; Bai, S.; Han, X.; Liang, J.; Meng, J.; Zhang, H. Research progress on the raw and modified montmorillonites as adsorbents for mycotoxins: A review. *Appl. Clay Sci.* **2018**, *163*, 299–311. [CrossRef]
6. Zellner, T.; Prasa, D.; Färber, E.; Hoffmann-Walbeck, P.; Genser, D.; Eyer, F. The use of activated charcoal to treat intoxications. *Dtsch. Aerztebl. Int.* **2019**, *116*, 311–317. [CrossRef]
7. Dizbay-Onat, M.; Vaidya, U.K.; Lungu, C.T. Preparation of industrial sisal fiber waste derived activated carbon by chemical activation and effects of carbonization parameters on surface characteristics. *Ind. Crop. Prod.* **2017**, *95*, 583–590. [CrossRef]
8. Copetti, M.V.; Iamanaka, B.T.; Pitt, J.I.; Taniwaki, M.H. Fungi and mycotoxins in cocoa: From farm to chocolate. *Int. J. Food Microbiol.* **2014**, *178*, 13–20. [CrossRef]
9. WHO. Mycotoxins. Available online: https://www.who.int/news-room/fact-sheets/detail/mycotoxins (accessed on 14 December 2022).
10. Wang, Y.; Liu, F.; Zhou, X.; Liu, M.; Zang, H.; Liu, X.; Shan, A.; Feng, X. Alleviation of Oral Exposure to Aflatoxin B1-Induced Renal Dysfunction, Oxidative Stress, and Cell Apoptosis in Mice Kidney by Curcumin. *Antioxidants* **2022**, *11*, 1082. [CrossRef]
11. Zhou, Z.; Li, R.; Ng, T.B.; Lai, Y.; Yang, J.; Ye, X. A new laccase of Lac 2 from the white rot fungus *Cerrena unicolor* 6884 and Lac 2-mediated degradation of aflatoxin B1. *Toxins* **2020**, *12*, 476. [CrossRef]
12. Hamad, G.M.; Omar, S.A.; Mostafa, A.G.M.; Cacciotti, I.; Saleh, S.M.; Allam, M.G.; El-Nogoumy, B.; Abou-Alella, S.A.E.; Mehany, T. Binding and removal of polycyclic aromatic hydrocarbons in cold smoked sausage and beef using probiotic strains. *Food Res. Int.* **2022**, *161*, 111793. [CrossRef]
13. Rathore, S.; Varshney, A.; Mohan, S.; Dahiya, P. An innovative approach of bioremediation in enzymatic degradation of xenobiotics. *Biotechnol. Genet. Eng. Rev.* **2022**, *38*, 1–32. [CrossRef]
14. Hill, C.; Guarner, F.; Reid, G.; Gibson, G.R.; Merenstein, D.J.; Pot, B.; Morelli, L.; Canani, R.B.; Flint, H.J.; Salminen, S.; et al. Expert consensus document: The international scientific association for probiotics and prebiotics consensus statement on the scope and appropriate use of the term probiotic. *Nat. Rev. Gastroenterol. Hepatol.* **2014**, *11*, 506–514. [CrossRef] [PubMed]
15. Pascari, X.; Ramos, A.J.; Marín, S.; Sanchís, V. Mycotoxins and beer. Impact of beer production process on mycotoxin contamination. A review. *Food Res. Int.* **2018**, *103*, 121–129. [CrossRef] [PubMed]
16. Díaz-Muñoz, C.; Van de Voorde, D.; Tuenter, E.; Lemarcq, V.; Van de Walle, D.; Soares Maio, J.P.; Mencía, A.; Hernandez, C.E.; Comasio, A.; Sioriki, E.; et al. An in-depth multiphasic analysis of the chocolate production chain, from bean to bar, demonstrates the superiority of *Saccharomyces cerevisiae* over *Hanseniaspora opuntiae* as functional starter culture during cocoa fermentation. *Food Microbiol.* **2023**, *109*, 104115. [CrossRef] [PubMed]
17. Parapouli, M.; Vasileiadis, A.; Afendra, A.S.; Hatziloukas, E. *Saccharomyces cerevisiae* and its industrial applications. *AIMS Microbiol.* **2020**, *6*, 1–31. [CrossRef]
18. Nathanail, A.V.; Gibson, B.; Han, L.; Peltonen, K.; Ollilainen, V.; Jestoi, M.; Laitila, A. The lager yeast Saccharomyces pastorianus removes and transforms Fusarium trichothecene mycotoxins during fermentation of brewer's wort. *Food Chem.* **2016**, *203*, 448–455. [CrossRef]
19. Dogi, C.; Cristofolini, A.; González Pereyra, M.L.; García, G.; Fochesato, A.; Merkis, C.; Dalcero, A.M.; Cavaglieri, L.R. Aflatoxins and *Saccharomyces cerevisiae*: Yeast modulates the intestinal effect of aflatoxins, while aflatoxin B1 influences yeast ultrastructure. *World Mycotoxin J.* **2017**, *10*, 171–181. [CrossRef]
20. Liew, W.P.P.; Nurul-Adilah, Z.; Than, L.T.L.; Mohd-Redzwan, S. The binding efficiency and interaction of *Lactobacillus casei* Shirota toward aflatoxin B1. *Front. Microbiol.* **2018**, *9*, 1503. [CrossRef]
21. Essia Ngang, J.J.; Yadang, G.; Sado Kamdem, S.L.; Kouebou, C.P.; Youte Fanche, S.A.; Tsochi Kougan, D.L.; Tsoungui, A.; Etoa, F.X. Antifungal properties of selected lactic acid bacteria and application in the biological control of ochratoxin A producing fungi during cocoa fermentation. *Biocontrol Sci. Technol.* **2015**, *25*, 245–259. [CrossRef]
22. Codex Alimentarius Commission. *Standard for Chocolate and Chocolate Products*; CODEX STAN; FAO/WHO: Rome, Italy, 2003.
23. Kabak, B. Aflatoxins and ochratoxin A in chocolate products in Turkey. *Food Addit. Contam. Part B* **2019**, *12*, 225–230. [CrossRef]
24. Copetti, M.V.; Iamanaka, B.T.; Pereira, J.L.; Lemes, D.P.; Nakano, F.; Taniwaki, M.H. Co-occurrence of ochratoxin a and aflatoxins in chocolate marketed in Brazil. *Food Control* **2012**, *26*, 36–41. [CrossRef]
25. Turcotte, A.M.; Scott, P.M.; Tague, B. Analysis of cocoa products for ochratoxin A and aflatoxins. *Mycotoxin Res.* **2013**, *29*, 193–201. [CrossRef] [PubMed]
26. Naz, N.; Kashif, A.; Kanwal, K.; Ajaz, H. Incidence of mycotoxins in local and branded samples of chocolates marketed in Pakistan. *J. Food Qual.* **2017**, *2017*, 1947871. [CrossRef]
27. Sartori, A.V.; Swensson de Mattos, J.; de Moraes, M.H.P.; da Nóbrega, A.W. Determination of Aflatoxins M1, M2, B1, B2, G1, and G2 and Ochratoxin A in UHT and Powdered Milk by Modified QuEChERS Method and Ultra-High-Performance Liquid Chromatography Tandem Mass Spectrometry. *Food Anal. Methods* **2015**, *8*, 2321–2330. [CrossRef]
28. Istiqomah, L.; Damayanti, E.; Arisnandhy, D.; Setyabudi, F.M.C.S.; Anwar, M. *Saccharomyces cerevisiae* B18 as antifungal and aflatoxin binder in vitro. *AIP Conf. Proc.* **2019**, *2099*, 020009. [CrossRef]
29. Rahaie, S.; Emam-Djomeh, Z.; Razavi, S.H.; Mazaheri, M. Immobilized *Saccharomyces cerevisiae* as a potential aflatoxin decontaminating agent in pistachio nuts. *Braz. J. Microbiol.* **2010**, *41*, 82–90. [CrossRef]
30. Palomino, M.M.; Allievi, M.C.; Gründling, A.; Sanchez-Rivas, C.; Ruzal, S.M. Osmotic stress adaptation in *Lactobacillus casei* BL23 leads to structural changes in the cell wall polymer lipoteichoic acid. *Microbiology* **2013**, *159*, 2416–2426. [CrossRef] [PubMed]

31. Hamad, G.; Ombarak, R.A.; Eskander, M.; Mehany, T.; Anees, F.R.; Elfayoumy, R.A.; Omar, S.A.; Lorenzo, J.M.; Abou-Alella, S.A.-E. Detection and inhibition of *Clostridium botulinum* in some Egyptian fish products by probiotics cell-free supernatants as bio-preservation agents. *LWT* **2022**, *163*, 113603. [CrossRef]

32. Aazami, M.H.; Nasri, M.H.F.; Mojtahedi, M.; Mohammadi, S.R. In vitro aflatoxin B1 binding by the cell wall and $(1{\rightarrow}3)$-β-D-glucan of Baker's yeast. *J. Food Prot.* **2018**, *81*, 670–676. [CrossRef]

33. Bueno, D.J.; Casale, C.H.; Pizzolitto, R.P.; Salvano, M.A.; Oliver, G. Physical adsorption of aflatoxin B1 by lactic acid bacteria and *Saccharomyces cerevisiae*: A theoretical model. *J. Food Prot.* **2007**, *70*, 2148–2154. [CrossRef]

34. De Mil, T.; Devreese, M.; De Baere, S.; Van Ranst, E.; Eeckhout, M.; De Backer, P.; Croubels, S. Characterization of 27 mycotoxin binders and the relation with in vitro zearalenone adsorption at a single concentration. *Toxins* **2015**, *7*, 21–33. [CrossRef]

35. Joannis-Cassan, C.; Tozlovanu, M.; Hadjeba-Medjdoub, K.; Ballet, N.; Pfohl-Leszkowicz, A. Binding of zearalenone, aflatoxin b 1, and ochratoxin a by yeast-based products: A method for quantification of adsorption performance. *J. Food Prot.* **2011**, *74*, 1175–1185. [CrossRef] [PubMed]

36. Ikegamai, T.; Yanagishita, H.; Kitamoto, D.; Haraya, K. Accelerated ethanol fermentation by *Saccharomyces cerevisiae* with addition of activated carbon. *Biotechnol. Lett.* **2000**, *22*, 1661–1665. [CrossRef]

37. Peltonen, K.; El-Nezami, H.; Haskard, C.; Ahokas, J.; Salminen, S. Aflatoxin B1 binding by dairy strains of lactic acid bacteria and bifidobacteria. *J. Dairy Sci.* **2001**, *84*, 2152–2156. [CrossRef] [PubMed]

38. Hernandez-Mendoza, A.; Guzman-De-Peña, D.; Garcia, H.S. Key role of teichoic acids on aflatoxin B1 binding by probiotic bacteria. *J. Appl. Microbiol.* **2009**, *107*, 395–403. [CrossRef]

39. Shetty, P.H.; Hald, B.; Jespersen, L. Surface binding of aflatoxin B1 by *Saccharomyces cerevisiae* strains with potential decontaminating abilities in indigenous fermented foods. *Int. J. Food Microbiol.* **2007**, *113*, 41–46. [CrossRef]

40. Gallo, A.; Masoero, F. In vitro models to evaluate the capacity of different sequestering agents to adsorb aflatoxins. *Ital. J. Anim. Sci.* **2009**, *9*, 109–116. [CrossRef]

41. Rasheed, U.; Ain, Q.U.; Yaseen, M.; Santra, S.; Yao, X.; Liu, B. Assessing the aflatoxins mitigation efficacy of blueberry pomace biosorbent in buffer, gastrointestinal fluids and modelwine. *Toxins* **2020**, *12*, 466. [CrossRef]

42. Tejada, C.N.; Almanza, D.; Villabona, A.; Colpas, F.; Granados, C. Caracterización de carbón activado sintetizado a baja temperatura a partir de cáscara de cacao (*Theobroma cacao*) para la adsorción de amoxicilina. *Ing. Compet.* **2017**, *19*, 45–54. [CrossRef]

43. Barrientos-Velázquez, A.L.; Arteaga, S.; Dixon, J.B.; Deng, Y. The effects of pH, pepsin, exchange cation, and vitamins on aflatoxin adsorption on smectite in simulated gastric fluids. *Appl. Clay Sci.* **2016**, *120*, 17–23. [CrossRef]

44. Kalagatur, N.K.; Karthick, K.; Allen, J.A.; Ghosh, O.S.N.; Chandranayaka, S.; Gupta, V.K.; Krishna, K.; Mudili, V. Application of activated carbon derived from seed shells of Jatropha curcas for decontamination of zearalenone mycotoxin. *Front. Pharmacol.* **2017**, *8*, 760. [CrossRef]

45. Vilela, A.; Schuller, D.; Mendes-Faia, A.; Côrte-Real, M. Redução da acidez volátil de vinhos por células de *Saccharomyces cerevisiae* imobilizadas em esferas de alginato-quitosano. *Enologia* **2012**, 38–42. Available online: https://www.researchgate.net/publication/282154567_Reducao_da_acidez_volatil_de_vinhos_por_celulas_de_Saccharomyces_cerevisiae_imobilizadas_em_esferas_de_alginato-quitosano (accessed on 29 November 2022).

46. De Rossi, A.; Rigueto, C.V.T.; Dettmer, A.; Colla, L.M.; Piccin, J.S. Synthesis, characterization, and application of *Saccharomyces cerevisiae*/alginate composites beads for adsorption of heavy metals. *J. Environ. Chem. Eng.* **2020**, *8*, 104009. [CrossRef]

47. Saha, D.; Heldt, C.L.; Gencoglu, M.F.; Vijayaragavan, K.S.; Chen, J.; Saksule, A. A study on the cytotoxicity of carbon-based materials. *Mater. Sci. Eng. C* **2016**, *68*, 101–108. [CrossRef] [PubMed]

48. Ocampo, R.O.; Usita, N.P. Utilization of Bamboo Charcoal as Additives in Cakes. *Asia Pac. J. Multidiscip. Res.* **2015**, *3*, 82–86.

49. El-Nezami, H.; Kankaanpää, P.; Salminen, S.; Ahokas, J. Physicochemical alterations enhance the ability of dairy strains of lactic acid bacteria to remove aflatoxin from contaminated media. *J. Food Prot.* **1998**, *61*, 466–468. [CrossRef]

50. Hamad, G.M.; Mohdaly, A.A.A.; El-Nogoumy, B.A.; Ramadan, M.F.; Hassan, S.A.; Zeitoun, A.M. Detoxification of Aflatoxin B1 and Ochratoxin A Using *Salvia farinacea* and *Azadirachta indica* Water Extract and Application in Meat Products. *Appl. Biochem. Biotechnol.* **2021**, *193*, 3098–3120. [CrossRef]

51. Tabari, D.G.; Kermanshahi, H.; Golian, A.; Heravi, R.M. In Vitro Binding Potentials of Bentonite, Yeast Cell Wall and Lactic Acid Bacteria for Aflatoxin B1 and Ochratoxin A. *Iran. J. Toxicol.* **2018**, *12*, 7–13. [CrossRef]

52. Hamad, G.M.; Zahran, E.; Hafez, E.E. The efficacy of bacterial and yeasts strains and their combination to bind aflatoxin B1 and B2 in artificially contaminated infants food. *J. Food Saf.* **2017**, *37*, 12365. [CrossRef]

53. Mirković, M.; Seratlić, S.; Kilcawley, K.; Mannion, D.; Mirković, N.; Radulović, Z. The sensory quality and volatile profile of dark chocolate enriched with encapsulated probiotic *Lactobacillus plantarum* bacteria. *Sensors* **2018**, *18*, 2570. [CrossRef]

54. Ryan, R.M.; Deci, E.L. *Self-Determination Theory: Basic Psychological Needs in Motivation, Development, and Wellness*; Guilford Publications: New York, NY, USA, 2017.

55. Ranadheera, C.S.; Evans, C.A.; Adams, M.C.; Baines, S.K. Probiotic viability and physico-chemical and sensory properties of plain and stirred fruit yogurts made from goat's milk. *Food Chem.* **2012**, *135*, 1411–1418. [CrossRef]

Review

The Growing Importance of Three-Dimensional Models and Microphysiological Systems in the Assessment of Mycotoxin Toxicity

Veronica Zingales [1,2,3,*], Maria Rosaria Esposito [2,3], Noemi Torriero [2,3], Mercedes Taroncher [1], Elisa Cimetta [2,3] and María-José Ruiz [1,*]

[1] Laboratory of Toxicology, Faculty of Pharmacy, University of Valencia, Av. Vicent Andrés Estellés s/n, 46100 Valencia, Spain; mercedes.taroncher@uv.es

[2] Department of Industrial Engineering (DII), University of Padua, Via Marzolo 9, 35131 Padova, Italy; mr.esposito@irpcds.org (M.R.E.); noemi.torriero@phd.unipd.it (N.T.); elisa.cimetta@unipd.it (E.C.)

[3] Fondazione Istituto di Ricerca Pediatrica Cittá Della Speranza (IRP)—Lab BIAMET, Corso Stati Uniti 4, 35127 Padova, Italy

[*] Correspondence: vezin@uv.es (V.Z.); m.jose.ruiz@uv.es (M.-J.R.)

Abstract: Current investigations in the field of toxicology mostly rely on 2D cell cultures and animal models. Although well-accepted, the traditional 2D cell-culture approach has evident drawbacks and is distant from the in vivo microenvironment. To overcome these limitations, increasing efforts have been made in the development of alternative models that can better recapitulate the in vivo architecture of tissues and organs. Even though the use of 3D cultures is gaining popularity, there are still open questions on their robustness and standardization. In this review, we discuss the current spheroid culture and organ-on-a-chip techniques as well as the main conceptual and technical considerations for the correct establishment of such models. For each system, the toxicological functional assays are then discussed, highlighting their major advantages, disadvantages, and limitations. Finally, a focus on the applications of 3D cell culture for mycotoxin toxicity assessments is provided. Given the known difficulties in defining the safety ranges of exposure for regulatory agency policies, we are confident that the application of alternative methods may greatly improve the overall risk assessment.

Keywords: alternative methods; 3D culture; spheroids; in vitro toxicology; mycotoxins

Key Contribution: This review is focused on the applications of 3D spheroid and organ-on-a-chip models for advanced toxicological studies, highlighting their advantages and critical aspects. Advice on the correct establishment of the models is also provided based on our laboratory experience.

Citation: Zingales, V.; Esposito, M.R.; Torriero, N.; Taroncher, M.; Cimetta, E.; Ruiz, M.-J. The Growing Importance of Three-Dimensional Models and Microphysiological Systems in the Assessment of Mycotoxin Toxicity. *Toxins* **2023**, *15*, 422. https://doi.org/10.3390/toxins15070422

Received: 22 May 2023
Revised: 21 June 2023
Accepted: 25 June 2023
Published: 29 June 2023

1. Introduction

Since the early 1900s, assessing the effects of toxic chemicals mainly relies on in vitro two-dimensional (2D) cell cultures, which poorly reflect in vivo conditions and are affected by several limitations [1,2]. In 2D cultures, cells lack both the in vivo tissue organization and key biological functions such as cell–cell and cell–matrix interactions, all contributing to decreased differentiation, non-physiological distribution of nutrients and growth factors from the medium, reduced resistance to xenobiotics, and rapid proliferation [3]. Due to their inadequate ability to elucidate complex biological processes, 2D cell-based systems often require additional follow-up experiments with animal models to better assess toxicity. However, animal experiments also show evident limitations: they are costly, require relatively large amounts of test substances and longer experimental times [4], and express and regulate genes and proteins differently from humans, limiting the straightforward translation of information [5,6]. Last but not least, laboratory procedures that often lead to animal suffering and their sacrifice are raising ever-growing concerns among the general public, such that various countries and organizations are strictly controlling animal studies due to ethical reasons [7]. Over the last decades, several alternative in vitro toxicity-testing strategies that better mimic the in vivo cell behavior and provide more predictive results have been developed for evaluating the hazards associated with the exposure to toxic substances. In 2007, the Toxicology in the 21st Century program, or Tox-21c, prompted by the National Research Council, dominated the discussion [8]. Since then, numerous conferences have addressed Tox-21c's call for a paradigm shift in toxicology, several American agencies formed a coalition to facilitate its implementation, and the U.S. Environmental Protection Agency (EPA) made the novel approach its official toxicity-testing strategy [9,10]. Similarly, the 1986 European Directive 86/609/EEC article 7 encourages the development and validation of alternative techniques that would provide the same level of information as animal experiments [11]. Most recently, animals' wellbeing and welfare in laboratories have been further regulated by Directive 2010/63/EU [12]. However, although Europe is heavily investing in the 3R principles—reduce, replace, and refine—there is a certain reluctance to fully embrace the alternative methods due to the remaining difficulties in standardization, quality control, and validation [4].

The call for alternative and more predictive methods has led to the development of novel advanced in vitro systems with greater physiological relevance.

2. Towards More Predictive Toxicology

It is increasingly recognized that cells grown in a 3D environment more closely resemble in vivo cell functions due to the improved cell–cell and cell–matrix interactions [13,14]. Indeed, the phenotype and physiological behavior of individual cells are strongly dependent on interactions with neighboring cells and proteins of the extracellular matrix (ECM) [15]. Moreover, 3D cultures have well-differentiated cells, show more realistic proliferation rates, reacquire functions lost to monolayers, and are more resistant to xenobiotics treatments, providing a more accurate representation of their effects [3,16]. Finally, 3D cell cultures have longer-term stability, making them an appropriate tool for chronic toxicity studies. Overall, 3D models outperform standard 2D monolayer cultures and provide researchers with tools to better analyze poorly understood phenomena. Despite all these evident advantages, 3D cultures tend to be more expensive and time-consuming, technically challenging, and low in throughput and increase the difficulty of interpreting the data and replicating the experiments.

Simplifying, 3D cultures can be divided in two main groups: spheroids and organoids. Whereas spheroids can be defined as a cluster of differentiated cells that aggregate, exhibiting some tissue-like structure, organoids are multicellular self-assembled constructs that mimic the corresponding in vivo organ in terms of cell types, structure, and function [17]. The lack of components of the in vivo organ, such as vasculature [18], prompted the development and integration of micro-physiological systems (organ-on-a-chip, OoC). These technologies, which combine in vitro models with perfusion and micro-engineered environments, enable the creation of controlled micro-environmental niches and patterned biomolecular signals. Since the first published OoC by Shuler et al. [19], the number of studies on this topic has increased constantly. Although useful to accelerate the preclinical assessment of substances on cells and tissue mimics, they present limitations due to the lack of the complex inter-organ crosstalk in the human body. To improve the available models, researchers have thus devised various "multi-organ-on-a-chip" devices, connecting multiple organs on a single chip via microfluidic channels, enabling interactions through the flow of culture medium [20–24]. Organ-chip models are obviously more difficult to use than other 3D culture techniques in terms of development, sample handling, and manipulation [25].

Given this overview of in vitro alternative methods and with the goal of identifying the optimal tradeoff between adherence to in vivo conditions, quality of the results, and ease of use, the next sections will focus on key in vitro approaches (spheroids and OoC) and their application in toxicological studies.

3. 3D Spheroids

Spheroids represent a suitable model for improved toxicology risk assessments over other 3D cell-culture methods thanks to their faster production, higher reproducibility, and presence of cell heterogeneity within the sphere [26].

The self assembly of a spheroid from gently pelleted cells generally follows three main steps: first, extracellular matrix arginine-glycine-aspartate (RGD) motifs bind with integrins forming loose aggregates; second, cell–cell interactions induce increased expression of the cadherin gene, resulting in the membrane accumulation of cadherin protein; finally, a compact spheroid forms as a result of the homophilic cadherin-to-cadherin interactions [27]. Figure 1 schematizes these steps and shows representative images of the formation of a spheroid using human neuroblastoma SK-N-AS cells.

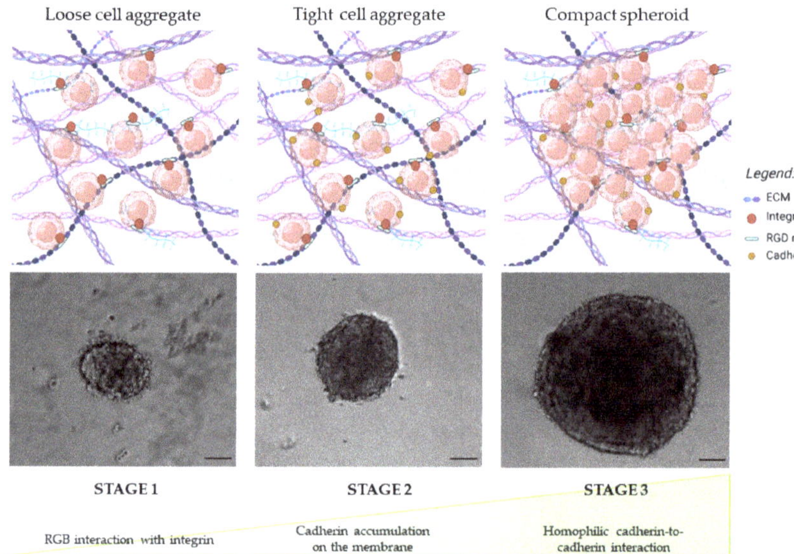

Figure 1. Representative optical images of SK-N-AS spheroids at different stages of growth over the course of 7 days, from the loose aggregate of cells to the formation of a compact spheroid. Scale bar = 100 μm (magnification 5×). This figure is original and unpublished.

While 2D cell cultures are almost exclusively based on plastic plates or flasks, the role of both synthetic and organic materials in the formation and maintenance of spheroids is key for their optimized use. Generally, the formation methods can be divided in two main groups: scaffold-based and scaffold-free [28].

Scaffold-based methods use materials providing external cell-anchoring systems mimicking the ECM and thus supporting cell growth [29,30]. Several scaffolding biomaterials are currently available and include synthetic hydrogels, e.g., poly (ethylene glycol) (PEG), poly (lactic-co-glycolic) acid (PLGA), and polydimethylsiloxane (PDMS), and natural protein-based hydrogels, such as collagen, alginate, matrigel, gelatin, and hyaluronic acid (HA) [28,31,32]. The ideal scaffold should be chosen according to several parameters, including the porosity and the pore size (100–500 μm) [33,34], which confer specific characteristics to the 3D culture, especially in terms of the diffusion of nutrients, oxygen, and metabolites [29,35]. Natural polymers present several advantages, including biocompatibility and natural cell-adhesive properties [36]. Nevertheless, the lack of strong mechanical properties and the batch-to-batch variation makes them not easily controllable [37]. Synthetic hydrogels overcome these limitations by providing high reproducibility, stability, and control over the biochemical and mechanical properties [38]. Comprehensive reviews of scaffold-based methods can be found in the literature [29,30,39,40].

Scaffold-free methods promote the self-aggregation of cells without adding biomaterials, use specialized culture surfaces, including ultra-low attachment (ULA) plates or hanging-drop microplates [36], and exploit factors such as magnetic and gravitational forces [29]. The hanging-drop method allows the production of high numbers of spheroids, but given its high operator dependency, it leads to heterogeneous sizes and morphologies [29,41,42]. In contrast, microwell plates characterized by low-adhesion micropatterned compartments combine high-throughput production with little plate-to-plate variation [29,43]. These systems are available in different formats and an example is represented by AggreWell^{TM} plates (STEMCELL), where each of the 24 or six wells contains a matrix of pyramid-shaped microwells enabling the production of thousands of uniform cellular aggregates within 24–48 h with simple centrifugation of the cell suspensions. The exact diameter of the aggregates will be determined by the microwell dimensions and the chosen cell-seeding concentration [40]. Aggregates can be recovered from the microwells by gentle pipetting and transferred into ULA plates or bioreactors for further culturing and testing. Despite the clear advantages, this method is not ideal when analyses of single spheroids are required (such as for microscopy imaging or cytotoxicity evaluation) because of the difficulty of diluting the obtained suspension of 3D structures to the single-spheroid level.

Single spheroids can be obtained via a liquid-overlay technique or using ULA plates. The first one uses substrates creating non-adherent surfaces that favor cell aggregation [44]. Agarose diluted in medium can be used to coat the plate and prevent cell adhesion, but the generated spheroids are typically non-uniform in size and the bottom concave agarose layer may hinder some applications. Since agarose solidifies within seconds to minutes, a critical step is that it must be kept at a temperature of around 60 °C during dispension to avoid irregular or insufficient coating due to cooling. In addition, a relatively large volume (50 μL) per surface area is needed to guarantee the formation of a concave surface [45]. Finally, agarose does not interact with tumor cells, leading to its inability to activate specific pathways. Alternative materials can be used, including HA, which can interact with the surface receptors of cancer cells [44].

The ULA plates are characterized by cell-repellant surfaces promoting cell–cell interaction and self-aggregation into spheroids without the need for coating [46]. Our group has recently shown a successful application of ULA (Corning$^{®}$, 7007, New York, NY, USA) for the formation, characterization, and evaluation of sterigmatocystin-induced cytotoxicity on neuroblastoma SH-SY5Y- and SK-N-DZ-derived spheroids [47]. The non-adhesive U-bottom surface of ULA 96-well plates allowed the formation of one centrally located spheroid per well, compatible with several applications and analyses. In particular, in the aforementioned work, the spheroids were employed for various tests, from cytotoxicity assays, such as thiazolyl blue tetrazolium bromide (MTT) and ATP assays, to immunofluorescence and Western blotting. Moreover, the round bottom of the wells directs the aggregation of cells at its center and facilitates the generation of a uniform spheroid upon centrifugation, an essential feature for the reproducibility of cytotoxicity experiments. Among the commercially available formats, we believe 96-well plates are the ideal choice for cytotoxicity experiments since they guarantee monitoring and manipulation of individual spheroids with high numbers of replicas for experimental conditions. Careful optimization of both the cell numbers and the centrifugation times needed for aggregation is necessary, as well as determination of the appropriate days of culture for spheroids to form and grow to the required size.

Dynamic culture in bioreactors brings several advantages over static methods for the generation of spheroids, including a homogeneous environment, better diffusion of oxygen and nutrients, and longer periods of cultivation [48,49]. The stirred-tank bioreactor (STR) allows an easy and large-scale production of spheroids using culture modalities that allow the control of several parameters, including oxygen and pH [48,50]. Their main disadvantage is related to the relatively high shear forces that could have a detrimental effect on the spheroids' shape and structure. Rotating wall vessel bioreactors (RWB) are also called microgravitational bioreactors and produce spheroids in a microgravity environment with a lower shear and turbulence compared to other bioreactor systems [48,51].

3.1. Considerations for Spheroid Handling

To use a spheroid model for advanced toxicology risk assessment, some critical considerations need to be further addressed. First, it is crucial to have one single spheroid per well; second, the spheroids must be uniform in shape and size to reduce variability in the readouts. Although we are aware of the difficulty of obtaining an absolute standardization of spheroid culturing, we will here discuss some key points that are often overlooked.

3.1.1. Morphology

Employing regular and well-rounded spheroids in in vitro assays ensures a higher reproducibility of the results [52,53]. For this reason, the evaluation of morphological parameters such as solidity and circularity is the starting step to limit the bias and choose the appropriate spheroid for a given biological application. According to Santo et al. [54], spheroids are considered regular in shape if their solidity values are higher than 0.90. Circularity (*Cir*) is used to calculate the sphericity index (SI), in turn indicating how close to a spherical geometry the construct is, according to Equation (1) [39].

$$\text{SI} = \sqrt{Cir} \tag{1}$$

Zanoni et al. [30] consider spheroids spherical when SI $\geq$ 0.90. The shape parameters may be estimated using AnaSP, a user-friendly software automatically analyzing brightfield images acquired with a standard optical microscope [55]. In our laboratories, we characterized spheroids generated from tumoral and non-tumoral cell lines. As shown in Figure 2, spheroids generated from bone-marrow-derived mesenchymal stem cells (BM-MSCs) can be considered regular and spherical in shape at all tested cell-seeding densities and culture times, with solidity values higher than 0.92 after 1 day of culture and SI ranging from 0.93 to 0.97. Regarding neuroblastoma SH-SY5Y cells, the best results were obtained for spheroids after 7 days of culture, with measured solidity values and SI over 0.85 and close to 0.90, independent of the cell-seeding concentration.

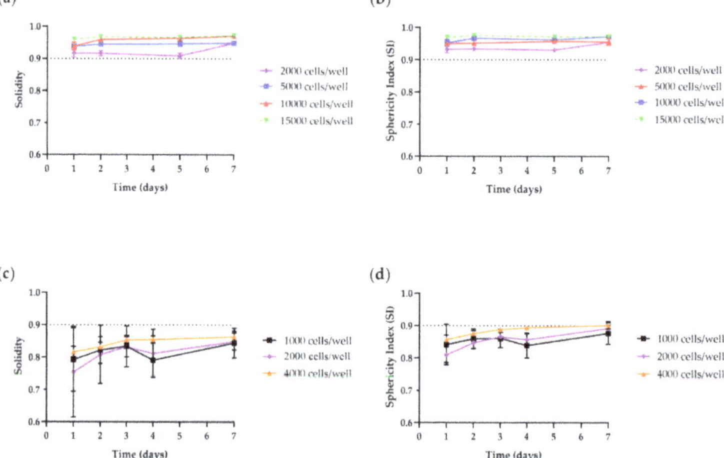

Figure 2. Shape parameters of BM-MSCs (**a,b**) and SH-SY5Y (**c,d**) spheroids throughout culture period in ULA 96-well round-bottom plates obtained using AnaSP software. (**a,c**) Solidity; (**b,d**) sphericity index (SI). Curves represent the mean $\pm$ SEM of 3 replicates for each cell density/well. This figure is original and unpublished.

3.1.2. Size and Time

The physiological state of spheroids strongly depends on their size and culture time. A diameter of 400–500 µm is considered appropriate for the spontaneous formation of gradients of oxygen and nutrients and of differential proliferation rates, all essential for relevant 3D experimental studies [46]. Smaller spheroids sized up to 200 µm allow one to obtain cell-to-cell and cell-to-matrix interactions, but they do not develop a stratified composition with viable cells in the rim and necrotic/apoptotic cells in the center [55]. The seeding cell number only partially correlates to the spheroid size, which is also affected by the compactness in a cell-line-dependent manner (Figure 3). Moreover, given the specific doubling times, different cell lines require different culture lengths to obtain spheroids of a defined diameter. Therefore, the optimal seeding cell density and time in culture need to be identified for each cell line. Figure 4 shows representative size measurements of spheroids generated from BM-MSCs and SH-SY5Y cells over time in culture. While SH SY5Y spheroids had an almost linear growth, BM-MSC spheroids showed a slight decrease, likely as a consequence of a reduction in the amount of cytoplasm and in cell volume [56].

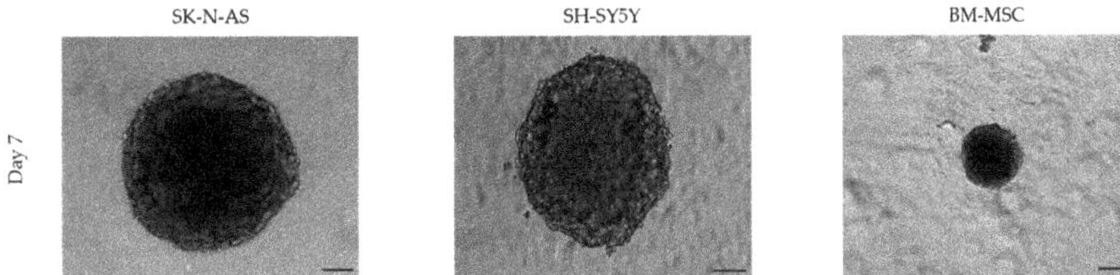

Figure 3. Optical images of spheroids generated from different cell lines on ULA 96-well round-bottom plates and seeding the same number of cells (2000 cells/well). Scale bar = 100 µm (magnification 5×). This figure is original and unpublished.

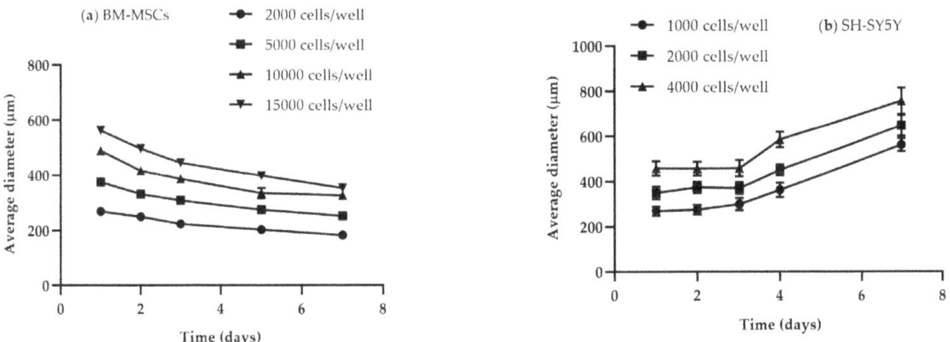

Figure 4. Assessments of spheroid growth on ULA 96-well round-bottom plates. Growth kinetics of (**a**) BM-MSCs and (**b**) SH-SY5Y spheroids evaluated over a period of 7 days. Curves represent the mean ± SEM of 3 replicates for each cell density/well. This figure is original and unpublished.

Another fundamental aspect is the assessment of the maintenance of the desired cell viability throughout the 3D constructs. Recently, several user-friendly kits have been developed to measure cell viability on 3D spheroids. Among these, the Live & Dead Viability/Cytotoxicity assay (Invitrogen®, L3224, Waltham, MA, USA) distinguishes live from dead cells in a population by exploiting the fluorescent properties of Calcein AM (acetomethoxycalcein) and ethidium homodimer-1 dyes (Figure 5). Calcein, as a permeable and liposoluble dye, when internalized in live cells is cleaved by intracellular esterases with the subsequent emission of green fluorescence. Ethidium homodimer-1, a cell-membrane-impermeable dye, confers red fluorescence to nucleic acids and is an indicator of cell death.

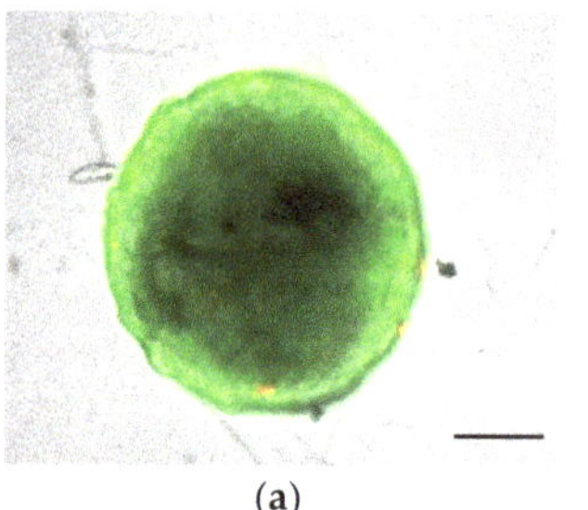 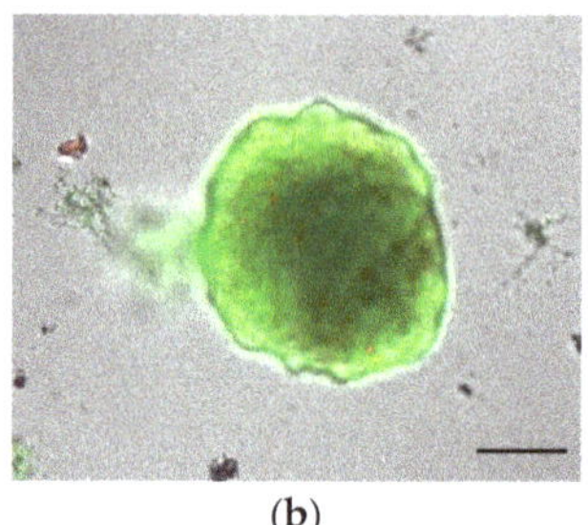

(**a**)   (**b**)

Figure 5. Assessments of spheroid viability on ULA 96-well round-bottom plates. Cell viability of BM-MSCs spheroids was evaluated using a Live & Dead assay after (**a**) 1 day and (**b**) 7 days of growth. Live cells are stained green and dead cells are stained red. Scale bar = 100 μm (magnification 5×). This figure is original and unpublished.

3.1.3. Extracellular Matrices

Not all cell lines spontaneously form spheroids using standard culture medium but require the addition of ECMs such as collagen or methylcellulose. For example, human umbilical vein endothelial cells (HUVECs) seeded in ULA 96-well round-bottom plates form loose aggregates and only upon the addition of ECM components to the growth medium yield a 3D spheroid-like morphology (Figure 6). In this case, the spheroid-generation conditions need to be optimized by testing various ECM components. For HUVECs, we found that collagen I worked best to form spheroids with a defined boundary, with an optimized concentration of 7.5 μg/mL, while higher concentrations led to a disrupted morphology (Figure 7). Considering that some ECMs may have poor biocompatibility, their use requires a validation of their potential toxicity. The cell viability of spheroids cultured in the presence of ECMs may be tested with a Live & Dead fluorescence staining assay (Figure 8).

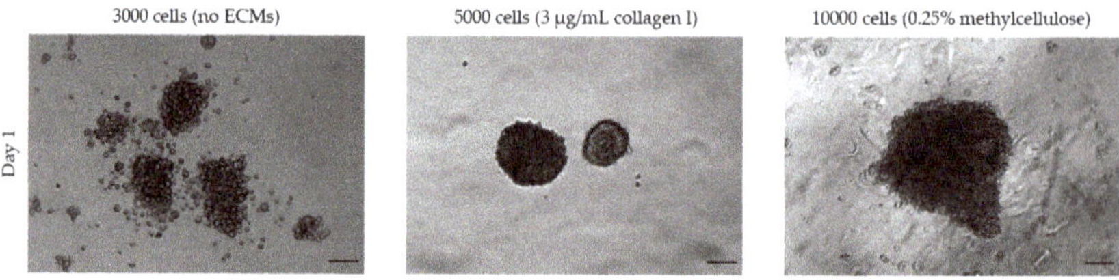

Figure 6. HUVEC cells seeded for spheroid formation with and without ECMs addition. Scale bar = 100 μm (magnification 5×). This figure is original and unpublished.

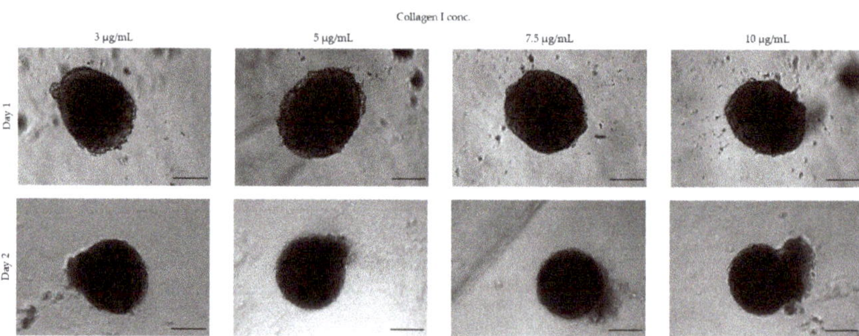

Figure 7. Standardizing collagen I concentration. A total of 10,000 HUVEC cells was seeded in medium supplemented with different concentrations of collagen I. Scale bar: 100 μm (magnification 5×). This figure is original and unpublished.

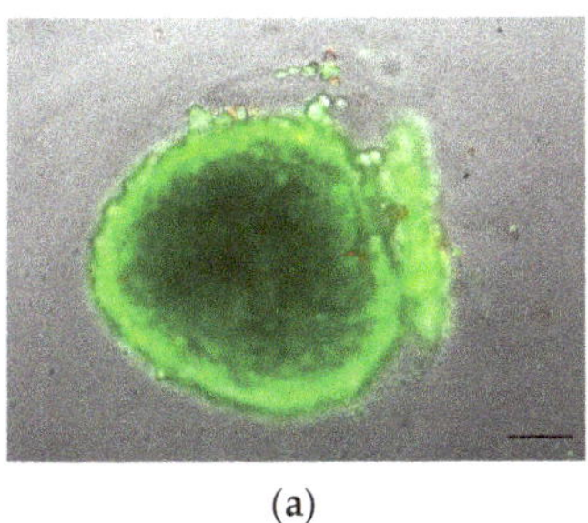 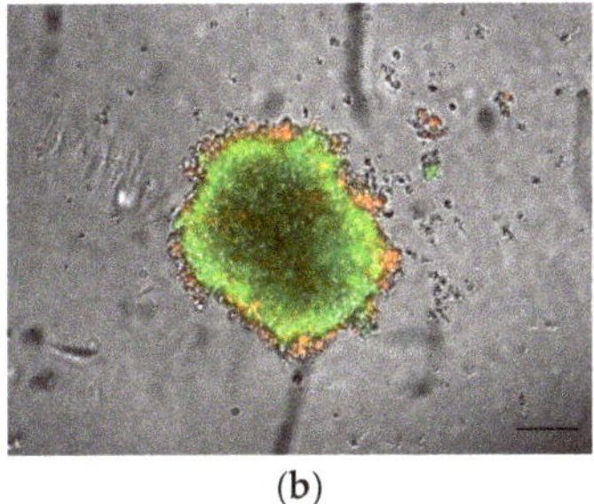

(a)   (b)

Figure 8. Assessments of HUVEC spheroid viability cultured in the presence of ECMs. A total of 10,000 HUVEC cells was seeded in medium supplemented with (**a**) 7.5 µg/mL collagen I and (**b**) 0.25% methylcellulose and stained with Live & Dead 1 day later. Scale bar: 100 µm (magnification 5×). This figure is original and unpublished.

3.2. Downstream Functional Assays on 3D Spheroid Model: Advantages and Critical Issues

As previously stated, 3D spheroids improve upon the canonical 2D culture techniques by enabling us to address previously unanswered research questions [36]. Of course, bridging the gap from 2D culture techniques to 3D innovative models is not free from obstacles and challenges. In addition to the previously discussed critical aspects such as the spheroids' formation, size, and shape, the type and method of acquisition of the viability/cytotoxicity tests following treatments with toxic substances must also be carefully considered to minimize biases and obtain reliable data.

After exposure to toxic substances, the first evidence of toxic effects is based on optical observations of potential alterations on the structural organization of the spheroids, assessed by measuring the integrity and viability of the spheroids over time [47]. Commercially available cell-viability assays can be adapted to the 3D model. An example is the CellTiter-Glo® Luminescent Cell Viability Assay (Promega®, G968B, Milan, Italy), a luminescent assay that measures ATP to determine the number of metabolically active cells. An additional viability test applicable to 3D spheroids is the MTT assay. Compared to monolayer cultures, the MTT assay for spheroids requires slight protocol modifications. In particular, at the end of the treatment, the spheroids must be transferred to a flat-bottom 96-well plate before the MTT solution is added. However, it should be noted that the spheroid transfer could not only disturb the culture, but also lead to the loss of cell material and impede a large-scale throughput [47,57].

Moreover, a limiting step when working with 3D cultures can be the reproducibility of data. To bypass this issue and provide major robustness to the results it is mandatory to have at least four to six spheroids per experimental condition.

Microscopy and imaging are among the most widespread analytical methods employed in cell biology. However, one of the biggest issues when 3D spheroids are used as a working model is the difficulty of staining and imaging the entire structure due to the dense network of cells in the core mass. Furthermore, the plates where spheroids are formed are not always fully compatible with microscopes, hindering high-resolution imaging and often requiring spheroids to be transferred to more suitable surfaces, with consequent disturbances to the culture conditions and the potential loss of cellular material [58]. A further challenge is to capture the full complexity of the spheroid structure. To do this, a series of *xy* images can be captured at fixed steps in the vertical direction using automated microscopes to obtain a *z* stack [58]. Moreover, since not all antibodies labeled with fluorescent dyes are able to diffuse into and stain the inner core of the spheroids, we recommend performing primary and secondary antibody incubations under constant and gentle shaking to improve upon the limited penetration capacity.

Spheroids are also adaptable to routine molecular and biochemistry applications. A recent study optimized the protocol to determine the protein profile in 3D cultures after mycotoxin treatments [47]. Surely, one aspect to consider for specific downstream applications is the overall number of viable cells available at the desired endpoint. A simple but not necessarily easy strategy is culturing high numbers of spheroids per condition to obtain the quantity and purity of RNA and proteins suitable for the analyses. After the careful collection of spheroids, the genes and proteins of interest can be detected according to standard PCR and blotting procedures.

Other assays that have been shown to be adaptable to the spheroid model are cell cycle and apoptosis analysis using flow cytometry. The crucial aspect to consider for this type of acquisition is the sensitivity of the cytometer, with a minimum cell count requirement of 500,000 cells, thus requiring large numbers of spheroids according to the initial seeding density, growth timing, and final cell number obtained [47]. The experimental procedure is based on the labeling and acquisition of single-cell suspensions; thus, a defined number of spheroids per experimental condition must be collected and trypsinized before staining and subsequent cytometer acquisition [59].

Angiogenesis is a multi-step process involving the parent vessel's extracellular matrix degradation, causing the migration and proliferation of endothelial cells to form a new vessel with a lumen and a layer of mural cells [60]. Being a complex phenomenon, reproducing it in vitro is a challenge not addressable with standard cell culture. Dr. Thomas Korff and Dr. Hellmut Augustin pioneered an endothelial cell spheroid-based 3D angiogenesis technique for in vitro studies [61]. Based on this system, in vitro angiogenesis assays were widely developed to investigate the putative angiogenesis-related toxic effects of compounds and/or genetic manipulations [62]. By culturing endothelial-cell spheroids with the hanging-drop method and embedding them into a collagen matrix, the ability to form capillary-like tubes can be investigated by counting the number and length of sprouts after treatment with toxins [63]. The spheroid-based sprouting assay is an advantageous tool that makes the study of angiogenesis in an in vivo-like microenvironment easier and more robust. Similarly, 3D cultures can be instrumental in analyzing with better soundness the general cell-migration phenomenon. For this purpose, different matrices have been tested to include the spheroids, such as gelatin, collagen, and matrigel [64–66]. Based on our experience, a gelatin coat guarantees a good experimental reproducibility and spheroids generated in ULA 96-well round-bottom plates can be transferred to gelatin-coated "migration" plates and monitored over time using an optical microscope. Within a few hours, tumor cells begin spreading from the spheroid over the coated surface [47]. Fundamental for correct data interpretation are the image-processing steps performed on several images for each sample. The complete image-analysis routine needs to be managed with different and sequential image-processing techniques with the aim of measuring the cells spreading on the gelatin coat. The Sobel mathematical model, based on the edge detection of optical greyscale images, provides a reliable measure of migrating cells. The edge of the spheroid masks detected at time 0 is employed to identify the inner regions from the surrounding migrated cells [47,67,68].

Regarding the assessment of genotoxicity associated with exposure to toxic substances, there are many limitations associated with 2D cultures, as extensively reviewed previously [69]. Moreover, the regulated genotoxicity in vitro testing methods have demonstrated low specificity and produced misleading false positives, resulting in the need for additional animal testing [70]. Recently, 3D in vitro models have had increasing application in the field of genotoxic assessment due to their reliability in reproducing the physiological environment and metabolism of chemicals [71]. An existing application tested genotoxic effects on 3D HepG2 cells and compared the effects in 2D monolayer cultures, showing that the spheroids had improved sensitivity in detecting genotoxic compounds evaluated with a comet assay [72]. Genotoxicity assessments, like other analyses described above, require at least 48–60 spheroids per condition to provide reliable results (Figure 9). Other advanced tools are the 3D skin comet and the reconstructed skin micronucleus (RSMN) assays [73], which detect DNA lesions and chromosomal damage, respectively, that could be due to exposure to a variety of compounds found in cosmetics, industrial chemicals, and household products. In the field of dermally applied chemical exposure, the 3D skin comet assay resulted in a 70–100% range of predictivity and good intra- and inter-laboratory reproducibility.

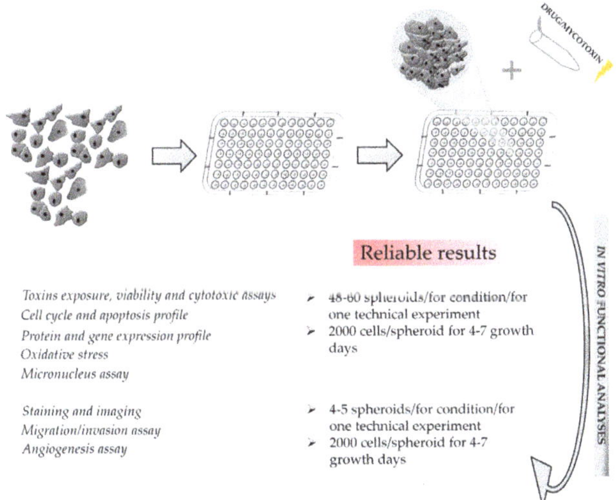

Figure 9. Summary of functional assays and possible approaches for the obtainment of reliable results in a spheroid model. This figure is original and unpublished.

In addition, novel genotoxicity testing enabled researchers to observe DNA damage through a high-throughput comet chip assay on metabolically competent HepaRG cells [74]. The optimized 3D culture system used 96- or 384-well ULA plates for evaluating the response to various direct-acting and indirect-acting genotoxicant/carcinogen exposure. The high-throughput comet chip platform on 3D spheroids proved to be a reliable in vitro approach improving the risk assessment for human genotoxic carcinogen exposure.

Another common effect associated with toxin exposure is the induction of oxidative stress. Currently, the role of reactive oxygen species (ROS) on decreased cell viability induced by toxic substances is mainly investigated using simplified 2D monolayers [72]. The use of a 3D spheroid model enables the capture of the complexity of these biological processes and cellular behaviors, better resembling in vivo conditions. ROS generation can be measured using an ROS-specific fluorescent dye such as 2′,7′-dichlorodihydrofluorescein diacetate (DCFH-DA), a cell-permeable probe used to detect intracellular ROS through plate-reader acquisition [47]. In a 3D glioblastoma model, the CellROX Green Reagent (Thermo Fisher Scientific, Waltham, MA, USA) was used to measure the release of intracellular ROS. CellROX, a DNA-binding cell-permeant dye, displays green photostable fluorescence if oxidation is active. This fluorescence followed the fluorescent probe's incorporation due to drug treatments. To provide further insights on the oxidative status of the spheroids, the authors also assessed glutathione (GSH) levels using a luminescence-based assay (Promega, Madison, WI, USA) as well as the expression of glutathione peroxidase4 (GXP4) and 8-Oxo-2′-deoxyguanosine (8-oxo-dG) using immunohistochemistry in sections from the spheroid tumors [75]. The expression of enzymes associated with ROS metabolism can also be measured via PCR and/or Western blot. These time-consuming methods require a skilled operator and, to reduce the data variability, multiple technical replicates per condition [47].

4. Organ-on-a-Chip (OoC) Systems

OoCs combine cell biology and microfluidic technology in a microdevice capable of recapitulating the dynamics, functionality, and (patho)physiology of human organs. Typically, a microfluidic-based device houses cell constructs embedded in engineered 3D microenvironments, with medium flow driven by pumps (syringe or peristaltic). Compared to conventional 2D cultures, cells in OoC systems show a cell polarization and cytoskeletal architecture more similar to those observed in vivo, in addition to maintaining cell viability and functionality for longer periods of time. OoCs also enable a fine regulation of the biological and physicochemical micro- or nano-environment thanks to a tight control of the flow of the culture medium. The flowing medium mimics the continuous supply of oxygen and nutrients, mirroring the in vivo physiological conditions while also allowing the delivery of drugs or other compounds of interest to the cells in culture. Another advantage derives from the use of small volumes of culture medium and test substances with great economic gain that should not be underestimated.

So far, OoCs have greatly advanced life science and medical research, playing an increasingly important role in preclinical trials and drug development, but also provide new broad opportunities for the toxicological field. With the ability to recapitulate in vivo physiology and study complex biological systems in a controlled in vitro environment, OoCs predict more accurately the acute and chronic effects induced by chemical and natural toxicants to which humans may be exposed. These advanced cell-culturing systems promote the improvement of toxicological hazard and risk assessment and overcome ethical limitations such as the use of animal testing for toxin risk assessment.

Although OoCs could play an important role in the revolution of toxicology, their application shows downsides, such as the low throughput, difficulty of standardization and validation, and most importantly, the inability to give ultimate answers on the adverse systemic effects at the level of the whole individual. The following sections provide an overview of the usability, compatibility, and assay ability aspects of OoC systems.

4.1. Considerations for OoC Handling

In OoC technology, the design, fabrication methods, and culture strategy strongly depend on the ultimate experimental purposes. All the choices relating to the development of an OoC model must depend on them, from the selection of materials to the cell source.

4.1.1. Material Selection

There are various materials that can be used for the fabrication of OoCs; however, each of them shows advantages and disadvantages. Here, we present only the most important and widely used materials in OoC design and development.

The most commonly used material is PDMS, a silicone polymer whose biocompatibility, elasticity, optical transparency, and gas permeability make it the ideal candidate for biological experiments [76]. However, its tendency to adsorb a wide range of chemicals limits its application in toxicity studies and must be carefully taken into account [77,78]. The absorption of biomolecules is reduced when using glass, whose great optical transparency also makes it the first choice for real-time imaging. However, glass chips are not suitable for long-term cell culture, since they are not gas-permeable [79]. Furthermore, in contrast to PDMS, which can be processed with soft lithography and micromolding techniques, glass is typically processed with the more expensive and time-consuming standard lithography [76,80].

Recently, plastic materials such as poly(methylmethacrylate) (PMMA), polycarbonate (PC), and polystyrene (PS) have been increasingly replacing the traditional PDMS- and glass-based chips due to their low cost and easy fabrication [81]. Furthermore, they show several interesting properties, such as: (i) high biocompatibility, enabling cell growth and adhesion; (ii) excellent optical transmittance, allowing high-quality fluorescent imaging; and (iii) resistance to small-molecule permeation and, therefore, approval by the Food and Drug Administration (FDA) [82,83]. Nevertheless, there are some limitations in their use, due to the low gas permeability, which has a negative impact on long-term experiments, and their incompatibility with most organic solvents [84,85].

Hydrogels are polymeric materials with a high water content showing many attractive properties for the OoC field, including biocompatibility, high permeability, biodegradability, elasticity, and low cytotoxicity [37]. In addition, their mechanical properties mimic some elements of ECMs, also protecting biological entities in long-term studies [86]. Due to their high biocompatibility and presence of cell-binding sites, hydrogels are mostly coated on the surface of OoC devices fabricated with a material with less-favorable cell-attachment properties [87,88]. However, hydrogels, especially natural ones, present some limitations due to the poor stability and batch-to-batch variability. Collagen is one of the most widely used hydrogels for bioengineered tissues, being the most common ECM component in the human body [89]. Techniques for hydrogel-based OoC fabrication include lithography, 3D printing, and molding [90].

Finally, as for conventional cell-culture platforms, all components of the OoC devices must be sterile, and the choice of sterilization method depends on the materials involved. An inappropriate sterilization may result in component damage, ultimately leading to system failure. The conventional autoclave sterilization methods cannot be used for materials with low thermal resistance, such as PMMA and PC plastics, and should be replaced with UV and/or ethanol treatments. However, materials that exhibit opacity and/or the ability to absorb UV rays are not suitable for UV sterilization, while ethanol is not compatible with materials that can absorb it, including PDMS [91].

Once sterilized, the device surfaces that will be in direct contact with cell cultures may require specific treatments to ensure biocompatibility and regulate cell attachment. If adhesion needs to be favored, ECM coating is often adopted (see above), while pluronic acid treatment, which prevents cell adhesion, can be used in those chips designed for 3D culture [3].

4.1.2. Selection of Cell Culture

Closely related to the scientific question is the selection of the appropriate cell source. Immortalized cells, primary cell cultures, and induced pluripotent stem cells (iPSCs) can be used. Cancer cells are widely employed in toxicological studies because they are highly proliferative, easy to culture, and well-characterized. Despite these advantages, immortalized cell lines show an altered genotype and phenotype, which limit their ability to reproduce physiological cell behaviors [92]. For these reasons, primary cells are a valid option to evaluate the toxicological effects induced by exposure to toxic substances under physiological conditions. However, they show several limitations, including poor accessibility due to the lack of donors, inter-donor variability, limited proliferative capacity, and the loss of tissue-specific functions when maintained in vitro, preventing their use in chronic toxicity studies [93,94].

The use of iPSCs is in demand in toxicological studies using OoC models. The main advantages of iPSCs are their capacity to differentiate in various cell lineages [95] and their adult origin, avoiding the ethical concerns associated with the use of embryonic tissues. In addition, like primary cells, iPSCs derived from donors with known disease phenotypes inherit the patient genotype, making them ideal to study toxic responses for susceptible groups, but less suitable for broader populations [95]. The major challenge in the use of iPSCs is their correct differentiation into specific cells or tissues. Due to their natural variability, standardization remains difficult, but protocols to maintain, differentiate, and mature iPSCs in vitro are constantly being developed and updated. In addition, direct on-chip culture techniques have also been performed using iPSC-derived intestinal organoids [96], proving that a synergy between iPSC culture and the OoC technology is expected to greatly progress research.

OoCs culturing monolayers of only one cell phenotype might underrepresent the in vivo complexity. For these reasons, new approaches include more complex 3D architectures with multicellular compartments [97]. When culture wells allow direct access (open at the top), 3D cultures previously formed with scaffold-free techniques can be directly seeded without the need for delicate injections through microchannels [98–101].

Single-organ systems are ideal to study the effects of toxicants on a specific target; however, a more comprehensive analysis would require a multi-organ system that would reproduce the correlations that one site has on the functionality of another one. However, obtaining true multi-organ systems is not as straightforward as the simple connection of two or more single-organ systems. For example, multi-OoCs require an optimized formulation of a common culture medium, capable of ensuring and maintaining the correct viability and physiology of each cell population. Mixing the culture media used for each cell type has been shown to generally produce good outcomes [102,103]. However, as the number of cell types in a system increases, optimizing a co-culture medium can become more challenging. In these cases, the compartmentalization of cells into semi-insulated culture chambers can be a viable option to circumvent the issue [99,104].

4.2. Downstream Function Assays on OoC Model: Advantages and Critical Issues

The functional assays that can be performed in OoC systems can generally be divided into two classes: on-chip and off-chip. The first comprises immunohistochemistry, trans-epithelial electric resistance (TEER), and migration and angiogenesis assays. Off-chip assays include high-performance liquid chromatography/mass spectrometry (HPLC/MS), gas chromatography/mass spectrometry (GC/MS), and enzyme-linked immunosorbent assays (ELISA) [105].

Immunohistochemical staining and other microscopy readouts are the most widespread methods used in OoC devices. By using the microchannel network of the OoC, it is possible to directly deliver the fluorescent stains or antibodies to assess cell viability or the expression of specific markers. Furthermore, the use of optically transparent materials ensures excellent imaging quality [106–109] while also enable on-line monitoring of cell behavior (if live cell stains are available, i.e., in reporter cell lines). Other on-chip assays (e.g., calcium-imaging, colorimetric, and luminescence assays) are still feasible but less frequently adopted due to the difficult optimization in a microfluidic environment.

The measurement of TEER values allows the evaluation of the integrity and permeability of any barrier tissue. This technique is mainly employed in OoCs where cells form a natural barrier between fluid compartments. TEER is easily measured by applying Ohm's law, but impedence spectoscopy, using microelectrodes integrated into the chip during fabrication [110,111] or before an experiment [112,113], provides a more accurate measure. Complex systems containing several cell-culture chambers, such as the one reported by Ramadan et al. [114], which measured the barrier integrity of human keratinocytes in co-culture with monocytes, require highly trained end-users.

Off-chip assays have the advantage of measuring many markers simultaneously. However, only approaches that require a few microliters of sample volume are compatible with microfluidic chips. Perhaps for these reasons, HPLC/MS and GC/MS are used by few in OoC studies [115–117]. To meet the needs of OoC systems, in recent years various microarray techniques and commercial assay kits have been developed for specific biomarkers using very low volumes (<5 µL) [91].

Due to the high sensitivity of HPLC/MS and GC/MS techniques and their ability to provide quantitative analysis, considerable efforts have been made by researchers to overcome the difficulties encountered in coupling OoC to off-chip mass spectrometers. However, with the continued development of microfabrication technology, coupling of microfluidic systems to MS has become more common [118]. It must be noted that producing a device with chromatographic separation for LC/MS systems remains challenging due to the high back pressure generated when pushing the mobile phases through a particle-packed channel. Recently, Chen et al. developed a reproducible, robust, and stable 3D bioprinting microfluid chip coupled to LC/MS, demonstrating the suitability of the device for drug analysis and straightforward quantification [119]. The field of development of new coupling strategies is still growing and oriented towards the use of miniaturized analyzers, which would greatly increase the possibility of on-site analysis [120].

Lytic, transcriptomic, proteomic, and cell-viability assays typically require the retrieval of the cell samples from the chip prior to analysis, thus increasing the operator dependency.

It is also essential that the chip be designed to enable sample retrieval without detrimental effects on their function and structure in order not to affect the results [121].

It should be noted that most publications reported performing the analysis only on a single chip, although it is well-accepted that a minimal number of replicas per condition as well as positive and negative controls is required to produce statistically relevant and reliable data.

5. Applications of Advanced In Vitro Models for Mycotoxin Assessment

There is promising evidence that the toxicology field could benefit from the application of alternative culture methods. An innovative approach could contribute to a more reliable toxicity evaluation of compounds and support regulatory agency policies on allowable exposure levels. To date, regulatory toxicology has only partially embraced alternative methods, and food toxicology in particular is a sector still largely outside of the 3R and Tox-21c proposed regulations [122]. Searching terms as "food" AND "toxicology" AND "alternative methods" in PubMed results in a bibliography of just 118 articles from 1985 to the present. Of these, very few address mycotoxins and their risk assessment. On the contrary, searching terms such as "mycotoxin" and "animal model" yield 1200 articles in the last 20 years only, highlighting how far this topic is from adopting alternative approaches.

Mycotoxins are toxic secondary metabolites produced by filamentous fungi frequently found as contaminants of food and feed. Recently, Royal DSM, a global science-based company in nutrition, health and sustainable living, released Biomin results for the 2020 World Mycotoxin Survey, identifying 65% of analyzed samples contaminated with at least one mycotoxin above the threshold levels [123]. Mycotoxins are objects of concern, as they are known to induce adverse health effects in humans and animals following the consumption of contaminated foodstuffs. To prevent and contain the negative effects on consumers and animals, regulatory limits and guidance values are stipulated in several countries for many mycotoxins [124]. So far, mycotoxin risk assessment mainly relies on studies evaluating their cytotoxicity using in vitro 2D cell models, which can overestimate or underestimate the cellular toxicity due to the lack of a 3D architecture. In fact, evidence is accumulating that the same degree of sensitivity of the 2D systems did not translate into 3D culture systems [57,125,126]. As described in the sections above, this discrepancy can be attributed to the presence of more pronounced intracellular junctions in 3D cell models, mimicking physiological barriers, as well as a dense ECM, which influences xenobiotic transport [127,128]. In agreement with the literature, 3D models exposed to the mycotoxins ochratoxins (OTs), citrinin (CIT), sterigmatocystin (STE), and fumonisin B1 (FB1) showed a remarkably different sensitivity compared to monolayer cells [47,129,130]. These data shed light on the need for a more precise evaluation of mycotoxin risk using more complex models.

A list of studies aimed at assessing mycotoxin toxicity using alternative methods is provided in Table 1. Aflatoxin B1 (AFB1) is the mycotoxin that is most frequently studied using alternative in vitro models. The already-known genotoxic potential of AFB1 has recently been further investigated using hepatic spheroids, which show a higher expression of metabolic enzymes than 2D monoculture and, thus, better represent the metabolic activity of the hepatic cells in vivo [131,132]. Likely because of the increased metabolic activity of the spheroids, 3D HepG2 spheroids demonstrated a greater efficacy than standard monolayer cells in detecting genotoxicity, showing a higher sensitivity to low concentrations of AFB1 [71,72]. These data confirm that the advanced features of the 3D model make it an ideal candidate to improve the state-of-the-art for AFB1 genotoxicological assessment. The acute toxicity of AFB1 was also assessed on a microphysiological system (MPS) that enabled the maintenance of the hepatic functionality of hepatocytes for at least 14 days. An analysis of the LDH activity showed a modest increase in LDH after exposure to 30 μM AFB1. The system has proven to be a valuable tool for evaluating the hepatotoxicity of toxicants that require bioactivation, such as AFB1. However, only one liver cell type was employed in the system, representing a limitation of the approach [106]. Based on the study by Ma et al., an even more reliable toxicity evaluation of AFB1 may be obtained using 3D co-culture spheroids. In fact, the authors observed an important difference in terms of metabolic viability and sensitivity of 3D co-culture spheroids compared not only to 2D cells, but also to 3D mono-type cell spheroids [133]. Specifically, AFB1 ($\leq$30 μg/mL) was shown to induce apoptosis in 2D HepG2 cells, but did not significantly affect 3D cell spheroids, especially those in triple co-culture, which exhibited a higher resistance. Combined with an analysis of gene expression, both metabolic activation and detoxification efficiency were higher in 3D than in 2D cells, explaining the different sensibility shown by the two models. Similarly, in another study, the authors observed different interactions associated with the co-exposure of AFB1 with cyclopiazonic acid (CPA) compared with those reported in the literature. Synergistic effects were obtained in co-cultured hepatocyte spheroids at concentrations close to real-life exposure levels (0.625 μg/mL AFB1 and 3.125 μg/mL CPA) [134], and the authors speculated that this could be because the 3D culture environment might confer metabolic processes that differ from those of adherent cells, leading to differences in interactions between 3D and 2D cells. Since it is widely accepted that the carcinogenic and toxic effects of AFB1 are dependent on its metabolic activation [135], 3D cell culture seems to be a complex and suitable model for a more precise evaluation of mycotoxin co-exposure.

Considering that the toxicity of mycotoxins can be influenced by other organs once in the bloodstream, Bovard et al. developed a microfluidic multi-organ chip that overcomes the inaccuracies associated with 3D models cultured under conventional static conditions [136]. The novel system consisted of a 3D organotypic bronchial model connected with liver spheroids acting as a metabolizing compartment. In normal human bronchial epithelial cells at the air–liquid interface, AFB1 cytotoxicity was delayed by co-culture with liver spheroids, suggesting that at least part of AFB1 was metabolized into the less-toxic parent compound aflatoxin Q1 (AFQ1) by the hepatic compartment. Similarly, Schimek et al. observed that AFB1 induced a greater decrease in functionality and viability in mono-cultured bronchial cultures compared with lung–liver co-cultures [137]. Taken together, these studies lay the first stone upon a new alternative and physiologically relevant approach to assess the potential toxicity of AFB1.

For many mycotoxins of concern, the mechanisms underlying their toxicity are still elusive due to the lack of adequate models that fully recapitulate human functions in vivo. Imaoka et al. used a 3D human kidney proximal tube microphysiological system (kidney MPS) to define the dose–response relationships of ochratoxin A (OTA)-induced nephropathy [138]. The LD_{50} values obtained by the authors (1.21 and 0.375 μM at 72 and 186 h of exposure, respectively) agreed with the clinically relevant toxic concentrations of OTA in urine (0.37 μM), suggesting that the kidney MPS may represent a good model to reflect chronic toxicity of OTA and support changes in its risk assessment.

For the mycotoxin deoxynivalenol (DON), a deeper understanding of its effects on intestinal barrier functions is needed, considering that it is rapidly and almost completely absorbed in the proximal small intestine. However, the effects of DON on intestinal stem cells have not yet been studied in vitro due to the lack of a model containing them. To overcome this issue, Hanyu et al. used intestinal organoids containing intestinal stem cells (enteroids) to evaluate DON toxicity on luminal and basolateral intestinal sides [139]. Enteroids consist of a central lumen lined by a villus-like epithelium and several crypt-like domains and allow mimicry of most features of the native intestine. In the study, DON was delivered to the enteroid lumen using a microinjection technique and its basolateral exposure was shown to affect intestinal stem cells more than luminal exposure at equal concentrations. Similar results were obtained using the native small intestine of mice exposed to DON orally. DON toxicity in enteroids was also reported by Li et al., who showed that acute exposure to DON suppresses intestinal-stem-cell-based enteroids' expansion [140]. These findings support the use of enteroids as a powerful alternative tool to test the effects of toxins on intestinal stem cells. Very recently, DON-mediated effects on intestinal barrier leakage were further confirmed using a three-layered 3D gut-on-a-chip Caco-2 cell-culture model that allowed the inclusion of intestinal flow, ECM, and compartmentalization in three compartments (apical gut tube with lumen, ECM, and basolateral compartment). Similar to the previous findings, the intestinal cells were found to be more sensitive to basolateral DON exposure [141]. Noteworthy, with the addition of the third dimension and the application of fluid flow, shear stress, and the integration of the ECM, this study took one step further towards a more physiological model solution.

Table 1. Studies assessing mycotoxin toxicity using alternative in vitro methods.

Mycotoxin	3D Model	Endpoint	Reference
AFB1	Hepatic spheroids (HepG2 cells)	DNA damage	[72]
	Hepatic spheroids (HepG2 and HepaRG cells)	Cytotoxicity, liver functionality, genotoxicity	[71]
	Human hepatocytes cultured in a MPS	LDH release	[106]
	Mono-type spheroids (HepG2 cells); co-cultured spheroids (HepG2 cells + EA.hy 926 cells); triple co-cultured spheroids (HepG2 cells + EA.hy 926 cells + LX-2 cells)	Cell viability, mitochondria, oxidative stress, cell membrane	[133]
	Normal human bronchial epithelial (NHBE) cells cultured at the air–liquid interface (ALI); lung/liver-on-a-chip (NHBE ALI + HepaRG spheroids)	Transepithelial electrical resistance (TEER), ATP content	[136]
	Lung/liver-on-a-chip (Bronchial MucilAir + HepaRG and HHSteCs spheroids)	Intracellular ATP levels, LDH release	[137]
	Paper-based 3D HepG2 culture	Hepatotoxicity at different oxygen tensions	[142]
AFB1, CPA	Triple co-cultured spheroids (HepG2 cells + EA.hy 926 cells + LX-2 cells)	Individual and combined cell viability, mitochondria, oxidative stress, metabolomic analysis	[134]
STE	Human neuroblastoma spheroids (SH-SY5Y and SK-N-DZ cells)	Cell viability, oxidative stress, apoptosis, DNA damage, migration	[47]
CIT, OTs,	Canine kidney spheroids (MDCK cells)	Individual and combined cytotoxicity	[129]
DON	Mouse enteroids	Intestinal barrier function	[139]
	Porcine enteroids	Intestinal stem cells activity	[140]
	3-layered 3D gut-on-a-chip Caco-2 cell culture	Intestinal barrier function	[141]
FB1	Rat hepatic spheroids	Cytotoxicity	[130]
	3D human esophageal epithelial cells (HEEC)	Cell viability	[143]
OTA	3D human kidney proximal tubule microphysiological system	Cytotoxicity, analysis of kidney injury biomarkers, OTA transport, detoxification, and bioactivation	[138]

AFB1: Aflatoxin B1, ALI: air–liquid interface, CIT: citrinin, CPA: cyclopiazonic acid, DON: deoxynivalenol, EA.hy 926: immortalized human vascular endothelial cells, FB1: fumonisin B1, HHSteCs: human hepatic dtellate cells, LDH: lactate dehydrogenase, LX-2: human hepatic stellate cell line, MPS: microphysiological system, NHBE: normal human bronchial epithelial, OTs: ochratoxins, OTA ochratoxin A, SK-N-DZ: human neuroblastoma MYCN-amplified cells, STE: sterigmatocystin.

6. Conclusions

Both 2D and 3D cell-culture methods allow us to obtain an advanced understanding of the cellular response to toxin exposure. However, 3D models have proven to have the potential to better recapitulate the in vivo architecture of natural tissues and organs. Thus, researchers working to test toxins, including mycotoxins, on 2D cell monolayers should seriously consider 3D cell-culturing alternatives. Although there is encouraging evidence that the toxicological field could benefit from the application of advanced and complex culture methods, it has only partially embraced them, especially in the field of regulatory toxicology. Several alternative methods for assessing toxicity endpoints, such as irritation, corrosion, and genotoxicity, have already been validated and accepted by the Organization for Economic Co-operation and Development (OECD) [144,145]; however, there is still a lack of alternative models for many other toxicity endpoints. One of the reasons for the slow adaptation to alternative methods could be due to the difficulties of standardization, quality control, and validation. Furthermore, some alternative methods, while representing a successful example of in vitro reproduction of the in vivo microenvironment, are technically challenging and lack in throughput. Indeed, as the complexity of the system increases, the design, fabrication, and other phenomena commonly ignored in conventional macroscale cell cultures, such as bubble formation, evaporation, and nutrient depletion, assume a key and fundamental role. One of the major benefits of alternative in vitro models is the great potential to increase the biological relevance of in vitro studies. The development of cheaper and more user-friendly alternative systems would facilitate the adoption of this technology by more laboratories. With a shift in methodology and an increased effort for alternative models, the techniques will be better understood, and more advanced methods will be developed and adopted for toxicology studies.

Author Contributions: Conceptualization, V.Z., M.R.E. and N.T.; methodology, V.Z. and N.T.; investigation, V.Z. and N.T.; data curation, V.Z.; writing—original draft preparation, V.Z., M.R.E. and N.T.; writing—review and editing, E.C. and M.-J.R.; visualization, M.T.; supervision, E.C. and M.-J.R.; project administration, E.C. and M.-J.R.; funding acquisition, E.C. and M.-J.R. All authors have read and agreed to the published version of the manuscript.

Funding: This work was funded by a Spanish Ministry of Science and Innovation grant (PID2020-11587RB-100), a Spanish Ministry of Science and Innovation PhD grant (PRE2021-096941), a Spanish Ministry of Universities post-doctoral grant "Ayudas Margarita Salas para la formación de doctores jóvenes (UP2021-044)", and an ERC Starting Grant "759467"—MICRONEX project (UERI17).

Institutional Review Board Statement: Not applicable.

Informed Consent Statement: Not applicable.

Data Availability Statement: Not applicable.

Conflicts of Interest: The authors declare no conflict of interest.

References

1. Wrzesinski, K.; Fey, S.J. From 2D to 3D—A New Dimension for Modelling the Effect of Natural Products on Human Tissue. *Curr. Pharm. Des.* **2015**, *21*, 5605–5616. [CrossRef] [PubMed]
2. Ferreira, L.P.; Gaspar, V.M.; Mano, J.F. Design of spherically structured 3D in vitro tumor models—Advances and prospects. *Acta Biomater.* **2018**, *75*, 11–34. [CrossRef]
3. Jensen, C.; Teng, Y. Is It Time to Start Transitioning From 2D to 3D Cell Culture? *Front. Mol. Biosci.* **2020**, *7*, 33. [CrossRef] [PubMed]
4. Hartung, T. From alternative methods to a new toxicology. *Eur. J. Pharm. Biopharm.* **2011**, *77*, 338–349. [CrossRef] [PubMed]
5. Plevkova, J.; Brozmanova, M.; Matloobi, A.; Poliacek, I.; Honetschlager, J.; Buday, T. Animal models of cough. *Respir. Physiol. Neurobiol.* **2021**, *290*, 103656. [CrossRef] [PubMed]
6. Milani-Nejad, N.; Janssen, P.M. Small and large animal models in cardiac contraction research: Advantages and disadvantages. *Pharmacol. Ther.* **2014**, *141*, 235–249. [CrossRef]
7. Loder, N. UK researchers call for limits on animal experiment 'red tape'. *Nature* **2000**, *405*, 725. [CrossRef]
8. National Research Council, Committee on Toxicity Testing and Assessment of Environmental Agents. *Toxicity Testing in the 21st Century: A Vision and a Strategy*; Academic Press: Cambridge, MA, USA, 2007; 178p.
9. Firestone, M.; Kavlock, R.; Zenick, H.; Kramer, M.; the, U.S. EPA Working Group on the Future of Toxicity Testing. The U.S. Environmental Protection Agency strategic plan for evaluating the toxicity of chemicals. *J. Toxicol. Environ. Health B Crit. Rev.* **2010**, *13*, 139–162. [CrossRef]
10. EPA. Available online: https://www.epa.gov/assessing-and-managing-chemicals-under-tsca/alternative-test-methods-and-strategies-reduce (accessed on 7 December 2022).
11. EEC. Council Directive 86/609/EEC of 24 November 1986 on the approximation of laws, regulations and administrative provisions of the Member States regarding the protection of animals used for experimental and other scientific purposes. *Off. J. Eur. Union.* 1986, L 358, pp. 1–29. Available online: http://data.europa.eu/eli/dir/1986/609/oj (accessed on 7 December 2022).

2. The European Parlament; Council of the European Union. Directive 2010/63/EU of the European Parliament and of the Council of 22 September 2010 on the protection of animals used for Scientific purposes. *Off. J. Eur. Union* **2010**, *276*, 33–79.

3. Gunness, P.; Mueller, D.; Shevchenko, V.; Heinzle, E.; Ingelman-Sundberg, M.; Noor, F. 3D organotypic cultures of human HepaRG cells: A tool for in vitro toxicity studies. *Toxicol. Sci.* **2013**, *133*, 67–78. [CrossRef]

4. Zhang, X.; Yang, S.T. High-throughput 3-D cell-based proliferation and cytotoxicity assays for drug screening and bioprocess development. *J. Biotechnol.* **2011**, *151*, 186–193. [CrossRef]

5. Abbott, A. Cell culture: Biology's new dimension. *Nature* **2003**, *424*, 870–872. [CrossRef]

6. Ramaiahgari, S.C.; den Braver, M.W.; Herpers, B.; Terpstra, V.; Commandeur, J.N.; van de Water, B.; Price, L.S. A 3D in vitro model of differentiated HepG2 cell spheroids with improved liver-like properties for repeated dose high-throughput toxicity studies. *Arch. Toxicol.* **2014**, *88*, 1083–1095. [CrossRef] [PubMed]

7. Augustyniak, J.; Bertero, A.; Coccini, T.; Baderna, D.; Buzanska, L.; Caloni, F. Organoids are promising tools for species-specific in vitro toxicological studies. *J. Appl. Toxicol.* **2019**, *39*, 1610–1622. [CrossRef]

8. Iakobachvili, N.; Peters, P.J. Humans in a Dish: The Potential of Organoids in Modeling Immunity and Infectious Diseases. *Front. Microbiol.* **2017**, *8*, 2402. [CrossRef]

9. Viravaidya, K.; Sin, A.; Shuler, M.L. Development of a microscale cell culture analog to probe naphthalene toxicity. *Biotechnol. Prog.* **2004**, *20*, 316–323. [CrossRef]

10. Marx, U.; Andersson, T.B.; Bahinski, A.; Beilmann, M.; Beken, S.; Cassee, F.R.; Cirit, M.; Daneshian, M.; Fitzpatrick, S.; Frey, O.; et al. Biology-inspired microphysiological system approaches to solve the prediction dilemma of substance testing. *ALTEX* **2016**, *33*, 272–321. [CrossRef]

11. Materne, E.M.; Maschmeyer, I.; Lorenz, A.K.; Horland, R.; Schimek, K.M.; Busek, M.; Sonntag, F.; Lauster, R.; Marx, U. The multi-organ chip—A microfluidic platform for long-term multi-tissue coculture. *J. Vis. Exp.* **2015**, *28*, e52526. [CrossRef]

12. Lee, S.H.; Sung, J.H. Microtechnology-Based Multi-Organ Models. *Bioengineering* **2017**, *4*, 46. [CrossRef] [PubMed]

13. Bein, A.; Shin, W.; Jalili-Firoozinezhad, S.; Park, M.H.; Sontheimer-Phelps, A.; Tovaglieri, A.; Chalkiadaki, A.; Kim, H.J.; Ingber, D.E. Microfluidic Organ-on-a-Chip Models of Human Intestine. *Cell. Mol. Gastroenterol. Hepatol.* **2018**, *5*, 659–668. [CrossRef] [PubMed]

14. Choe, A.; Ha, S.K.; Choi, I.; Choi, N.; Sung, J.H. Microfluidic Gut-liver chip for reproducing the first pass metabolism. *Biomed. Microdevices* **2017**, *19*, 4. [CrossRef] [PubMed]

15. Sontheimer-Phelps, A.; Hassell, B.A.; Ingber, D.E. Modelling cancer in microfluidic human organs-on-chips. *Nat. Rev. Cancer* **2019**, *19*, 65–81. [CrossRef]

16. Riffle, S.; Hegde, R.S. Modeling tumor cell adaptations to hypoxia in multicellular tumor spheroids. *J. Exp. Clin. Cancer Res.* **2017**, *36*, 102. [CrossRef] [PubMed]

17. Kouroupis, D.; Correa, D. Increased Mesenchymal Stem Cell Functionalization in Three-Dimensional Manufacturing Settings for Enhanced Therapeutic Applications. *Front. Bioeng. Biotechnol.* **2021**, *9*, 621748. [CrossRef]

18. Han, S.J.; Kwon, S.; Kim, K.S. Challenges of applying multicellular tumor spheroids in preclinical phase. *Cancer Cell Int.* **2021**, *21*, 152. [CrossRef]

19. Pinto, B.; Henriques, A.C.; Silva, P.M.A.; Bousbaa, H. Three-Dimensional Spheroids as In Vitro Preclinical Models for Cancer Research. *Pharmaceutics* **2020**, *12*, 1186. [CrossRef] [PubMed]

20. Zanoni, M.; Piccinini, F.; Arienti, C.; Zamagni, A.; Santi, S.; Polico, R.; Bevilacqua, A.; Tesei, A. 3D tumor spheroid models for in vitro therapeutic screening: A systematic approach to enhance the biological relevance of data obtained. *Sci. Rep.* **2016**, *6*, 19103. [CrossRef]

21. Li, Y.; Kumacheva, E. Hydrogel microenvironments for cancer spheroid growth and drug screening. *Sci. Adv.* **2018**, *4*, eaas8998. [CrossRef]

22. Lee, K.H.; Kim, T.H. Recent Advances in Multicellular Tumor Spheroid Generation for Drug Screening. *Biosensors* **2021**, *11*, 445. [CrossRef]

23. Ikada, Y. Challenges in tissue engineering. *J. R. Soc. Interface* **2006**, *3*, 589–601. [CrossRef]

24. Bova, L.; Maggiotto, F.; Micheli, S.; Giomo, M.; Sgarbossa, P.; Gagliano, O.; Falcone, D.; Cimetta, E. A Porous Gelatin Methacrylate-Based Material for 3D Cell-Laden Constructs. *Macromol. Biosci.* **2023**, *23*, e2200357. [CrossRef] [PubMed]

35. Alghuwainem, A.; Alshareeda, A.T.; Alsowayan, B. Scaffold-Free 3-D Cell Sheet Technique Bridges the Gap between 2-D Cell Culture and Animal Models. *Int. J. Mol. Sci.* **2019**, *20*, 4926. [CrossRef]

36. Langhans, S.A. Three-Dimensional in Vitro Cell Culture Models in Drug Discovery and Drug Repositioning. *Front. Pharmacol.* **2018**, *9*, 6. [CrossRef] [PubMed]

37. Catoira, M.C.; Fusaro, L.; Di Francesco, D.; Ramella, M.; Boccafoschi, F. Overview of natural hydrogels for regenerative medicine applications. *J. Mater. Sci. Mater. Med.* **2019**, *30*, 115. [CrossRef] [PubMed]

38. Jubelin, C.; Munoz-Garcia, J.; Griscom, L.; Cochonneau, D.; Ollivier, E.; Heymann, M.F.; Vallette, F.M.; Oliver, L.; Heymann, D. Three-dimensional in vitro culture models in oncology research. *Cell Biosci.* **2022**, *12*, 155. [CrossRef]

39. Kelm, J.M.; Timmins, N.E.; Brown, C.J.; Fussenegger, M.; Nielsen, L.K. Method for generation of homogeneous multicellular tumor spheroids applicable to a wide variety of cell types. *Biotechnol. Bioeng.* **2003**, *83*, 173–180. [CrossRef] [PubMed]

40. Razian, G.; Yu, Y.; Ungrin, M. Production of large numbers of size-controlled tumor spheroids using microwell plates. *J. Vis. Exp.* **2013**, *81*, e50665. [CrossRef]

41. Habanjar, O.; Diab-Assaf, M.; Caldefie-Chezet, F.; Delort, L. 3D Cell Culture Systems: Tumor Application, Advantages, and Disadvantages. *Int. J. Mol. Sci.* **2021**, *22*, 12200. [CrossRef]
42. Bartosh, T.J.; Ylostalo, J.H. Preparation of anti-inflammatory mesenchymal stem/precursor cells (MSCs) through sphere formation using hanging-drop culture technique. *Curr. Protoc. Stem Cell Biol.* **2014**, *28*, 2B.6.1–2B.6.23. [CrossRef]
43. Fang, Y.; Eglen, R.M. Three-Dimensional Cell Cultures in Drug Discovery and Development. *SLAS Discov.* **2017**, *22*, 456–472 [CrossRef]
44. Ryu, N.E.; Lee, S.H.; Park, H. Spheroid Culture System Methods and Applications for Mesenchymal Stem Cells. *Cells* **2019**, *8*, 1620. [CrossRef] [PubMed]
45. Friedrich, J.; Seidel, C.; Ebner, R.; Kunz-Schughart, L.A. Spheroid-based drug screen: Considerations and practical approach. *Nat. Protoc.* **2009**, *4*, 309–324. [CrossRef]
46. Vinci, M.; Gowan, S.; Boxall, F.; Patterson, L.; Zimmermann, M.; Court, W.; Lomas, C.; Mendiola, M.; Hardisson, D.; Eccles, S.A. Advances in establishment and analysis of three-dimensional tumor spheroid-based functional assays for target validation and drug evaluation. *BMC Biol.* **2012**, *10*, 29. [CrossRef]
47. Zingales, V.; Torriero, N.; Zanella, L.; Fernandez-Franzon, M.; Ruiz, M.J.; Esposito, M.R.; Cimetta, E. Development of an in vitro neuroblastoma 3D model and its application for sterigmatocystin-induced cytotoxicity testing. *Food Chem. Toxicol.* **2021**, *157*, 112605. [CrossRef]
48. Fuentes, P.; Torres, M.J.; Arancibia, R.; Aulestia, F.; Vergara, M.; Carrion, F.; Osses, N.; Altamirano, C. Dynamic Culture of Mesenchymal Stromal/Stem Cell Spheroids and Secretion of Paracrine Factors. *Front. Bioeng. Biotechnol.* **2022**, *10*, 916229 [CrossRef] [PubMed]
49. Niibe, K.; Ohori-Morita, Y.; Zhang, M.; Mabuchi, Y.; Matsuzaki, Y.; Egusa, H. A Shaking-Culture Method for Generating Bone Marrow Derived Mesenchymal Stromal/Stem Cell-Spheroids with Enhanced Multipotency in vitro. *Front. Bioeng. Biotechnol.* **2020**, *8*, 590332. [CrossRef] [PubMed]
50. Petry, F.; Salzig, D. Large-Scale Production of Size-Adjusted β-Cell Spheroids in a Fully Controlled Stirred-Tank Reactor. *Processes* **2022**, *10*, 861. [CrossRef]
51. Phelan, M.A.; Gianforcaro, A.L.; Gerstenhaber, J.A.; Lelkes, P.I. An Air Bubble-Isolating Rotating Wall Vessel Bioreactor for Improved Spheroid/Organoid Formation. *Tissue Eng. Part C Methods* **2019**, *25*, 479–488. [CrossRef]
52. Thoma, C.R.; Zimmermann, M.; Agarkova, I.; Kelm, J.M.; Krek, W. 3D cell culture systems modeling tumor growth determinants in cancer target discovery. *Adv. Drug Deliv. Rev.* **2014**, *69–70*, 29–41. [CrossRef]
53. Weiswald, L.B.; Bellet, D.; Dangles-Marie, V. Spherical cancer models in tumor biology. *Neoplasia* **2015**, *17*, 1–15. [CrossRef]
54. Santo, V.E.; Estrada, M.F.; Rebelo, S.P.; Abreu, S.; Silva, I.; Pinto, C.; Veloso, S.C.; Serra, A.T.; Boghaert, E.; Alves, P.M.; et al. Adaptable stirred-tank culture strategies for large scale production of multicellular spheroid-based tumor cell models. *J. Biotechnol.* **2016**, *221*, 118–129. [CrossRef] [PubMed]
55. Kunz-Schughart, L.A. Multicellular tumor spheroids: Intermediates between monolayer culture and in vivo tumor. *Cell Biol. Int.* **1999**, *23*, 157–161. [CrossRef] [PubMed]
56. Bartosh, T.J.; Ylostalo, J.H.; Mohammadipoor, A.; Bazhanov, N.; Coble, K.; Claypool, K.; Lee, R.H.; Choi, H.; Prockop, D.J. Aggregation of human mesenchymal stromal cells (MSCs) into 3D spheroids enhances their antiinflammatory properties. *Proc. Natl. Acad. Sci. USA* **2010**, *107*, 13724–13729. [CrossRef]
57. Salehi, F.; Behboudi, H.; Kavoosi, G.; Ardestani, S.K. Monitoring ZEO apoptotic potential in 2D and 3D cell cultures and associated spectroscopic evidence on mode of interaction with DNA. *Sci. Rep.* **2017**, *7*, 2553. [CrossRef]
58. Booij, T.H.; Price, L.S.; Danen, E.H.J. 3D Cell-Based Assays for Drug Screens: Challenges in Imaging, Image Analysis, and High-Content Analysis. *SLAS Discov.* **2019**, *24*, 615–627. [CrossRef]
59. Vermes, I.; Haanen, C.; Steffens-Nakken, H.; Reutelingsperger, C. A novel assay for apoptosis. Flow cytometric detection of phosphatidylserine expression on early apoptotic cells using fluorescein labelled Annexin V. *J. Immunol. Methods* **1995**, *184*, 39–51 [CrossRef] [PubMed]
60. Carmeliet, P. Mechanisms of angiogenesis and arteriogenesis. *Nat. Med.* **2000**, *6*, 389–395. [CrossRef]
61. Korff, T.; Augustin, H.G. Tensional forces in fibrillar extracellular matrices control directional capillary sprouting. *J. Cell Sci.* **1999**, *112 Pt 19*, 3249–3258. [CrossRef]
62. Heiss, M.; Hellstrom, M.; Kalen, M.; May, T.; Weber, H.; Hecker, M.; Augustin, H.G.; Korff, T. Endothelial cell spheroids as a versatile tool to study angiogenesis in vitro. *FASEB J.* **2015**, *29*, 3076–3084. [CrossRef]
63. Tetzlaff, F.; Fischer, A. Human Endothelial Cell Spheroid-based Sprouting Angiogenesis Assay in Collagen. *Bio-Protocol* **2018**, *8*, e2995. [CrossRef]
64. Nazari, S.S. Generation of 3D Tumor Spheroids with Encapsulating Basement Membranes for Invasion Studies. *Curr. Protoc. Cell Biol.* **2020**, *87*, e105. [CrossRef]
65. Carey, S.P.; Martin, K.E.; Reinhart-King, C.A. Three-dimensional collagen matrix induces a mechanosensitive invasive epithelial phenotype. *Sci. Rep.* **2017**, *7*, 42088. [CrossRef]
66. Lee, J.Y.; Chang, J.K.; Dominguez, A.A.; Lee, H.P.; Nam, S.; Chang, J.; Varma, S.; Qi, L.S.; West, R.B.; Chaudhuri, O. YAP-independent mechanotransduction drives breast cancer progression. *Nat. Commun.* **2019**, *10*, 1848. [CrossRef]
67. Kanopoulos, N.; Vasanthavada, N.; Baker, R.L. Design of an image edge detection filter using the Sobel operator. *IEEE J. Solid State Circuits* **1988**, *23*, 358–367. [CrossRef]

68. Chan, T.F.; Vese, L.A. Active contours without edges. *IEEE Trans. Image Process.* **2001**, *10*, 266–277. [CrossRef]
69. Evans, S.J.; Clift, M.J.; Singh, N.; de Oliveira Mallia, J.; Burgum, M.; Wills, J.W.; Wilkinson, T.S.; Jenkins, G.J.; Doak, S.H. Critical review of the current and future challenges associated with advanced in vitro systems towards the study of nanoparticle (secondary) genotoxicity. *Mutagenesis* **2017**, *32*, 233–241. [CrossRef]
70. Corvi, R.; Madia, F. In vitro genotoxicity testing-Can the performance be enhanced? *Food Chem. Toxicol.* **2017**, *106*, 600–608. [CrossRef]
71. Conway, G.E.; Shah, U.K.; Llewellyn, S.; Cervena, T.; Evans, S.J.; Al Ali, A.S.; Jenkins, G.J.; Clift, M.J.D.; Doak, S.H. Adaptation of the in vitro micronucleus assay for genotoxicity testing using 3D liver models supporting longer-term exposure durations. *Mutagenesis* **2020**, *35*, 319–330. [CrossRef]
72. Stampar, M.; Tomc, J.; Filipic, M.; Zegura, B. Development of in vitro 3D cell model from hepatocellular carcinoma (HepG2) cell line and its application for genotoxicity testing. *Arch. Toxicol.* **2019**, *93*, 3321–3333. [CrossRef] [PubMed]
73. Reisinger, K.; Blatz, V.; Brinkmann, J.; Downs, T.R.; Fischer, A.; Henkler, F.; Hoffmann, S.; Krul, C.; Liebsch, M.; Luch, A.; et al. Validation of the 3D Skin Comet assay using full thickness skin models: Transferability and reproducibility. *Mutat. Res. Genet. Toxicol. Environ. Mutagen.* **2018**, *827*, 27–41. [CrossRef] [PubMed]
74. Seo, J.E.; He, X.; Muskhelishvili, L.; Malhi, P.; Mei, N.; Manjanatha, M.; Bryant, M.; Zhou, T.; Robison, T.; Guo, X. Evaluation of an in vitro three-dimensional HepaRG spheroid model for genotoxicity testing using the high-throughput CometChip platform. *ALTEX* **2022**, *39*, 583–604. [CrossRef]
75. Shaw, P.; Kumar, N.; Privat-Maldonado, A.; Smits, E.; Bogaerts, A. Cold Atmospheric Plasma Increases Temozolomide Sensitivity of Three-Dimensional Glioblastoma Spheroids via Oxidative Stress-Mediated DNA Damage. *Cancers* **2021**, *13*, 1780. [CrossRef]
76. McDonald, J.C.; Whitesides, G.M. Poly(dimethylsiloxane) as a material for fabricating microfluidic devices. *Acc. Chem. Res.* **2002**, *35*, 491–499. [CrossRef] [PubMed]
77. Toepke, M.W.; Beebe, D.J. PDMS absorption of small molecules and consequences in microfluidic applications. *Lab Chip* **2006**, *6*, 1484–1486. [CrossRef]
78. Wang, J.D.; Douville, N.J.; Takayama, S.; ElSayed, M. Quantitative analysis of molecular absorption into PDMS microfluidic channels. *Ann. Biomed. Eng.* **2012**, *40*, 1862–1873. [CrossRef]
79. Iliescu, C.; Taylor, H.; Avram, M.; Miao, J.; Franssila, S. A practical guide for the fabrication of microfluidic devices using glass and silicon. *Biomicrofluidics* **2012**, *6*, 016505. [CrossRef]
80. Schulze, T.; Mattern, K.; Fruh, E.; Hecht, L.; Rustenbeck, I.; Dietzel, A. A 3D microfluidic perfusion system made from glass for multiparametric analysis of stimulus-secretioncoupling in pancreatic islets. *Biomed. Microdevices* **2017**, *19*, 47. [CrossRef]
81. Sackmann, E.K.; Fulton, A.L.; Beebe, D.J. The present and future role of microfluidics in biomedical research. *Nature* **2014**, *507*, 181–189. [CrossRef] [PubMed]
82. Lee, S.; Lim, J.; Yu, J.; Ahn, J.; Lee, Y.; Jeon, N.L. Engineering tumor vasculature on an injection-molded plastic array 3D culture (IMPACT) platform. *Lab Chip* **2019**, *19*, 2071–2080. [CrossRef]
83. Mottet, G.; Perez-Toralla, K.; Tulukcuoglu, E.; Bidard, F.C.; Pierga, J.Y.; Draskovic, I.; Londono-Vallejo, A.; Descroix, S.; Malaquin, L.; Louis Viovy, J. A three dimensional thermoplastic microfluidic chip for robust cell capture and high resolution imaging. *Biomicrofluidics* **2014**, *8*, 024109. [CrossRef] [PubMed]
84. Ren, K.; Zhou, J.; Wu, H. Materials for microfluidic chip fabrication. *Acc. Chem. Res.* **2013**, *46*, 2396–2406. [CrossRef]
85. Gencturk, E.; Mutlu, S.; Ulgen, K.O. Advances in microfluidic devices made from thermoplastics used in cell biology and analyses. *Biomicrofluidics* **2017**, *11*, 051502. [CrossRef] [PubMed]
86. Seliktar, D. Designing cell-compatible hydrogels for biomedical applications. *Science* **2012**, *336*, 1124–1128. [CrossRef] [PubMed]
87. Zhang, X.; Li, L.; Luo, C. Gel integration for microfluidic applications. *Lab Chip* **2016**, *16*, 1757–1776. [CrossRef]
88. Annabi, N.; Selimovic, S.; Acevedo Cox, J.P.; Ribas, J.; Afshar Bakooshli, M.; Heintze, D.; Weiss, A.S.; Cropek, D.; Khademhosseini, A. Hydrogel-coated microfluidic channels for cardiomyocyte culture. *Lab Chip* **2013**, *13*, 3569–3577. [CrossRef]
89. Antoine, E.E.; Vlachos, P.P.; Rylander, M.N. Review of collagen I hydrogels for bioengineered tissue microenvironments: Characterization of mechanics, structure, and transport. *Tissue Eng. Part B Rev.* **2014**, *20*, 683–696. [CrossRef]
90. Goy, C.B.; Chaile, R.E.; Madrid, R.E. Microfluidics and hydrogel: A powerful combination. *React. Funct. Polym.* **2019**, *145*, 104314. [CrossRef]
91. Leung, C.M.; De Haan, P.; Ronaldson-Bouchard, K.; Kim, G.-A.; Ko, J.; Rho, H.S.; Chen, Z.; Habibovic, P.; Jeon, N.L.; Takayama, S. A guide to the organ-on-a-chip. *Nat. Rev. Methods Prim.* **2022**, *2*, 33. [CrossRef]
92. Gillet, J.P.; Varma, S.; Gottesman, M.M. The clinical relevance of cancer cell lines. *J. Natl. Cancer Inst.* **2013**, *105*, 452–458. [CrossRef] [PubMed]
93. Cristofalo, V.J.; Lorenzini, A.; Allen, R.G.; Torres, C.; Tresini, M. Replicative senescence: A critical review. *Mech. Ageing Dev.* **2004**, *125*, 827–848. [CrossRef]
94. Diederichs, S.; Tuan, R.S. Functional comparison of human-induced pluripotent stem cell-derived mesenchymal cells and bone marrow-derived mesenchymal stromal cells from the same donor. *Stem Cells Dev.* **2014**, *23*, 1594–1610. [CrossRef]
95. Wnorowski, A.; Yang, H.; Wu, J.C. Progress, obstacles, and limitations in the use of stem cells in organ-on-a-chip models. *Adv. Drug Deliv. Rev.* **2019**, *140*, 3–11. [CrossRef] [PubMed]
96. Naumovska, E.; Aalderink, G.; Wong Valencia, C.; Kosim, K.; Nicolas, A.; Brown, S.; Vulto, P.; Erdmann, K.S.; Kurek, D. Direct On-Chip Differentiation of Intestinal Tubules from Induced Pluripotent Stem Cells. *Int. J. Mol. Sci.* **2020**, *21*, 4964. [CrossRef]

97. Punt, A.; Bouwmeester, H.; Blaauboer, B.J.; Coecke, S.; Hakkert, B.; Hendriks, D.F.G.; Jennings, P.; Kramer, N.I.; Neuhoff, S.; Masereeuw, R.; et al. New approach methodologies (NAMs) for human-relevant biokinetics predictions. Meeting the paradigm shift in toxicology towards an animal-free chemical risk assessment. *ALTEX* **2020**, *37*, 607–622. [CrossRef] [PubMed]

98. Lasli, S.; Kim, H.J.; Lee, K.; Suurmond, C.E.; Goudie, M.; Bandaru, P.; Sun, W.; Zhang, S.; Zhang, N.; Ahadian, S.; et al. A Human Liver-on-a-Chip Platform for Modeling Nonalcoholic Fatty Liver Disease. *Adv. Biosyst.* **2019**, *3*, e1900104. [CrossRef]

99. Jang, K.J.; Otieno, M.A.; Ronxhi, J.; Lim, H.K.; Ewart, L.; Kodella, K.R.; Petropolis, D.B.; Kulkarni, G.; Rubins, J.E.; Conegliano, D.; et al. Reproducing human and cross-species drug toxicities using a Liver-Chip. *Sci. Transl. Med.* **2019**, *11*, eaax5516. [CrossRef] [PubMed]

100. Kostrzewski, T.; Cornforth, T.; Snow, S.A.; Ouro-Gnao, L.; Rowe, C.; Large, E.M.; Hughes, D.J. Three-dimensional perfused human in vitro model of non-alcoholic fatty liver disease. *World J. Gastroenterol.* **2017**, *23*, 204–215. [CrossRef]

101. Tao, T.P.; Brandmair, K.; Gerlach, S.; Przibilla, J.; Genies, C.; Jacques-Jamin, C.; Schepky, A.; Marx, U.; Hewitt, N.J.; Maschmeyer, I.; et al. Demonstration of the first-pass metabolism in the skin of the hair dye, 4-amino-2-hydroxytoluene, using the Chip2 skin-liver microphysiological model. *J. Appl. Toxicol.* **2021**, *41*, 1553–1567. [CrossRef]

102. Yang, F.; Cohen, R.N.; Brey, E.M. Optimization of Co-Culture Conditions for a Human Vascularized Adipose Tissue Model. *Bioengineering* **2020**, *7*, 114. [CrossRef]

103. Chang, S.Y.; Weber, E.J.; Sidorenko, V.S.; Chapron, A.; Yeung, C.K.; Gao, C.; Mao, Q.; Shen, D.; Wang, J.; Rosenquist, T.A.; et al. Human liver-kidney model elucidates the mechanisms of aristolochic acid nephrotoxicity. *JCI Insight* **2017**, *2*, e95978. [CrossRef] [PubMed]

104. Zhang, C.; Zhao, Z.; Abdul Rahim, N.A.; van Noort, D.; Yu, H. Towards a human-on-chip: Culturing multiple cell types on a chip with compartmentalized microenvironments. *Lab Chip* **2009**, *9*, 3185–3192. [CrossRef]

105. Junaid, A.; Mashaghi, A.; Hankemeier, T.; Vulto, P. An end-user perspective on Organ-on-a-Chip: Assays and usability aspects. *Curr. Opin. Biomed. Eng.* **2017**, *1*, 15–22. [CrossRef]

106. Chang, S.Y.; Voellinger, J.L.; Van Ness, K.P.; Chapron, B.; Shaffer, R.M.; Neumann, T.; White, C.C.; Kavanagh, T.J.; Kelly, E.J.; Eaton, D.L. Characterization of rat or human hepatocytes cultured in microphysiological systems (MPS) to identify hepatotoxicity. *Toxicol. In Vitro* **2017**, *40*, 170–183. [CrossRef] [PubMed]

107. Wevers, N.R.; van Vught, R.; Wilschut, K.J.; Nicolas, A.; Chiang, C.; Lanz, H.L.; Trietsch, S.J.; Joore, J.; Vulto, P. High-throughput compound evaluation on 3D networks of neurons and glia in a microfluidic platform. *Sci. Rep.* **2016**, *6*, 38856. [CrossRef] [PubMed]

108. Shen, J.; Cai, C.; Yu, Z.; Pang, Y.; Zhou, Y.; Qian, L.; Wei, W.; Huang, Y. A microfluidic live cell assay to study anthrax toxin induced cell lethality assisted by conditioned medium. *Sci. Rep.* **2015**, *5*, 8651. [CrossRef]

109. Sobrino, A.; Phan, D.T.; Datta, R.; Wang, X.; Hachey, S.J.; Romero-Lopez, M.; Gratton, E.; Lee, A.P.; George, S.C.; Hughes, C.C. 3D microtumors in vitro supported by perfused vascular networks. *Sci. Rep.* **2016**, *6*, 31589. [CrossRef] [PubMed]

110. Henry, O.Y.F.; Villenave, R.; Cronce, M.J.; Leineweber, W.D.; Benz, M.A.; Ingber, D.E. Organs-on-chips with integrated electrodes for trans-epithelial electrical resistance (TEER) measurements of human epithelial barrier function. *Lab Chip* **2017**, *17*, 2264–2271. [CrossRef]

111. van der Helm, M.W.; Odijk, M.; Frimat, J.P.; van der Meer, A.D.; Eijkel, J.C.T.; van den Berg, A.; Segerink, L.I. Direct quantification of transendothelial electrical resistance in organs-on-chips. *Biosens. Bioelectron.* **2016**, *85*, 924–929. [CrossRef]

112. Elbakary, B.; Badhan, R.K.S. A dynamic perfusion based blood-brain barrier model for cytotoxicity testing and drug permeation. *Sci. Rep.* **2020**, *10*, 3788. [CrossRef]

113. Yalcin, Y.D.; Luttge, R. Electrical monitoring approaches in 3-dimensional cell culture systems: Toward label-free, high spatiotemporal resolution, and high-content data collection in vitro. *Organs-on-a-Chip* **2021**, *3*, 100006. [CrossRef]

114. Ramadan, Q.; Ting, F.C. In vitro micro-physiological immune-competent model of the human skin. *Lab Chip* **2016**, *16*, 1899–1908. [CrossRef] [PubMed]

115. Zeller, P.; Legendre, A.; Jacques, S.; Fleury, M.J.; Gilard, F.; Tcherkez, G.; Leclerc, E. Hepatocytes cocultured with Sertoli cells in bioreactor favors Sertoli barrier tightness in rat. *J. Appl. Toxicol.* **2017**, *37*, 287–295. [CrossRef]

116. Shah, P.; Fritz, J.V.; Glaab, E.; Desai, M.S.; Greenhalgh, K.; Frachet, A.; Niegowska, M.; Estes, M.; Jager, C.; Seguin-Devaux, C.; et al. A microfluidics-based in vitro model of the gastrointestinal human-microbe interface. *Nat. Commun.* **2016**, *7*, 11535. [CrossRef] [PubMed]

117. Lee, H.; Kim, D.S.; Ha, S.K.; Choi, I.; Lee, J.M.; Sung, J.H. A pumpless multi-organ-on-a-chip (MOC) combined with a pharmacokinetic-pharmacodynamic (PK-PD) model. *Biotechnol. Bioeng.* **2017**, *114*, 432–443. [CrossRef]

118. Wang, X.; Yi, L.; Mukhitov, N.; Schrell, A.M.; Dhumpa, R.; Roper, M.G. Microfluidics-to-mass spectrometry: A review of coupling methods and applications. *J. Chromatogr. A* **2015**, *1382*, 98–116. [CrossRef]

119. Chen, P.-C.; Zhang, W.-Z.; Chen, W.-R.; Jair, Y.-C.; Wu, Y.-H.; Liu, Y.-H.; Chen, P.-Z.; Chen, L.-Y.; Chen, P.-S. Engineering an integrated system with a high pressure polymeric microfluidic chip coupled to liquid chromatography-mass spectrometry (LC-MS) for the analysis of abused drugs. *Sens. Actuators B Chem.* **2022**, *350*, 130888. [CrossRef]

120. Varuni, K.; Nivetha, V.; Gandhimathi, R. A Role of Microfluidic Chip Technology in Analytical Method Development: An Overview. *NeuroQuantology* **2022**, *20*, 2125.

121. Vit, F.F.; Nunes, R.; Wu, Y.T.; Prado Soares, M.C.; Godoi, N.; Fujiwara, E.; Carvalho, H.F.; Gaziola de la Torre, L. A modular, reversible sealing, and reusable microfluidic device for drug screening. *Anal. Chim. Acta* **2021**, *1185*, 339068. [CrossRef]

22. Hartung, T. Food for thought . . . on alternative methods for chemical safety testing. *ALTEX* **2010**, *27*, 3–14. [CrossRef]
23. Biomin. The Global Mycotoxin Threat. 2021. Available online: https://www.biomin.net/science-hub/world-mycotoxin-survey-impact-2021/ (accessed on 14 December 2022).
24. Bennett, J.W.; Klich, M. Mycotoxins. *Clin. Microbiol. Rev.* **2003**, *16*, 497–516. [CrossRef]
25. Baek, N.; Seo, O.W.; Kim, M.; Hulme, J.; An, S.S. Monitoring the effects of doxorubicin on 3D-spheroid tumor cells in real-time. *OncoTargets Ther.* **2016**, *9*, 7207–7218. [CrossRef]
26. De Simone, U.; Roccio, M.; Gribaldo, L.; Spinillo, A.; Caloni, F.; Coccini, T. Human 3D Cultures as Models for Evaluating Magnetic Nanoparticle CNS Cytotoxicity after Short- and Repeated Long-Term Exposure. *Int. J. Mol. Sci.* **2018**, *19*, 1993. [CrossRef]
27. Goodman, T.T.; Ng, C.P.; Pun, S.H. 3-D tissue culture systems for the evaluation and optimization of nanoparticle-based drug carriers. *Bioconjug. Chem.* **2008**, *19*, 1951–1959. [CrossRef]
28. Lee, J.; Lilly, G.D.; Doty, R.C.; Podsiadlo, P.; Kotov, N.A. In vitro toxicity testing of nanoparticles in 3D cell culture. *Small* **2009**, *5*, 1213–1221. [CrossRef] [PubMed]
29. Csenki, Z.; Garai, E.; Faisal, Z.; Csepregi, R.; Garai, K.; Sipos, D.K.; Szabo, I.; Koszegi, T.; Czeh, A.; Czompoly, T.; et al. The individual and combined effects of ochratoxin A with citrinin and their metabolites (ochratoxin B, ochratoxin C, and dihydrocitrinone) on 2D/3D cell cultures, and zebrafish embryo models. *Food Chem. Toxicol.* **2021**, *158*, 112674. [CrossRef] [PubMed]
30. Kim, B.; Ejaz, S.; Chekarova, I.; Sukura, A.; Ashraf, M.; Lim, C.W. Cytotoxicity of fumonisin B(1) in spheroid and monolayer cultures of rat hepatocytes. *Drug Chem. Toxicol.* **2008**, *31*, 339–352. [CrossRef]
31. Bell, C.C.; Hendriks, D.F.; Moro, S.M.; Ellis, E.; Walsh, J.; Renblom, A.; Fredriksson Puigvert, L.; Dankers, A.C.; Jacobs, F.; Snoeys, J.; et al. Characterization of primary human hepatocyte spheroids as a model system for drug-induced liver injury, liver function and disease. *Sci. Rep.* **2016**, *6*, 25187. [CrossRef] [PubMed]
32. Shah, U.K.; Mallia, J.O.; Singh, N.; Chapman, K.E.; Doak, S.H.; Jenkins, G.J.S. Reprint of: A three-dimensional in vitro HepG2 cells liver spheroid model for genotoxicity studies. *Mutat. Res. Genet. Toxicol. Environ. Mutagen.* **2018**, *834*, 35–41. [CrossRef]
33. Ma, X.; Sun, J.; Ye, Y.; Ji, J.; Sun, X. Application of triple co-cultured cell spheroid model for exploring hepatotoxicity and metabolic pathway of AFB1. *Sci. Total. Environ.* **2022**, *807*, 150840. [CrossRef]
34. Ma, X.; Ye, Y.; Sun, J.; Ji, J.; Wang, J.S.; Sun, X. Coexposure of Cyclopiazonic Acid with Aflatoxin B1 Involved in Disrupting Amino Acid Metabolism and Redox Homeostasis Causing Synergistic Toxic Effects in Hepatocyte Spheroids. *J. Agric. Food Chem.* **2022**, *70*, 5166–5176. [CrossRef]
35. Neal, G.E.; Eaton, D.L.; Judah, D.J.; Verma, A. Metabolism and toxicity of aflatoxins M1 and B1 in human-derived in vitro systems. *Toxicol. Appl. Pharmacol.* **1998**, *151*, 152–158. [CrossRef]
36. Bovard, D.; Sandoz, A.; Luettich, K.; Frentzel, S.; Iskandar, A.; Marescotti, D.; Trivedi, K.; Guedj, E.; Dutertre, Q.; Peitsch, M.C.; et al. A lung/liver-on-a-chip platform for acute and chronic toxicity studies. *Lab Chip* **2018**, *18*, 3814–3829. [CrossRef] [PubMed]
37. Schimek, K.; Frentzel, S.; Luettich, K.; Bovard, D.; Rutschle, I.; Boden, L.; Rambo, F.; Erfurth, H.; Dehne, E.M.; Winter, A.; et al. Human multi-organ chip co-culture of bronchial lung culture and liver spheroids for substance exposure studies. *Sci. Rep.* **2020**, *10*, 7865. [CrossRef]
38. Imaoka, T.; Yang, J.; Wang, L.; McDonald, M.G.; Afsharinejad, Z.; Bammler, T.K.; Van Ness, K.; Yeung, C.K.; Rettie, A.E.; Himmelfarb, J.; et al. Microphysiological system modeling of ochratoxin A-associated nephrotoxicity. *Toxicology* **2020**, *444*, 152582. [CrossRef] [PubMed]
39. Hanyu, H.; Yokoi, Y.; Nakamura, K.; Ayabe, T.; Tanaka, K.; Uno, K.; Miyajima, K.; Saito, Y.; Iwatsuki, K.; Shimizu, M.; et al. Mycotoxin Deoxynivalenol Has Different Impacts on Intestinal Barrier and Stem Cells by Its Route of Exposure. *Toxins* **2020**, *12*, 610. [CrossRef]
40. Li, X.G.; Zhu, M.; Chen, M.X.; Fan, H.B.; Fu, H.L.; Zhou, J.Y.; Zhai, Z.Y.; Gao, C.Q.; Yan, H.C.; Wang, X.Q. Acute exposure to deoxynivalenol inhibits porcine enteroid activity via suppression of the Wnt/beta-catenin pathway. *Toxicol. Lett.* **2019**, *305*, 19–31. [CrossRef]
41. Poschl, F.; Hoher, T.; Pirklbauer, S.; Wolinski, H.; Lienhart, L.; Ressler, M.; Riederer, M. Dose and route dependent effects of the mycotoxin deoxynivalenol in a 3D gut-on-a-chip model with flow. *Toxicol. In Vitro* **2023**, *88*, 105563. [CrossRef]
42. DiProspero, T.J.; Dalrymple, E.; Lockett, M.R. Physiologically relevant oxygen tensions differentially regulate hepatotoxic responses in HepG2 cells. *Toxicol. In Vitro* **2021**, *74*, 105156. [CrossRef] [PubMed]
43. Yu, S.; Jia, B.; Liu, N.; Yu, D.; Zhang, S.; Wu, A. Fumonisin B1 triggers carcinogenesis via HDAC/PI3K/Akt signalling pathway in human esophageal epithelial cells. *Sci. Total. Environ.* **2021**, *787*, 147405. [CrossRef]
44. Hardwick, R.N.; Betts, C.J.; Whritenour, J.; Sura, R.; Thamsen, M.; Kaufman, E.H.; Fabre, K. Drug-induced skin toxicity: Gaps in preclinical testing cascade as opportunities for complex in vitro models and assays. *Lab Chip* **2020**, *20*, 199–214. [CrossRef]
45. Gordon, S.; Daneshian, M.; Bouwstra, J.; Caloni, F.; Constant, S.; Davies, D.E.; Dandekar, G.; Guzman, C.A.; Fabian, E.; Haltner, E.; et al. Non-animal models of epithelial barriers (skin, intestine and lung) in research, industrial applications and regulatory toxicology. *ALTEX* **2015**, *32*, 327–378. [CrossRef] [PubMed]

Article

The Occurrence and Health Risk Assessment of Aflatoxin M1 in Raw Cow Milk Collected from Tunisia during a Hot Lactating Season

Khouloud Ben Hassouna [1,2], Jalila Ben Salah-Abbès [1], Kamel Chaieb [3], Samir Abbès [1,4], Emilia Ferrer [5], Francisco J. Martí-Quijal [5], Noelia Pallarés [5,*] and Houda Berrada [5]

[1] Laboratory of Genetic, Biodiversity and Bio-Resources Valorisation, University of Monastir, Monastir 5000, Tunisia; khouloudhsn20@gmail.com (K.B.H.); jalila.bensalah@yahoo.fr (J.B.S.-A.); abb_samir@yahoo.fr (S.A.)

[2] Laboratory of Analysis, Treatment and Valorization of Environmental Pollutants and Products, Faculty of Pharmacy, Monastir University, Monastir 5000, Tunisia

[3] Department of Biochemistry, Faculty of Science, King Abdulaziz University, Jeddah 21589, Saudi Arabia; chaieb_mo@yahoo.fr

[4] High Institute of Biotechnology of Béja, University of Jendouba, Jendouba 8189, Tunisia

[5] Nutrition and Food Science Area, Preventive Medicine and Public Health, Food Science, Toxicology and Forensic Medicine Department, Faculty of Pharmacy, Universitat de València, Burjassot, 46100 València, Spain; emilia.ferrer@uv.es (E.F.); francisco.j.marti@uv.es (F.J.M.-Q.); houda.berrada@uv.es (H.B.)

* Correspondence: noelia.pallares@uv.es; Tel.: +34-963544117

Citation: Hassouna, K.B.; Salah-Abbès, J.B.; Chaieb, K.; Abbès, S.; Ferrer, E.; Martí-Quijal, F.J.; Pallarés, N.; Berrada, H. The Occurrence and Health Risk Assessment of Aflatoxin M1 in Raw Cow Milk Collected from Tunisia during a Hot Lactating Season. *Toxins* **2023**, *15*, 518. https://doi.org/10.3390/toxins15090518

Received: 4 July 2023
Revised: 28 July 2023
Accepted: 22 August 2023
Published: 24 August 2023

Abstract: Milk is a staple food that is essential for human nutrition because of its high nutrient content and health benefits. However, it is susceptible to being contaminated by Aflatoxin M1 (AFM1), which is a toxic metabolite of Aflatoxin B1 (AFB1) presented in cow feeds. This research investigated AFM1 in Tunisian raw cow milk samples. A total of 122 samples were collected at random from two different regions in 2022 (Beja and Mahdia). AFM1 was extracted from milk using the QuEChERS method and contamination amounts were determined using liquid chromatography (HPLC) coupled with fluorescence detection (FD). Good recoveries were shown with intra-day and inter-day precisions of 97 and 103%, respectively, and detection and quantification levels of 0.003 and 0.01 µg/L, respectively. AFM1 was found in 97.54% of the samples, with amounts varying from values below the LOQ to 197.37 µg/L. Lower AFM1 was observed in Mahdia (mean: 39.37 µg/L), respectively. In positive samples, all AFM1 concentrations exceeded the EU maximum permitted level (0.050 µg/L) for AFM1 in milk. In Tunisia, a maximum permitted level for AFM1 in milk and milk products has not been established. The risk assessment of AFM1 was also determined. Briefly, the estimated intake amount of AFM1 by Tunisian adults through raw cow milk consumption was 0.032 µg/kg body weight/day. The Margin of Exposure (MOE) values obtained were lower than 10,000. According to the findings controls as well as the establishment of regulations for AFM1 in milk are required in Tunisia.

Keywords: aflatoxin M1; raw cow milk samples; Tunisia; HPLC-FD; risk assessment

Key Contribution: According to the findings of this survey; AFM1 is commonly found in the majority of raw cow milk collected in Tunisia; and all positive samples had concentrations that exceeded the EU maximum tolerable limit. Furthermore; the risk assessment of AFM1 through milk consumption was determined; suggesting some health concerns for Tunisian consumers.

---

1. Introduction

Milk is essential for human nutrition due to its high nutrient content and health benefits [1]. Humans consume milk either raw (unprocessed) or processed (condensed, pasteurized, powdered, liquid, heat-treated, or UHT-treated) [2]. Cow milk is the most frequently consumed by humans. When comparing cow milk and other animal milk types

---

to human milk, it was previously thought to be the ideal food for children, who represent the majority of consumers [3].

In fact, cow milk accounts for 83% of world milk production, followed by buffalo milk with 13%, goat milk with 2%, sheep milk with 1%, and camel milk with 0.3% [4]. Furthermore, it accounts for 77% of African animal milk production [4]. The dairy sector is an important component of Tunisia's agricultural and agro-alimentary sectors. Furthermore, in Tunisia, raw cow milk is used for making special and traditional products such as cheese and gouta. Tunisia produced an average of 1,100,000 tons of fresh cow milk per year in the period of 2010–2014 [5]. Later, in 2017, Tunisia produced approximately 1424 million liters of cow milk [6]. Its consumption by the Tunisian population is about 110 L/cap/year. This consumption is higher than the global average (100 L per capita per year) and the African average (17 L/cap/year). The cow milk collected represents more than 62.6% of the national milk produced, and the Tunisian industry dairy obtain about 89.6% of the fresh milk from collection centers [6]. Tunisia, in North Africa, is surrounded by the Mediterranean Sea, which has a warm climate with high relative humidity. These conditions are favorable to the proliferation of toxigenic molds and mycotoxins production. In fact, it is essential that milk be free of toxic compounds that can be harmful to humans, particularly children, who are more susceptible to the action of toxic compounds. In fact, the quality of milk, in terms of toxic contaminants, has a direct relation with the type and quality of animal feed. Nonetheless, milk can be indirectly contaminated by mycotoxins such as aflatoxins (AFs) and ochratoxin A (OTA) at relatively high levels, as well as fumonisins (FBs), zearalenone (ZEA), and deoxynivalenol (DON) at trace levels, by ingesting contaminated feed [7]. Mycotoxins are a group of highly toxic secondary metabolites produced by *Aspergillus*, *Fusarium*, *Penicillium*, *Claviceps*, and *Alternaria* fungal species [8]. Mycotoxins can cause health risks ranging from allergic reactions to death, both in humans and animals. In effect, their toxicity is determined by the concentration and type of mycotoxins consumed. They are responsible for chronic effects, namely immunotoxicity, genotoxicity, hepatotoxicity, nephrotoxicity, teratogenicity, cytotoxicity, and neurotoxicity [9]. Furthermore, the International Agency for Research on Cancer has classified AFs as carcinogenic to humans and animals [10].

The majority of studies on mycotoxins' presence in milk have focused on aflatoxin M1 (AFM1) analysis. AFM1 is a hepatic carcinogenic mycotoxin, and its presence in milk poses a danger for humans [11]. In fact, AFM1 is a monohydroxilated product of aflatoxin B1 (AFB1) biotransformation in the cow liver, and is detected in milk within 12–24 h of the first ingestion of AFB1 presented in animal feed [12]. It is excreted in milk at rates ranging from 1% to 6% of the dietary consumption [13]. The rate of biotransformation varies between animals, and other factors such as diet, rate of ingestion, digestion rate, animal health, and the biotransformation capacity of the liver [14]. AFM1 was classified as a Group 1 carcinogen by the International Agency for Research on Cancer [10]. Milk contamination by mycotoxins is not only a public health problem, but it also costs farmers financially due to the negative effects of them on animal production [15]. The presence of AFM1 in milk is a worldwide issue. Once it is secreted into milk, it cannot be removed by technological treatments such as pasteurization, sterilization, or drying [16]. In order to protect humans, the European Union established 0.05 µg/L as the maximum permitted level (MPL) of AFM1 in milk [17]. Similarly, the Food and Drug Administration (FDA) in the United States has established a maximum permitted level of 0.50 µg/L for AFM1 in milk [18]. Tunisia, unfortunately, does not have a maximum limit for mycotoxins in milk and milk products.

For the detection of mycotoxins in milk, many sensitive, selective, and reliable analytical techniques are used, such as enzyme-linked immunosorbent assay (ELISA) and liquid chromatography coupled with a fluorescence or ultraviolet detector (HPLC-FD/HPLC-UV), thin-layer chromatography (TLC), LC-MS/MS, and ultra-performance liquid chromatography coupled with tandem mass spectrometry (UHPLC-MS/MS) [19–21].

Despite the high consumption of cow milk by Tunisian people and the dangers that mycotoxins pose to humans and animals, to the best of our knowledge, only one study about mycotoxins presence in milk was carried out in Tunisia [22], and no data on the risk assessment of mycotoxin consumption was available in this study. Thus, it is essential to investigate the occurrence of mycotoxins in milk to furnish information about human exposure. Taking those considerations into account, the objectives of the present study are to determine and quantify AFM1 in raw cow milk collected from two Tunisian regions, Beja (the continental region) and Mahdia (the coastal region), using HPLC-FD, as well as to determine the daily exposure of Tunisians consumers to AFM1 through cow milk consumption.

2. Results

2.1. AFM1 Levels in Raw Cow Milk

One hundred and twenty-two samples of raw cow milk produced in the Tunisian provinces of Mahdia (n = 60) and Beja (n = 62) were analyzed by HPLC-FD to evaluate the occurrence of AFM1. The AFM1 occurrence, concentration range, and mean concentration are presented in Table 1. In fact, only positive samples above the LOD (0.003 µg/L) were considered.

Table 1. Incidence of the studied AFM1 (µg/L) in raw cow milk samples from tow Tunisian regions.

Region	Positive Samples		<LOD	<LOQ	Minimum Concentration (µg/L)	Maximum Concentration (µg/L)	Mean Concentration (µg/L)	Lower Bound Scenario (µg/L)	Upper Bound Scenario (µg/L)
	n	**%**							
Mahdia (60)	58	96.67	2	48	<LOQ	99.36	39.37	6.56	6.57
Beja (62)	61	98.39	1	53	<LOQ	197.37	64.27	8.29	8.30
Total (122)	119	97.54	3	101	<LOQ	197.37	50.44	7.44	7.45

Only positive samples (≥LOD: 0.003 µg/L) were considered for calculations. Lower bound scenario (LB): the value of 0 was attributed when AFM1 was not detected or detected below the LOQ. Upper bound scenario (UB): the value of LOD was attributed to AFM1 that was not detected, and the value of LOQ was attributed to AFM1 detected at levels below the LOQ. Values are expressed as means of three assays and significant vs. control assay: $p < 0.05$. LOD: limit of detection: 0.003 µg/L. LOQ: limit of quantification: 0.01µg/L.

This study exposed high contamination levels of AFM1 in the raw cow milk collected in Tunisia. AFM1 was found in 119 out of 122 (97.54%) of analyzed milk samples, at concentrations ranging between values lower than the LOQ and 197.37 µg/L, with a mean value of 50.44 µg/L. However, it is important to highlight that only 18 of 119 positive samples were contaminated at levels above LOQ. Furthermore, AFM1 was not detected in 3 of 122 samples. Moreover, 18 of 122 analyzed samples exceeded the maximum permitted level (MPL) set for AFM1 by the EU (0.05µg/L) [17] and the FDA 0.5 µg/L [18].

2.2. Regional Differences in AFM1 Contamination Levels

Based on the milk sampling region, the levels of AFM1 in samples collected from two different regions in Tunisia were compared (Tables 2 and 3). The highest AFM1 maximum concentration was found in the continental region (Beja), at 197.37 µg/L, compared to 99.36 µg/L in the coastal region (Mahdia). Moreover, the mean value of AFM1 in milk samples from the continental region (64.27 µg/L) was about 1.6 times higher than in milk samples from the littoral region (39.73 µg/L). AFM1 levels of some samples, eight from the Beja region and ten from Mahdia, were above the thresholds set by the EU and FDA [17,18].

Table 2. Meteorological data during sampling period of 2022 in Tunisia.

Region (Period)	Average of Precipitation (mm)	Average of Temperatures (°C)	Average of Humidity (%)
Littoral region (March–June)	3.75	21.25	71.5
Continental region (March–June)	10.5	21.25	71

Data from the National Institute of Meteorology (Tunisia).

Table 3. Geographic data related to the sampling regions in Tunisia.

Sampling Region	Geographic Position	Altitude (m)	Latitude	Longitude	Bioclimatic Zone
Mahdia (Coastal)	Center East	6	36807 N	10_22051 E	Semi-Arid Inferior
Beja (Continental north)	Northwest	93	36_33004 N	9_26035 W	Sub-Humid

2.3. Estimation of Daily Intake

Based on the findings of this study and the amount of milk consumed by the Tunisian population (approximately 110 L/person/year [23]), the estimated daily intake of AFM1 (EDI) for a Tunisian adult weighing 70 kg was estimated to be 0.032 µg/kg body weight/day, considering the mean concentration calculated under Lower Bound scenario (LB) and Upper Bound scenario (UB).

2.4. AFM1 Risk Characterization

The characterization of AFM1 risk from dairy product consumption in the Tunisian population was performed using a deterministic approach. The Margin of Exposure (MOE) was calculated. Considering the LB and UB scenarios the MOE value for raw milk was 12.54.

3. Discussion

3.1. AFM1 Presence in Raw Cow Milk

Mycotoxins are a worldwide issue that harms both animals and humans. Similar to the rest of the world, the Tunisian people consume a great quantity of cow milk. The determination of AFM1 (a carcinogenic mycotoxin) levels in raw cow milk samples from Tunisia, as well as the investigation of the risk assessment associated with this contamination, piqued our interest in this study. In the context of this study, there are other studies on AFM1 occurrence in raw cow milk from countries over the world highlighting their presence. To the extent of our knowledge, only one study on mycotoxins in milk has been published in Tunisia [22].

In this survey, AFM1 contamination levels in raw milk were found in 97.54 percent of analyzed samples. Similar to our results, AFM1 was found in raw cow milk samples with very high occurrences in various other North African countries. In Egypt, Kamel et al. [24] used UPLC-MS/MS to analyze raw cow milk samples and found AFM1 in all of them (100%) at quantities ranging from 0.14 to 8.54 µg/L. Later, Abdallah et al. [25] found AFM1 in 20 of 20 samples taken from different local markets in Assiut Governorate in Upper Egypt, with concentrations ranging from 0.02 to 0.19 µg/L. In a previous Sudanese investigation, Ali et al. [26] also revealed AFM1 in all raw milk samples tested with concentrations ranging from 0.1 to 2.52 µg/L. In 2023, Hamed et al. [27] used HPLC-FD to analyze raw cow milk from Egypt, and detected AFM1 in 88% of samples with a maximum concentration of 0.122.

More recently, in 2023, AFM1 was detected in all of the raw cow milk samples from Ethiopia [28]. However, no AFM1 occurrence was detected in raw cow milk from Portugal in 2023 [29]. North African counties, including Tunisia, which are surrounded by the Atlantic Ocean, the Mediterranean Sea, and the Red Sea, have a humid and warm climate that may encourage fungal development and mycotoxin production. In fact, the high temperatures associated with climate change induce fungi apparition and mycotoxin production in animal feeds [30]. Recently, The Food and Agriculture Organization (FAO) [31] have published a paper on climate change and food safety, notably analyzing the effects of climate change on mycotoxins. In the sampling period, the temperature (21 °C) was high (Table 2). Aflatoxins (AFs), including AFB1, are mycotoxins produced in animal feed by fungal genera such as *Aspergillus flavus*, *Aspergillus parasiticus*, and *Aspergillus nomius* [32]. These fungi can grow at temperatures ranging from 19 to 35 °C [33], producing AFB1 in Tunisian animal feed [34]. In comparison to our findings, AFM1 contamination has been recorded in other studies with zero or low prevalence in raw cow milk samples. In North Africa, lower incidences were reported. In 2015, Redouane et al. [35] reported that an only one out of 22 (4.5%) raw cow milk sample from Algeria was contaminated by AFM1. Also in Egypt, a study revealed AFM1 in only 18% of raw cow milk samples taken from Cairo provinces [36], and later Zakaria et al. [37] found AFM1 in 49% of raw milk from the Aswan province. In an Algerian study, the contamination of raw cow milk was shown to have an incidence of 46.43% [38].

The concentrations found in this work ranged from values below LOQ and 197.37 µg/L. In accordance with our results, Esam et al. [39] found high concentrations of AFM1, ranging between 10 and 110 µg/L in all raw cow milk samples analyzed from Egypt. Added to that, Mukhtar et al. [40] detected the presence of AFM1 in the overall fresh raw cow milk samples collected from Nigeria at concentrations ranging from 22.33 to 119.99 µg/L.

The high AFM1 concentration in milk is usually explained by the contamination of cow feeds by AFB1. According to Table 2, the sampling period was a dry one during which farmers were not able to feed their animals (kept in local dairy farms) with green feed. Therefore, the main feed supplements are grains (corn and soya), and concentrated feed is a mixed of barley, wheat, corn, and soya, which are susceptible to mycotoxins and fungal contamination due to poor transportation and storage conditions. Tunisians typically store those feeds in non-hermetic bags made by jute and woven polypropylene. These bags have inadequate construction because they allow air to enter, exposing the stored feed to fungus spores [41].

The maximum concentration of AFM1 (197.37 µg/L) in Tunisian raw milk was higher than that reported in other countries around the world, including Cyprus (0.017 µg/L; [42]), Turkey (0.034 µg/L; [43]), Italy (0.026 µg/kg; [44]), Pakistan (0.021 µg/L; [45]), Brazil (0.045 µg/L; [46]), China (0.036 µg/L; [47]), Lebanon (0.440 µg/L; [48]), and Spain (1.36 µg/L; [49]), and lower than that reported in Kenya (273.8 µg/L; [50]) and Iran (240 µg/L; [51]).

Those different results indicate that milk contamination levels with AFM1 vary between countries. These differences are due to different techniques for extracting and detecting mycotoxins, the type and quality of forage and feed, the kind of cow diet, the geographic location, the climatic variation, the sampling seasons, the genetic variation in dairy cows, the farming systems, and the feed storage [51–53].

In terms of regulations, all raw milk samples had AFM1 concentrations that exceeded the MPL (0.050 µg/L for the EU and 0.5 µg/L for the FDA). In the same way, the percentage of AFM1 contamination levels of samples exceeding regulatory limits have been reported for cow milk from north Africa, such as Egypt (100%; [24]) and Sudan (100%; [26]); however, Mohammedi-Ameur et al. [38] found that only one sample (1.19%) of Algerian raw milk had an AFM1 concentration above the MPL. Cow milk contamination levels (%) that exceeded regulatory limits have also been reported in other African countries such as Burundi (100%; [54]), Kenya (63%; [55]), Zimbabwe (79.2%; [56]), and Tanzania (83.8%; [57]), as well as in other countries over the world such as Iran (83.7%; [58]) and Indonesia (68.4%; [59]).The high quantities of AFM1 reported in this study can be attributed to the nature of the milk examined, which was fresh and untreated. In a Moroccan study,

15.2% of pasteurized milk samples were contaminated with AFM1 at lower levels ranging from 0.010 to 0.077 µg/L while in ultra-high temperature treated (UHT) milk, only 9.5% of samples were contaminated with AFM1 at concentrations ranging from 0.013 to 0.048 µg/L [60]. AFM1 was found in three different milk products from Punjab, India, with varying levels of contamination (percentage of positive samples/maximum concentration), including raw milk (80%, 4.185 µg/L), pasteurized milk (41%, 2.330 µg/L), and ultra-high-temperature milk (47%, 2.585 µg/L) [61]. Inadequate conditions, contamination of the milking instrument and/or manipulation of raw milk may also contribute to mycotoxin contamination of milk [62].

3.2. Regional Differences in AFM1 Concentrations

The current study reported considerable differences in AFM1 contamination levels in raw milk samples from two different regions in Tunisia, with the continental region having the highest contamination levels with a mean concentration of 64.27 µg/L and all the positive samples having concentrations above the MPL (0.05 µg/L). However, Abbès et al. [22] found contradictory findings when analyzing raw cow milk from the continental region, Beja, in Tunisia using ELISA technique. They detected moderate contamination levels of AFM1, with a mean concentration of 13.62 µg/L, and only 5 (4.4%) had concentrations above 0.05 µg/L. The differences in concentrations observed could be attributed to geographic and climatic variations [63]. In fact, high levels of AFM1 in milk samples were found in continental regions with more precipitation, including rain (Table 2), which increases humidity and favors the growth of aflatoxigenic fungi and the accumulation of aflatoxins, in particular AFB1, in dairy animal feed, which was then transferred to milk in the form of AFM1 [64]. These variations could also be caused by other factors such as animal species, milking time, the level of mycotoxins in feed intake, and the volume of milk produced by the animal mammal in such region. In fact, due to limited resources for this survey, we could not collect background information on farming practices and cow feeds (green feed, dry feed, and concentrate) provided at the different regions. So far, we have no information about differences in AFM1 contamination.

3.3. Estimated Daily Consumption of AFM1

Humans are exposed daily to AFM1 through the consumption of contaminated milk. Exposure assessment, as one component of risk assessment methodology, combines mycotoxin levels in food with consumption habits, providing valuable information for risk management if mycotoxins affect food safety and health on an individual or population level [65]. The estimated exposure of humans to mycotoxins, in particular AFM1, is calculated by combining contamination and consumption data. In this context and, for the first time in Tunisia, the determination of the EDI of AFM1 was carried out to evaluate the risk associated with the intake of this mycotoxin through milk consumption by Tunisian adults (70 kg). In Tunisia, milk consumption was estimated at 110 L/person/year. This consumption is compared with neighboring countries. It was higher than 72 L/person/year in Morocco and lower than 140 L/person/year in Algeria [60]. In this survey, considering the mean of total samples, EDI values obtained were 0.032 µg/kg body weight/day in both the LB and UB scenarios. The great finding of EDI can be explained by the Tunisian population's higher intake of raw cow milk, due to its low cost, accessibility, and taste. It can also be explained by the high prevalence of AFM1 in the raw milk samples collected, given that this toxin is not removed or degraded by the industry's typical technological processes, such as heat treatments, drying, and/or milk fermentation [60].

Although various studies have reported the presence of AFM1 in milk [66], few data on the risk assessment of this mycotoxin for milk consumers are still accessible, and no data are available in Tunisia. In comparison to data from other countries, the EDI (0.032 µg/kg body weight/day) of AFM1 taken from cow raw milk consumption reported in this research was the highest. In comparison to our reported data, negligible EDI values were reported in other studies. Bilandžić et al. [67] found EDI values from mean raw milk intake ranging from 0.000017 to 0.00282 μg/kg bw/day in a Croatian study. Further, Mohammedi-Ameur et al. [38] found an EDI of 0.0003425 μg/kg bw/day in raw milk from Algeria. Also, the research by Gizachew et al. [68] reported an EDI of 0.0007891 μg/kg bw/day in raw milk from Ethiopia. Recently, Daou et al. [48] found an EDI of 0.0001126 μg/kg bw/day for raw milk in Lebanon. In Ghana, the EDI ranged between 0.000006 and 0.000203 μg/kg bw/day for infants, toddlers, children, teenagers, and adults [69]. Due to the lack of consumption data for various age groups in this study, we were able to estimate exposure and health risks only for adults.

3.4. Risk Characterization of AFM1 Exposure

Regarding the MOE, it can be used to describe the health risk posed by carcinogenic and genotoxic substances found in food [70]. When the value of the MOE is ≥ than 10,000, it is assumed that there is a low risk to public health [66]. In our study, the MOE values measured in both scenarios were less than 10,000 (12.54 in both the LB and UB scenarios). In any case, the ingestion of raw cow milk contaminated with AFM1 can result in a public health problem (liver cancer). Similar to our results, Kortei et al. [69] found MOE values ranging from 197 to 6666.7, which was below 10,000 and caused a problem for public health by consuming raw cow milk. Further, Zebib et al. [28] revealed that the MOE detected in the studied Ethiopian regions was less than 10,000 in adult populations, indicating a possible public health risk due to the high AFM1 exposure from raw milk consumption. Contrary to our findings, the MOE values for Serbian consumers' exposure to AFM1 were significantly higher than 10,000 in toddlers and other children, indicating that there is no health risk associated with AFM1 exposure through consuming milk [71]. The differences in results reported in this work and other studies could be attributed to a variety of factors, including milk consumption, nutritional habits, the number of analyzed samples, the milk sampling period, and the method used for AFM1 extraction and analysis (ELISA kits, HPLC, or LC-MS/MS, for example). That said, attention should be paid to the health risks associated with exposure to mycotoxins, including AFM1, through consumption of raw cow milk.

The high frequencies and concentrations of AFM1 in raw milk samples in the current investigation indicate that lactating cows were exposed to high levels of dietary AFB1. Thus, preventive approaches for reducing AFB1 concentrations in feed as well as good storage practices and hygiene in dairy farms should be explored to protect animal health. Furthermore, to protect consumers from possible health risks associated with AFM1 exposure, more extensive and periodic control of AFM1 concentration in milk is required, as is establishing regulations for mycotoxins in Tunisia.

4. Conclusions

In the current survey, AFM1 was frequently identified in raw cow milk from two regions of Tunisia (the continental and coastal regions) with high levels of contamination. The amounts of AFM1 in 18 of 119 positive samples exceeded the MRL. According to the toxicological parameters of EDI and MOE calculated in this study, where the MOE was less than 10,000, there was a public health problem, particularly hepatocellular carcinoma (HCC), due to the high exposure of the Tunisian population to AFM1 by consuming raw cow milk. To avoid the proliferation of fungi and the production of aflatoxins, farmers must be informed of the optimal storage conditions for animal feed as well as milk control. Strict regulation and a tolerable maximum for AFB1 and AFM1 should be established in Tunisia to reduce possible risks to health and economic losses. Further investigation of AFM1 in milk and its risk assessment in consumers, particularly children (the majority of milk consumers), is needed.

5. Materials and Methods

5.1. Reagents and Chemicals

The solvents used in this work, methanol (MeOH) and acetonitrile (ACN) (HPLC grade), were acquired from Merck (Darmstadt, Germany). The deionized water (resistivity > 18 M cm1) was obtained in the laboratory from a Milli-Q SP® Reagent Water System (Millipore Corporation, Bedford, MA, USA). The formic acid (95%) was purchased from Sigma Aldrich (St. Louis, MO, USA). The Nylon filters (0.45-μm pore size) were supplied by Scharlau (Barcelona, Spain). The syringe nylon filters (13 mm diameter and 0.22 m pore size) were supplied by Membrane Solutions (Plano, TX, USA). The derivatization reagent (pyridine hydrobromide perbromide, tech.90%) was acquired from Thermo Scientific (Kandel, Germany). The glacial acetic acid (99%) was supplied by Fisher Scientific (Leics, UK). The hexane (95%) was provided by VWR International bvba (Leuven, Belgium). The tri-sodium citrate dihydrate was purchased from VWR International BVBA (Leuven, Belgium). The disodium hydrogen citrate sesquihydrate (99%) was delivered from Alfa Aesar GmbH and Co. KG (Karlsruhe, Germany). The sodium chloride (99.5%) was supplied by Fisher Scientific (Loughborough, UK). The magnesium sulfate anhydrous was supplied by Thermo Scientific (Kandel, Germany). The standards for Aflatoxin M1 were purchased from Sigma Aldrich (Madrid, Spain). All solutions were kept in amber vials under secure conditions at −20 °C. Prior to injection into the LC-FD system, these stock solutions were diluted to obtain adequate working concentrations.

5.2. Study Region and Samples Collection

A total of 122 raw cow milk samples were randomly collected from two different geographical and agro-climatic conditions regions in Tunisia: the continental region (n = 62), defined as Beja, characterized by a more arid climate, and the coastal region (n = 60), represented by the location of Mahdia, characterized by a more semi-arid climate. Table 3 shows the geographic information for the coastal and continental areas. A total of 40 mL of fresh samples were collected directly from dairy cows in farmers' houses and dairy farms during a routine midday milking procedure. Samples were stored in 50 mL Falcon tubes at −20 °C until analysis without being treated in any way beforehand. All samples were transported to Valencia, Spain, in polystyrene cartons containing dry ice. Then, they were kept in the laboratory at −20 °C until the analysis.

5.3. AFM1 Extraction from Milk

A QuEChERS extraction procedure was employed to extract AFM1 from milk. To summarize, 500 μL of milk sample were placed in a 15-mL Falcon tube. Then, 2 mL of hexane were added to the sample. The mixture was vortexed for 30 s before being centrifuged for 5 min at 5000 rpm in an Eppendorf Centrifuge 5810R (Eppendorf, Hamburg, Germany). Following the removal of the hexane, 2.5 mL of deionized water and 2.5 mL of ACN were added to the tube and shaken for 1 min. Subsequently, a mix of salts (2 g of $MgSO_4$, 0.3 g of NaCl, 0.25 g of DSHCSH, and 0.5 g of TSCDH) was added. After vortexing for 1 min and centrifugation for 5 min at 3500 rpm, 2 mL of the upper phase was filtered through a 13 mm/0.22 μm nylon filter. Subsequently, 900 μL of the filtrate were dried at 40 °C under a nitrogen stream using a multi-sample Turbo-vap LV Evaporator (Zymark, Hoptkinton, USA). Finally, and before HPLC analysis, the final residue was reconstituted with 300 μL of methanol.

5.4. AFM1 Determination by HPLC-FD

A JASCO Lc-Net II/ADC coupled to a FP-2020 plus JASCO detector was used for the determination of AFM1. The chromatographic separation was achieved with a liquid purple ODS reverse-phase column C18 (5 μm, column LC 150 × 4.6 mm, Analisis vinicos). The mobile phases consisted of acetonitrile (A), 0.1% acetic acid water (B), and methanol (C). The gradient program began with a proportion of 15% for eluent A and 60% for eluent B. Then, mobile phase A changed to 50% and 40% of mobile phase B in 14 min and remained the same proportion until 30 min. Then, the column was cleaned and readjusted to initial conditions over the next 5 min. The following instrument parameters were set: injection volume of 10 µL, flow rate of 0.8 mL/min, excitation wavelength of 365 nm, and emission wavelength of 455 nm.

The methodology proposed for AFM1 determination was optimized in terms of recoveries, matrix effects, linearity, and limits of detection (LOD), and quantification (LOQ). All analytical parameters obtained were in accordance with the criteria established by European Commission Decision 2002/657/EC. In this sense, good recoveries were observed at 100 µg/L, with intra-day and inter-day precision between 97 and 103%, respectively. SSE (%) obtained evidenced no matrix effects. Calibrations curves constructed by spiking blank raw milk extract samples and solvent at levels comprised between <LOQ and 250 µg/L revealed good linearity, with correlation coefficients (R^2) between 0.990 and 0.999. Finally, the LOD and LOQ were 0.003 µg/L and 0.01 µg/L, respectively. Matrix-matched calibration curves were used for effective quantification of milk samples.

5.5. Estimation of the AFM1 Daily Intake (EDI)

The EDI was determined by the AFM1 mean concentration found in analyzed milk samples, the daily intake of milk by the population, and the average body weight of an adult consumer. The EDI was calculated using the following formula [72] and expressed in µg/kg per body weight (bw) per day [66].

$$EDI = \frac{CAFM1 \times ADC}{BW}$$

where CAFM1 is the mean AFM1 concentration in milk samples (µg/L), ADC is the average daily consumption of milk (L/day), and BW is the body weight of an individual consumer (kg).

To consider those concentration values that were below the LOD or LOQ (left-censored data), an exposure assessment was performed in two different scenarios [73,74]: the lower bound scenario (LB), in which the value of 0 was attributed when AFM1 was not detected or detected below the LOQ, and the Upper bound scenario (UB), in which the value of LOD was attributed to AFM1 that was not detected, and the value of LOQ was attributed to AFM1 detected at levels below the LOQ.

5.6. Risk Characterization Margin of Exposure (MOE)

The MOE was determined by dividing the Benchmark dose lower limit of 10% (BMDL 10) for AM1, which was 0.4 µg/kg bw/day (AFM1 potency in male Fischer rats based on a 2-year study) by EDI [66]. The benchmark is an estimate of the lowest dose that is 95% certain to cause no more than 10% cancer incidence; it is the dose that produces a minor but discernible reaction [28].

The main finding of the risk assessment of aflatoxins, including aflatoxin M1, was liver carcinogenicity. In fact, a MOE value below 10,000 indicates that exposure to these mycotoxins may increase the risk of developing hepatocellular carcinoma (HCC), which is a serious public health concern [66].

5.7. Statistical Study

The differences between AFM1 occurrence levels of the samples were evaluated using a randomized block experiment. Furthermore, significance level between sampling periods was elucidated using Duncan multiple comparison test. A multi sample ANOVA test was used also for determination of the significance (p value < 0.05).

Author Contributions: K.B.H.: Methodology, writing, original draft; J.B.S.-A.: data curation, writing; K.C.: review and editing; S.A.: overall project supervision, writing, and editing. N.P.: methodology, writing; H.B.: funding acquisition, supervision, writing and editing. E.F.: funding acquisition, writing and editing. F.J.M.-Q.: review and editing. All authors have read and agreed to the published version of the manuscript.

Funding: The authors gratefully acknowledge the technical and financial support from the Tunisian Ministry of Higher Education and Scientific research. This work has been supported by a scholarship offered by the University of Jendouba, Tunisia, by the Spanish Ministry of Science and Innovation project (PID2020-115871RB), and by the Generalitat Valenciana project (AICO/2021/037).

Institutional Review Board Statement: Not applicable.

Informed Consent Statement: Not applicable.

Data Availability Statement: Not applicable.

Acknowledgments: This work has been supported by a scholarship offered by the University of Jendouba, Tunisia, by the Spanish Ministry of Science and Innovation project (PID2020-115871RB), and by the Generalitat Valenciana project (AICO/2021/037). Khouloud Ben Hassouna would like to express his gratitude to the Department of Preventive Medicine and Public Health, Food Science, Toxicology, and Forensic Medicine, Faculty of Pharmacy, Valencia, Spain, for their technical assistance. Samir Abbès wants to express his gratitude to the Tunisian Ministry of Higher Education and Scientific Research for the support it has provided. Noelia Pallarés wishes to thank the post-PhD program of the University of Valencia for the requalification of the Spanish University System from the Ministry of Universities of the Government of Spain, modality "Margarita Salas" (MS21-045), financed by the European Union, Next Generation EU.

Conflicts of Interest: The authors declare no conflict of interest.

References

1. Ben Hassouna, K.; Ben Salah-Abbès, J.; Chaieb, K.; Abbès, S. Mycotoxins occurrence in milk and cereals in North African countries—A review. *Crit. Rev. Toxicol.* **2023**, *52*, 619–635. [CrossRef] [PubMed]
2. Becker-Algeri, T.A.; Castagnaro, D.; de Bortoli, K.; de Souza, C.; Drunkler, D.A.; Badiale-Furlong, E. Mycotoxins in Bovine Milk and Dairy Products: A Review. *J. Food Sci.* **2016**, *81*, R544–R552. [CrossRef]
3. Pietrzak-Fiećko, R.; Kamelska-Sadowska, A.M. The Comparison of Nutritional Value of Human Milk with other Mammals' Milk. *Nutrients* **2020**, *12*, 1404. [CrossRef]
4. FAO. Gateway to Dairy Production and Products. Milk and Milk Products. 2017. Available online: http://www.fao.org/dairy-production-products/production/dairy-animals/en/ (accessed on 1 December 2021).
5. Salah, M.; Boudiche, S.; Amara, S.; Ameur, M.; Bornaz, S. Stratégie de développement de la filière lait biologique en Tunisie à travers la chaine de valeur. *J. New Sci. Vol. Spécial Conférence IABC* **2016**, *10*, 1263–1275.
6. Givlait. Suivi Journalier de la Filière Laitière. 2018. Available online: http://www.givlait.com.tn/presentation-de-la-filiere-lait.html (accessed on 3 February 2023).
7. Flores-Flores, M.E.; González-Peñas, E. An LC–MS/MS method for multi-mycotoxin quantification in cow milk. *Food Chem.* **2017**, *218*, 378–385. [CrossRef]
8. Alshannaq, A.; Yu, J.-H. Occurrence, Toxicity, and Analysis of Major Mycotoxins in Food. *Int. J. Environ. Res. Public. Health* **2017**, *14*, 632. [CrossRef]
9. Omotayo, O.P.; Omotayo, A.O.; Mwanza, M.; Babalola, O.O. Prevalence of Mycotoxins and Their Consequences on Human Health. *Toxicol. Res.* **2019**, *35*, 1–7. [CrossRef] [PubMed]
10. IARC. *Mycotoxins and Human Health. Improving Public Health through Mycotoxin Control*; International Agency for Research on Cancer: Lyon, France, 2012; pp. 87–104.
11. Naeimipour, F.; Aghajani, J.; Kojuri, S.; Ayoubi, S. Useful approaches for reducing aflatoxin M1 content in milk and dairy products. *Biomed. Biotechnol. Res. J. (BBRJ)* **2018**, *2*, 94–99. [CrossRef]
12. Frazzoli, C.; Gherardi, P.; Saxena, N.; Belluzzi, G.; Mantovani, A. The Hotspot for (Global) One Health in Primary Food Production: Aflatoxin M1 in Dairy Products. *Front. Public. Health* **2017**, *4*, 294. [CrossRef] [PubMed]
13. Gonçalves, B.L.; Gonçalves, J.L.; Rosim, R.E.; Cappato, L.P.; Cruz, A.G.; Oliveira, C.A.F.; Corassin, C.H. Effects of different sources of Saccharomyces cerevisiae biomass on milk production, composition, and aflatoxin M1 excretion in milk from dairy cows fed aflatoxin B1. *J. Dairy Sci.* **2017**, *100*, 5701–5708. [CrossRef]
14. Iqbal, S.Z.; Asi, M.R.; Jinap, S. Variation of aflatoxin M1 contamination in milk and milk products collected during winter and summer seasons. *Food Control* **2013**, *34*, 714–718. [CrossRef]

15. Kemboi, D.C.; Antonissen, G.; Ochieng, P.E.; Croubels, S.; Okoth, S.; Kangethe, E.K.; Faas, J.; Lindahl, J.F.; Gathumbi, J.K. A Review of the Impact of Mycotoxins on Dairy Cattle Health: Challenges for Food Safety and Dairy Production in Sub-Saharan Africa. *Toxins* **2020**, *12*, 222. [CrossRef] [PubMed]
16. Campagnollo, F.B.; Ganev, K.C.; Khaneghah, A.M.; Portela, J.B.; Cruz, A.G.; Granato, D.; Corassin, C.H.; Oliveira, C.A.F.; Sant'Ana, A.S. The occurrence and effect of unit operations for dairy products processing on the fate of aflatoxin M1: A review. *Food Control* **2016**, *68*, 310–329. [CrossRef]
17. European Commission, E. Commission regulation (EC) no 1881/2006 of 19 December 2006 settingmaximumlevels for certain contaminants in foodstuffs. *Off. J. Eur. Union L* **2006**, *364*, 5–24.
18. FDA. *Sec. 527.400 Whole Milk, Low Fat Milk, Skim Milk-Aflatoxin M1 (CPG 7106.210). FDA Compliance Policy Guides*; FDA: Washington, DC, USA, 1996; p. 219.
19. Zhou, Y.; Xiong, S.; Zhang, K.; Feng, L.; Chen, X.; Wu, Y.; Huang, X.; Xiong, Y. Quantum bead-based fluorescence-linked immunosorbent assay for ultrasensitive detection of aflatoxin M1 in pasteurized milk, yogurt, and milk powder. *J. Dairy Sci.* **2019**, *102*, 3985–3993. [CrossRef]
20. Sani, A.M.; Nikpooyan, H. Determination of aflatoxin M1 in milk by high-performance liquid chromatography in Mashhad (north east of Iran). *Toxicol. Ind. Health* **2013**, *29*, 334–338. [CrossRef] [PubMed]
21. Mehta, R.; Shetty, S.A.; Young, M.F.; Ryan, P.B.; Rangiah, K. Quantification of aflatoxin and ochratoxin contamination in animal milk using UHPLC-MS/SRM method: A small-scale study. *J. Food Sci. Technol.* **2021**, *58*, 3453–3464. [CrossRef]
22. Abbès, S.; Salah-Abbès, J.B.; Bouraoui, Y.; Oueslati, S.; Oueslati, R. Natural occurrence of aflatoxins (B1 and M1) in feed, plasma and raw milk of lactating dairy cows in Beja, Tunisia, using ELISA. *Food Addit. Contam. Part B* **2012**, *5*, 11–15. [CrossRef]
23. INS (Institut National de Statistique)-Tunisia. Results of a National Survey on Cereal Products Consumption and Waste by Tunisian Consumers from Different Areas and Social Conditions. Internal Document. Available online: http://www.ins.nat.tn/indexfr.php (accessed on 18 November 2018).
24. Kamal, R.M.; Mansour, M.A.; Elalfy, M.M.; Abdelfatah, E.; Galala, W. Quantitative detection of aflatoxin M1, ochratoxin and zearalenone in fresh raw milk of cow, buffalo, sheep and goat by UPLC XEVO–TQ in Dakahli Governorate, Egypt. *J. Vet. Med. Health* **2019**, *3*, 1–5. [CrossRef]
25. Abdallah, M.F.; Girgin, G.; Baydar, T. Mycotoxin Detection in Maize, Commercial Feed, and Raw Dairy Milk Samples from Assiut City, Egypt. *Vet. Sci.* **2019**, *6*, 57. [CrossRef]
26. Ali, M.A.; El Zubeir, I.E.; Fadel Elseed, A.M. Aflatoxin M1 in raw and imported powdered milk sold in Khartoum state, Sudan. *Food Addit. Contam. Part B Surveill.* **2014**, *7*, 208–212. [CrossRef]
27. Hamad, G.M.; El-Makarem, H.S.A.; Allam, M.G.; El Okle, O.S.; El-Toukhy, M.I.; Mehany, T.; El-Halmouch, Y.; Abushaala, M.M.F.; Saad, M.S.; Korma, S.A.; et al. Evaluation of the Adsorption Efficacy of Bentonite on Aflatoxin M1 Levels in Contaminated Milk. *Toxins* **2023**, *15*, 107. [CrossRef]
28. Zebib, H.; Abate, D.; Woldegiorgis, A.Z. Exposure and Health Risk Assessment of Aflatoxin M1 in Raw Milk and Cottage Cheese in Adults in Ethiopia. *Foods* **2023**, *12*, 817. [CrossRef]
29. Leite, M.; Freitas, A.; Barbosa, J.; Ramos, F. Mycotoxins in Raw Bovine Milk: UHPLC-QTrap-MS/MS Method as a Biosafety Control Tool. *Toxins* **2023**, *15*, 173. [CrossRef]
30. Liu, C.; Van der Fels-Klerx, H. Quantitative modeling of climate change impacts on mycotoxins in cereals: A review. *Toxins* **2021**, *13*, 276. [CrossRef]
31. FAO. *Climate Change: Unpacking the Burden on Food Safety*; FAO—Food and Agriculture Organization of the United Nations: Rome, Italy, 2020. [CrossRef]
32. Granados-Chinchilla, F.; Redondo-Solano, M.; Jaikel-Víquez, D. Mycotoxin Contamination of Beverages Obtained from Tropical Crops. *Beverages* **2018**, *4*, 83. [CrossRef]
33. Asghar, M.A.; Ahmed, A.; Asghar, M.A. Influence of temperature and environmental conditions on aflatoxin contamination in maize collected from different regions of Pakistan during 2016–2019. *J. Stored Prod. Res.* **2020**, *88*, 101637. [CrossRef]
34. Juan, C.; Oueslati, S.; Mañes, J.; Berrada, H. Multimycotoxin Determination in Tunisian Farm Animal Feed. *J. Food Sci.* **2019**, *84*, 3885–3893. [CrossRef] [PubMed]
35. Redouane-Salah, S.; Morgavi, D.P.; Arhab, R.; Messaï, A.; Boudra, H. Presence of aflatoxin M1 in raw, reconstituted, and powdered milk samples collected in Algeria. *Environ. Monit. Assess.* **2015**, *187*, 375. [CrossRef]
36. Ismaiel, A.A.; Tharwat, N.A.; Sayed, M.A.; Gameh, S.A. Two-year survey on the seasonal incidence of aflatoxin M1 in traditional dairy products in Egypt. *J. Food Sci. Technol.* **2020**, *57*, 2182–2189. [CrossRef]
37. Zakaria, A.M.; Amin, Y.A.; Khalil, O.S.F.; Abdelhiee, E.Y.; Elkamshishi, M.M. Rapid detection of aflatoxin M1 residues in market milk in Aswan Province, Egypt and effect of probiotics on its residues concentration. *J. Adv. Vet. Anim. Res.* **2019**, *6*, 197–201. [CrossRef]
38. Mohammedi-Ameur, S.; Dahmane, M.; Brera, C.; Kardjadj, M.; Ben-Mahdi, M.H. Occurrence and seasonal variation of aflatoxin M(1) in raw cow milk collected from different regions of Algeria. *Vet. World* **2020**, *13*, 433–439. [CrossRef] [PubMed]
39. Esam, R.M.; Hafez, R.S.; Khafaga, N.I.M.; Fahim, K.M.; Ibrahim Ahmed, L. Assessment of aflatoxin M(1) and B(1) in some dairy products with referring to the analytical performances of enzyme-linked immunosorbent assay in comparison to high-performance liquid chromatography. *Vet. World* **2022**, *15*, 91–101. [CrossRef]

50. Mukhtar, F.; Umar, S.; Bukar, A.; Muhammad, H. Occurrence and level of Aflatoxin M1 in fresh raw cow milk within Zaria metropolis. *Bayero J. Pure Appl. Sci.* **2022**, *13*, 107–111.
51. Ayeni, K.; Atanda, O.; Krska, R.; Ezekiel, C. Present status and future perspectives of grain drying and storage practices as a means to reduce mycotoxin exposure in Nigeria. *Food Control* **2021**, *126*, 108074. [CrossRef]
52. Tuncay, C.; Oniz, A. Determination of Aflatoxin M1 in Raw Milk: A Review of an Island Sample with the HPLC Method. *Progress. Nutr.* **2022**, *23*, e2021185. [CrossRef]
53. Widiastuti, R.; Anastasia, Y. Aflatoxin M1 in fresh dairy milk from small individual farms in Indonesia. *J. Ilmu Ternak Dan Vet.* **2018**, *23*, 143. [CrossRef]
54. Armorini, S.; Altafini, A.; Zaghini, A.; Roncada, P. Occurrence of aflatoxin M1 in conventional and organic milk offered for sale in Italy. *Mycotoxin Res.* **2016**, *32*, 237–246. [CrossRef] [PubMed]
55. Iqbal, S.Z.; Waqas, M.; Latif, S. Incidence of Aflatoxin M1 in Milk and Milk Products from Punjab, Pakistan, and Estimation of Dietary Intake. *Dairy* **2022**, *3*, 577–586. [CrossRef]
56. Venâncio, R.L.; Ludovico, A.; de Santana, E.H.W.; de Toledo, E.A.; de Almeida Rego, F.C.; Dos Santos, J.S. Occurrence and seasonality of aflatoxin M1 in milk in two different climate zones. *J. Sci. Food Agric.* **2019**, *99*, 3203–3206. [CrossRef] [PubMed]
57. Jianglin, X.; Peng, L.; Zhou, H.; Lin, B.; Yan, P.; Wu, W.; Liu, Y.; Wu, L.; Qiu, Y. Prevalence of aflatoxin M1 in raw milk and three types of liquid milk products in central-south China. *Food Control* **2019**, *108*, 106840. [CrossRef]
58. Daou, R.; Afif, C.; Joubrane, K.; Rabbaa, L.; Maroun, R.; Ismail, A.; el Khoury, A. Occurrence of aflatoxin M1 in raw, pasteurized, UHT cows' milk, and dairy products in Lebanon. *Food Control* **2020**, *111*, 107055. [CrossRef]
59. Rodríguez-Blanco, M.; Ramos, A.J.; Prim, M.; Sanchis, V.; Marín, S. Usefulness of the analytical control of aflatoxins in feedstuffs for dairy cows for the prevention of aflatoxin M(1) in milk. *Mycotoxin Res.* **2020**, *36*, 11–22. [CrossRef]
60. Kang'ethe, E.K.; Sirma, A.J.; Murithi, G.; Mburugu-Mosoti, C.K.; Ouko, E.O.; Korhonen, H.J.; Nduhiu, G.J.; Mungatu, J.K.; Joutsjoki, V.; Lindfors, E.; et al. Occurrence of mycotoxins in food, feed, and milk in two counties from different agro-ecological zones and with historical outbreak of aflatoxins and fumonisins poisonings in Kenya. *Food Qual. Saf.* **2017**, *1*, 161–170. [CrossRef]
61. Hajmohammadi, M.; Valizadeh, R.; Naserian, A.; Nourozi, M.; Rocha, R.; Oliveira, C. Composition and occurrence of aflatoxin M 1 in cow's milk samples from Razavi Khorasan Province, Iran. *Int. J. Dairy Technol.* **2019**, *73*, 40–45. [CrossRef]
62. Sahin, H.Z.; Celik, M.; Kotay, S.; Kabak, B. Aflatoxins in dairy cow feed, raw milk and milk products from Turkey. *Food Addit. Contam. Part B Surveill.* **2016**, *9*, 152–158. [CrossRef]
63. Sumon, A.H.; Islam, F.; Mohanto, N.C.; Kathak, R.R.; Molla, N.H.; Rana, S.; Degen, G.H.; Ali, N. The Presence of Aflatoxin M1 in Milk and Milk Products in Bangladesh. *Toxins* **2021**, *13*, 440. [CrossRef]
64. Udomkun, P.; Mutegi, C.; Wossen, T.; Atehnkeng, J.; Nabahungu, N.L.; Njukwe, E.; Vanlauwe, B.; Bandyopadhyay, R. Occurrence of aflatoxin in agricultural produce from local markets in Burundi and Eastern Democratic Republic of Congo. *Food Sci. Nutr.* **2018**, *6*, 2227–2238. [CrossRef] [PubMed]
65. Kiarie, G.; Dominguez-Salas, P.; Kang'Ethe, S.; Grace, D.; Lindahl, J. Aflatoxin exposure among young children in urban low-income areas of Nairobi and association with child growth. *Afr. J. Food Agric. Nutr. Dev.* **2016**, *16*, 10967–10990. [CrossRef]
66. Choga, R.; Chiriseri, B.; Pfukenyi, D.M. Detection and levels of aflatoxin m_1 in raw milk of dairy cows from selected small scale and commercial farms in harare, Zimbabwe. *Zimb. Vet. J.* **2016**, *34*, 1–6.
67. Mohammed, S.; Munissi, J.J.E.; Nyandoro, S.S. Aflatoxin M1 in raw milk and aflatoxin B1 in feed from household cows in Singida, Tanzania. *Food Addit. Contam. Part. B* **2016**, *9*, 85–90. [CrossRef]
68. Koutamehr, M.E.; Akbari, H.; Akbari, S.; Hassanzadazar, H. Aflatoxin M1 level in raw milk samples of Maragheh, Bonab and Malekan cities, East Azerbaijan province, Iran. Studia Universitatis "Vasile Goldis" Arad. *Ser. Stiintele Vietii (Life Sci. Ser.)* **2017**, *27*, 85–89.
69. Maryam, R.; Widiyanti, P.M.; Dalilah, D. The Occurrence of Aflatoxin M1 in Fresh Milk and Its Possible Effects to Public Health. In Proceedings of the 1st International Conference for Health Research–BRIN (ICHR), Jakarta, Indonesia, 23–24 November 2022; pp. 541–549. [CrossRef]
70. Mannani, N.; Tabarani, A.; El Adlouni, C.; Abdennebi, E.H.; Zinedine, A. Aflatoxin M1 in pasteurized and UHT milk marked in Morocco. *Food Control* **2021**, *124*, 107893. [CrossRef]
71. Patyal, A.; Gill, J.P.S.; Bedi, J.S.; Aulakh, R.S. Occurrence of Aflatoxin M1 in raw, pasteurized and UHT milk from Punjab, India. *Curr. Sci.* **2020**, *118*, 79–86. [CrossRef]
72. Huang, L.C.; Zheng, N.; Zheng, B.Q.; Wen, F.; Cheng, J.B.; Han, R.W.; Xu, X.M.; Li, S.L.; Wang, J.Q. Simultaneous determination of aflatoxin M1, ochratoxin A, zearalenone and α-zearalenol in milk by UHPLC–MS/MS. *Food Chemistry* **2014**, *146*, 242–249. [CrossRef]
73. Rama, A.; Latifi, F.; Bajraktari, D.; Ramadani, N. Assessment of aflatoxin M1 levels in pasteurized and UHT milk consumed in Prishtina, Kosovo. *Food Control* **2015**, *57*, 351–354. [CrossRef]
74. Patyal, A.; Gill, J.P.S.; Bedi, J.S.; Aulakh, R.S. Potential risk factors associated with the occurrence of aflatoxin M1 in raw milk produced under different farm conditions. *J. Environ. Sci. Health B* **2020**, *55*, 827–834. [CrossRef] [PubMed]
75. Wild, C.P.; Miller, J.D.; Groopman, J.D. *Mycotoxin Control in Low-and Middle-Income Countries*; WHO: Geneva, Switzerland, 2015.
76. Schrenk, D.; Bignami, M.; Bodin, L.; Chipman, J.K.; Del Mazo, J.; Grasl-Kraupp, B.; Hogstrand, C.; Hoogenboom, L.R.; Leblanc, J.C.; Nebbia, C.S.; et al. Risk assessment of aflatoxins in food. *Efsa J.* **2020**, *18*, e06040. [CrossRef]

67. Bilandžić, N.; Varga, I.; Varenina, I.; Solomun Kolanović, B.; Božić Luburić, Đ.; Đokić, M.; Sedak, M.; Cvetnić, L.; Cvetnić, Ž. Seasonal Occurrence of Aflatoxin M1 in Raw Milk during a Five-Year Period in Croatia: Dietary Exposure and Risk Assessmen *Foods* **2022**, *11*, 1959. [CrossRef]
68. Gizachew, D.; Szonyi, B.; Tegegne, A.; Hanson, J.; Grace, D. Aflatoxin contamination of milk and dairy feeds in the Greater Addis Ababa milk shed, Ethiopia. *Food Control* **2016**, *59*, 773–779. [CrossRef]
69. Kortei, N.K.; Annan, T.; Kyei-Baffour, V.; Essuman, E.K.; Boakye, A.A.; Tettey, C.O.; Boadi, N.O. Exposure assessment and cancer risk characterization of aflatoxin M1 (AFM1) through ingestion of raw cow milk in southern Ghana. *Toxicol. Rep.* **2022**, *9*, 1189–1197. [CrossRef] [PubMed]
70. Njombwa, C.A.; Moreira, V.; Williams, C.; Aryana, K.; Matumba, L. Aflatoxin M(1) in raw cow milk and associated hepatocellular carcinoma risk among dairy farming households in Malawi. *Mycotoxin Res.* **2021**, *37*, 89–96. [CrossRef]
71. Milićević, D.R.; Milešević, J.; Gurinović, M.; Janković, S.; Đinović-Stojanović, J.; Zeković, M.; Glibetić, M. Dietary Exposure and Risk Assessment of Aflatoxin M1 for Children Aged 1 to 9 Years Old in Serbia. *Nutrients* **2021**, *13*, 4450. [CrossRef] [PubMed]
72. Mahdjoubi, C.K.; Arroyo-Manzanares, N.; Hamini-Kadar, N.; García-Campaña, A.M.; Mebrouk, K.; Gámiz-Gracia, L. Multi Mycotoxin Occurrence and Exposure Assessment Approach in Foodstuffs from Algeria. *Toxins* **2020**, *12*, 194. [CrossRef] [PubMed]
73. EFSA. Management of left-censored data in dietary exposure assessment of chemical substances. *EFSA J.* **2010**, *8*, 1557. [CrossRef]
74. Pallarés, N.; Carballo, D.; Ferrer, E.; Fernández-Franzón, M.; Berrada, H. Mycotoxin Dietary Exposure Assessment through Fruit Juices Consumption in Children and Adult Population. *Toxins* **2019**, *11*, 684. [CrossRef]

MDPI
St. Alban-Anlage 66
4052 Basel
Switzerland
www.mdpi.com

Toxins Editorial Office
E-mail: toxins@mdpi.com
www.mdpi.com/journal/toxins